Interpolation of Operators and Singular Integrals

PURE AND APPLIED MATHEMATICS

A Program of Monographs, Textbooks, and Lecture Notes

MONOGRAPHS AND TEXTBOOKS IN PURE AND APPLIED MATHEMATICS

1. *K. Yano,* Integral Formulas in Riemannian Geometry (1970)
2. *S. Kobayashi,* Hyperbolic Manifolds and Holomorphic Mappings (1970)
3. *V. S. Vladimirov,* Equations of Mathematical Physics (A. Jeffrey, editor; A. Littlewood, translator) (1970)
4. *B. N. Pshenichnyi,* Necessary Conditions for an Extremum (L. Neustadt, translation editor; K. Makowski, translator) (1971)
5. *L. Narici, E. Beckenstein, and G. Bachman,* Functional Analysis and Valuation Theory (1971)
6. *D. S. Passman,* Infinite Group Rings (1971)
7. *L. Dornhoff,* Group Representation Theory (in two parts). Part A: Ordinary Representation Theory. Part B: Modular Representation Theory (1971, 1972)
8. *W. Boothby and G. L. Weiss (eds.),* Symmetric Spaces: Short Courses Presented at Washington University (1972)
9. *Y. Matsushima,* Differentiable Manifolds (E. T. Kobayashi, translator) (1972)
10. *L. E. Ward, Jr.,* Topology: An Outline for a First Course (1972) *(out of print)*
11. *A. Babakhanian,* Cohomological Methods in Group Theory (1972)
12. *R. Gilmer,* Multiplicative Ideal Theory (1972)
13. *J. Yeh,* Stochastic Processes and the Wiener Integral (1973) *(out of print)*
14. *J. Barros-Neto,* Introduction to the Theory of Distributions (1973) *(out of print)*
15. *R. Larsen,* Functional Analysis: An Introduction (1973)
16. *K. Yano and S. Ishihara,* Tangent and Cotangent Bundles: Differential Geometry (1973)
17. *C. Procesi,* Rings with Polynomial Identities (1973)
18. *R. Hermann,* Geometry, Physics, and Systems (1973)
19. *N. R. Wallach,* Harmonic Analysis on Homogeneous Spaces (1973)
20. *J. Dieudonné,* Introduction to the Theory of Formal Groups (1973)
21. *I. Vaisman,* Cohomology and Differential Forms (1973)
22. *B.-Y. Chen,* Geometry of Submanifolds (1973)
23. *M. Marcus,* Finite Dimensional Multilinear Algebra (in two parts) (1973, 1975)
24. *R. Larsen,* Banach Algebras: An Introduction (1973)
25. *R. O. Kujala and A. L. Vitter (eds.),* Value Distribution Theory: Part A; Part B. Deficit and Bezout Estimates by Wilhelm Stoll (1973)
26. *K. B. Stolarsky,* Algebraic Numbers and Diophantine Approximation (1974)
27. *A. R. Magid,* The Separable Galois Theory of Commutative Rings (1974)
28. *B. R. McDonald,* Finite Rings with Identity (1974)
29. *J. Satake,* Linear Algebra (S. Koh, T. Akiba, and S. Ihara, translators) (1975)

30. *J. S. Golan,* Localization of Noncommutative Rings (1975)
31. *G. Klambauer,* Mathematical Analysis (1975)
32. *M. K. Agoston,* Algebraic Topology: A First Course (1976)
33. *K. R. Goodearl,* Ring Theory: Nonsingular Rings and Modules (1976)
34. *L. E. Mansfield,* Linear Algebra with Geometric Applications (1976)
35. *N. J. Pullman,* Matrix Theory and Its Applications: Selected Topics (1976)
36. *B. R. McDonald,* Geometric Algebra Over Local Rings (1976)
37. *C. W. Groetsch,* Generalized Inverses of Linear Operators: Representation and Approximation (1977)
38. *J. E. Kuczkowski and J. L. Gersting,* Abstract Algebra: A First Look (1977)
39. *C. O. Christenson and W. L. Voxman,* Aspects of Topology (1977)
40. *M. Nagata,* Field Theory (1977)
41. *R. L. Long,* Algebraic Number Theory (1977)
42. *W. F. Pfeffer,* Integrals and Measures (1977)
43. *R. L. Wheeden and A. Zygmund,* Measure and Integral: An Introduction to Real Analysis (1977)
44. *J. H. Curtiss,* Introduction to Functions of a Complex Variable (1978)
45. *K. Hrbacek and T. Jech,* Introduction to Set Theory (1978)
46. *W. S. Massey,* Homology and Cohomology Theory (1978)
47. *M. Marcus,* Introduction to Modern Algebra (1978)
48. *E. C. Young,* Vector and Tensor Analysis (1978)
49. *S. B. Nadler, Jr.,* Hyperspaces of Sets (1978)
50. *S. K. Sehgal,* Topics in Group Rings (1978)
51. *A. C. M. van Rooij,* Non-Archimedean Functional Analysis (1978)
52. *L. Corwin and R. Szczarba,* Calculus in Vector Spaces (1979)
53. *C. Sadosky,* Interpolation of Operators and Singular Integrals: An Introduction to Harmonic Analysis (1979)

Other Volumes in Preparation

Interpolation of Operators and Singular Integrals

An Introduction to Harmonic Analysis

Cora Sadosky

Universidad Central de Venezuela
Caracas, Venezuela

MARCEL DEKKER, INC.   New York and Basel

Library of Congress Cataloging in Publication Data

Sadosky, Cora.
Interpolation of operators and singular integrals.

(Monographs and textbooks in pure and applied mathematics; 53)
Bibliography: p. 363
Includes index.
1. Harmonic analysis. 2. Operator theory. 3. Interpolation. 4. Integrals, Singular. I. Title.
QA403.S2 515.2433 79-19595
ISBN 0-8247-6883-3

MARCEL DEKKER, INC.

270 Madison Avenue, New York, New York 10016

Current printing (last digit):

10 9 8 7 6 5 4 3 2 1

PRINTED IN THE UNITED STATES OF AMERICA

To Daniel J. Goldstein,
who through fourteen years of life in common
allowed me to enjoy a privilege unique for a
woman: to be able to work like a man,
in solidarity,

to Cora and Manuel, my parents, and
to Corasol, my child,
in deep love.

PREFACE

This book is an introduction to harmonic analysis on Euclidean spaces, aiming at the study of singular integrals. Thus it provides a basis for the study of topics such as differentiability properties of functions of several variables and the applications to partial differential equations, and for some recent developments in classical harmonic analysis. In particular, it leads to the more advanced treatises of Stein [5] and Stein and Weiss [6]. While certain topics had to be excluded, some of those which are presented here are not found in the existing introductory texts on harmonic analysis, as the Hardy-Littlewood theory of maximal functions and some of its modern applications, the Marcinkiewicz interpolation theorem, the class of functions of bounded mean oscillation, and ergodic theorems.

I hope that the book will be accesible to a wide audience that includes graduate students first approaching the subject. For this purpose, I tried to make the text as self-contained as possible, and most proofs are given in great detail, thereby making the development understandable for a beginner. The paragraphs marked with an asterisk are more technical and its reading can be omitted without altering the comprehension of the remaining text.

The reader should be familiar with the basic concepts of integration theory and normed linear spaces, as presented in a book like Royden's Real Analysis. Working knowledge of the elementary theory of functions of one complex variable is also desirable. The basic results used throughout the book are summarized in Chapter 0.

Chapters 1, 2 and 3 lay the groundwork for Fourier analysis. Chapter 1 deals with convolution units (or approximations of the identity) and the group algebra of $\mathbb{R}^n$, Chapter 2 with Fourier transforms of integrable functions and finite measures in $\mathbb{R}^n$, and Chapter 3 with Fourier transforms of square integrable functions, inversion theory for the Fourier integrals and harmonic functions of several variables. Exercises are supplied in the text, both to test the reader's understanding and complete some of the points. One of the differences between this and other introductory presentations is that Chapter 2 is mainly devoted to the study of the Fourier theory of finite measures and its applications to convergence theorems in probability theory. Although the setting is $\mathbb{R}^n$, the presentation makes clear the possibility of extensions to the abstract case of locally compact abelian groups. In particular, the proof of the Bochner theorem is based on the consideration of positive functionals, and is independent of the Euclidean structure of $\mathbb{R}^n$.

The following three chapters are deeper in nature and more technically involved. Some of the material included there appears in a text for the first time.

Chapter 4 deals with interpolation of operators on L^p spaces. The theorems of M. Riesz and Marcinkiewicz are presented in a way that leads to the abstract complex and real methods of interpolation of Calderón and Lions-Peetre respectively. Appendix B gives an exposition of the former.

Chapter 5 develops the theory of maximal functions as it is now widely used, giving the theory of differentiation of integrals and ergodic theorems as typical applications. Within this framework the space of functions of bounded mean oscillation, BMO, is introduced and some of its properties are studied.

Chapter 6 deals with the Calderón-Zygmund theory of singular integrals, including its extension to BMO. In Section 6 some generalizations to $L^p(\mathbb{R}^n, d\mu)$, for weighted measures $d\mu$, are given, with new proofs. This includes a new and sharper version of the Helson-Szegö theorem in prediction theory. Appendix A gives an exposition, without proofs, about the connections between singular integrals and the theory of partial differential equations.

All chapters conclude with a detailed Reference section and there is a Bibliography at the end of the book, containing the general references for the whole text. The end of a proof is marked by a ∇.

This book is the outgrowth of a volume of Lecture Notes written at the Universidad Central de Venezuela (Publ. Mat., U.C. V., Segunda Serie, Fasc. 1) in 1976. They correspond to material presented in courses taught at the Universidad Central de Venezuela in 1975, 1977 and 1978. The origin of those notes is much older, starting with a course taught at the Universidad de le República, Montevideo, Uruguay, in 1970, and a three months lecture series given in 1973 at the Universidad del Sur, Bahía Blanca, Argentina.

The initial inspiration on the treatment of this subject comes from magnificent courses given by E. M. Stein, G. Weiss and A. P. Calderón, which I attended as a graduate student at the University of Chicago and the University of Buenos Aires. The overall influence is that of Professor A. Zygmund who taught me how beautiful singular integrals are and induced the will to try to share with others the pleasure of their beauty. For this, for his helpful comments on the manuscript, and for his constant encouragement and support I am greatly indebted and want to express here my deepest gratitude.

This book would not exist without the friendship and help, both mathematical and personal, given me by Mischa Cotlar. Specifically, he read the entire manuscript and made very many valuable suggestions. Most of what is good and all which is original in this presentation comes from his comments.

A person in my situation--I have been forced out of mathematics for many years and have been moving from one country to another in the last decade--fortunately has many people to thank. Even without mentioning them all, I want to express my thanks to all those who helped me through this endeavor. Also my appreciation goes to the Universidad Central de Venezuela, where I have been a professor during most of the writing of this book, and to the Institute for Advanced Study, where I was able to complete it in the best possible atmosphere.

My dear friend, the late Professor E. T. Oklander from Bahía Blanca, Argentina, read a good part of the original manuscript, specially the Appendix on interpolation theory.

I am happy to acknowledge the support received from Professors G. Weiss and E. Taft, editor of this series, during the final stages of this work.

J. Rogawski helped me with my English, Lucía Flores and Beatriz Molina of Caracas, Venezuela, skillfully typed the original manuscript, and Irene Gaskill, from the Institute for Advanced Study, did an impressive and beautiful job with the final version. To all of them, I am greatly indebted.

My final thanks go to my family, and especially to my daughter Corasol, who was unbelievably patient and cooperative to me while enduring my writing of this book.

Cora Sadosky
Princeton, 1979

CONTENTS

Interpolation of Operators and Singular Integrals

Chapter 0

PRELIMINARIES

Measure theory, integration and the theory of normed linear spaces underlie all of harmonic analysis. We assume the reader is familiar with the basic notions and techniques of these theories. Royden's book Real Analysis [1] can be used as a reference.

In this preliminary chapter we summarize some results and fix the notation which will be used throughout. The body of the text begins with Chapter 1.

1. SOME DEFINITIONS FROM MEASURE THEORY

Let $\mathbb{R}^n$, $n \geq 1$, denote the n-dimensional euclidean space and $x = (x_1, \ldots, x_n)$, $t = (t_1, \ldots, t_n), \ldots$ be elements of $\mathbb{R}^n$. The scalar product of elements in $\mathbb{R}^n$ is denoted by $x.t = x_1t_1 + \ldots + x_nt_n$ and, in particular, $|x| = (x.x)^{1/2} = (x_1^2 + \ldots + x_n^2)^{1/2}$ is the distance of x to the origin.

A multiindex $\alpha = (\alpha_1, \ldots, \alpha_n)$ is an n-tuple of nonnegative integers and $|\alpha| = \alpha_1 + \ldots + \alpha_n$, $x^\alpha = x_1^{\alpha_1} \ldots x_n^{\alpha_n}$, $\partial^\alpha = \partial^{|\alpha|}/\partial x_1^{\alpha_1} \ldots \partial x_n^{\alpha_n}$.

The characteristic function of a set A is denoted by χ_A, hence $\chi_A(x) = 1$ if $x \in A$ and zero otherwise. A^c will always denote the complement of A.

An interval $I = \Pi_{k=1}^n I_k$ in $\mathbb{R}^n$ is the product of n intervals $I_1, \ldots, I_n$ in $\mathbb{R}$, so that I is a parallelepiped of sides parallel to the axes.

A translation in $\mathbb{R}^n$ will be denoted by $\tau_h x = x + h$ for $h \in \mathbb{R}^n$. Then, $\tau_h A = \{\tau_h x : x \in A\}$. Also if f is a given function, $\tau_h f(x) = f(x - h)$ for all x. If $f = \chi_A$ and $g = \chi_B$, where $B = \tau_h A$, then $g = \tau_h f$. A dilation in $\mathbb{R}^n$ will be denoted by $\delta_a x = ax$, $\delta_a f(x) = f(\delta_a x) = f(ax)$, $a > 0$.

In the space $\mathbb{R}^n$, $n \geq 1$, the Lebesgue measure is denoted by $dx = dx_1 \ldots dx_n$. In what follows, $|A|$ stands for the Lebesgue measure of the set $A \subset \mathbb{R}^n$ and $\int f dx$ stands for $\int_{\mathbb{R}^n} f(x_1, \ldots, x_n) dx_1 \ldots dx_n$.

A set N is called negligible if there exists a Borel set B such that $N \subset B$ and $|B| = 0$. A set A is Lebesgue measurable if $A = B + N$ for B a Borel set and N, negligible. A function f is Lebesgue measurable in $\mathbb{R}^n$ if the set $\{x : f(x) > \alpha\}$ is measurable for all real α. All the sets and functions we deal with will be measurable, unless otherwise stated. A property is said to be satisfied "almost everywhere" or a.e. in a set A if it is satisfied in $A - N$ for some negligible set N.

The Lebesgue measure is distinguished because it is, up to constant multiples, the unique translation invariant measure on $\mathbb{R}^n$: if $B = \tau_h A$ then $|B| = |A|$ for any $h \in \mathbb{R}^n$. This property implies that if f is (Lebesgue) integrable then $\tau_h f$ is also integrable and

$$\int_{\mathbb{R}^n} \tau_h f(x) dx = \int_{\mathbb{R}^n} f(x - h) dx = \int_{\mathbb{R}^n} f(x) dx \tag{1.1}$$

for all $h \in \mathbb{R}^n$.

In (1.1) it is essential that the integral be taken on the whole space $\mathbb{R}^n$ and not over a subset, since we can easily produce examples of integrable f for which $\int_0^1 \tau_h f(x)dx \neq \int_0^1 f(x)dx$.

We say that f is <u>integrable</u> and write $f \in L = L(\mathbb{R}^n) = L^1(\mathbb{R}^n)$ if f is Lebesgue integrable in $\mathbb{R}^n$. A function which is integrable over every compact subset of $\mathbb{R}^n$ is called <u>locally integrable</u> and the space of such functions is denoted by $L_{loc}(\mathbb{R}^n)$. It contains all bounded functions even though L does not. For example, $f(x) \equiv 1$ lies in L_{loc} but not in L. Note that the integral $\int f(x)dx$ is defined for all positive f, but $f \in L$ only if $\int f < \infty$.

Let $\Phi(t)$, $t \geq 0$, be a nonnegative nondecreasing function. The class of functions $f(x)$, $x \in X \subset \mathbb{R}^n$, such that

$$\int_X \Phi(|f(x)|)dx < \infty$$

will be denoted by $\Phi_X(L)$. If no ambiguity arises we shall simply write $\Phi(L)$. The meaning of $\Phi_{loc}(L)$ is obvious. It is clear that $\Phi(s) \leq \Psi(s)$, $s \geq 0$, implies $\Phi_{loc}(L) \supset \Psi_{loc}(L)$. The class L^p, $p > 0$, given by $\Phi(t) = t^p$, is particularly important, but the class $L^p(\log^+ L)^q$ of functions f such that

$$\int_X |f|^p(\log^+|f|)^q dx < \infty$$

is also of interest. (By $\log^+ t$ we mean the function equal to $\log t$ for $t > 1$ and to 0 for $0 \leq t \leq 1$.) The class $L \log^+ L$ is called the Zygmund class.

Sometimes we will work on a finite interval instead of $\mathbb{R}^n$. The theory is easier in this case because all subsets have finite measure. For instance, the interval $[0, 2\pi)$ arises in the study of periodic functions and Fourier series. $\mathbb{T} \sim \mathbb{R} \pmod{2\pi}$ denotes the unit circle and is identified with $[0, 2\pi)$.

$\mathbb{Z}$ will denote the set of all integers.

Along with the Lebesgue measure we consider other (σ-additive) positive measures in $\mathbb{R}^n$. We will always assume that every measure μ is a Borel measure, i.e., that $\mu(A)$ is defined for every Borel set $A \subset \mathbb{R}^n$ and that $\mu(A) < \infty$ for all bounded sets A. A Radon measure is any Borel measure that is regular, i.e.

$$\begin{aligned} \mu(A) &= \inf\{\mu(0) : A \subset 0 \text{ open}\} \\ &= \sup\{\mu(F) : F \subset A \text{ closed}\} \end{aligned} \tag{1.2}$$

In $\mathbb{R}^n$, every Borel measure is regular and so Radon and Borel measures coincide in $\mathbb{R}^n$. We shall consider also finite real measures μ given by $\mu = \mu_1 - \mu_2$, μ_1 and μ_2 positive Borel measures such that $\mu_i(\mathbb{R}^n) < \infty$ for $i = 1, 2$. The measures we deal with, besides the Lebesgue measure, are usually finite. But if not, they will at least be σ-finite, since all Borel measures are σ-finite on $\mathbb{R}^n$. The set of finite Borel measures in $\mathbb{R}^n$ is denoted by $\mathcal{M} = \mathcal{M}(\mathbb{R}^n)$.

Given a positive Borel measure μ, a set N is μ-negligible if $N \subset B$ where B is a Borel set and $\mu(B) = 0$. A set A is μ-measurable if $A = B + N$, B a Borel set and N μ-negligible. For any Borel measure μ in $\mathbb{R}^n$, all the Borel sets are μ-measurable, but two different Borel measures may have different negligible sets.

By δ_a we denote the Dirac measure at the point a. By definition, $\delta_a(A) = 1$ if $a \in A$ and $\delta_a(A) = 0$ if $a \notin A$. We write $\delta_0 = \delta$.

Sometimes we shall also consider measures μ on abstract spaces $\mathcal{X}$. Those μ will always be assumed σ-additive and σ-finite and, if $\mathcal{X}$ is a topological space, μ will be assumed regular as in (1.2). A pair $(\mathcal{X}, \mu)$ will be then called a measure space.

In particular, we consider measures in $\mathbb{Z}$, the discrete line. The analog of the Lebesgue measure on $\mathbb{Z}$ is that measure which gives measure one to each point of $\mathbb{Z}$. A function defined on $\mathbb{Z}$ is given by a sequence $f = \{f(n)\}$, $n \in \mathbb{Z}$, and its integral is $\Sigma_{-\infty}^{\infty} f(n)$.

Given two measure spaces $(\mathscr{X}, \mu)$ and $(\mathscr{Y}, \nu)$, $\mu \geq 0$, $\nu \geq 0$, the product measure $\mu \otimes \nu$ is defined on $\mathscr{X} \times \mathscr{Y}$ (e.g., if $\mathscr{X} = \mathscr{Y} = R$, $\mu = \nu$ = Lebesgue measure in $\mathbb{R}$, then $\mu \otimes \nu$ in the Lebesgue measure $dx_1 \cdot dx_2$ in $\mathbb{R}^2 = \mathbb{R} \times \mathbb{R}$). <u>Fubini's theorem</u> asserts that if $F(x, y)$ is $\mu \otimes \nu$-integrable then

$$\iint_{\mathscr{X} \times \mathscr{Y}} F(x, y) d\mu \otimes \nu = \int_{\mathscr{X}} \left(\int_{\mathscr{Y}} F(x, y) d\nu\right) d\mu = \int_{\mathscr{Y}} \left(\int_{\mathscr{X}} F(x, y) d\mu\right) d\nu \tag{1.3}$$

Furthermore, if $F(x, y)$ is only $\mu \otimes \nu$-measurable but $F(x, y) \geq 0$, the Fubini-Tonelli theorem asserts that (1.3) holds, even if both sides are infinite. From this we obtain the following criterion that we use repeatedly:

<u>Proposition 1.1.</u> Let $F(x, y)$ be $\mu \otimes \nu$-measurable. If one of the iterated integrals of $|F(x, y)|$, $\int(\int |F(x, y)| d\mu) d\nu$ or $\int(\int |F(x, y)| d\nu) d\mu$ is finite, then F is $\mu \otimes \nu$-integrable and (1.3) holds.

A measure ν is said to be <u>absolutely continuous with respect to a positive Borel measure</u> μ if $\mu(A) = 0$ implies $\nu(A) = 0$ for each Borel set A. In particular, if μ is the Lebesgue measure, ν is called an <u>absolutely continuous measure</u>. The <u>Radon-Nikodym theorem</u> asserts that if ν is absolutely continuous with respect to μ, then there exists a μ-integrable function h such that

$$\nu(A) = \int_A h d\mu \quad \text{for all measurable sets } A$$

The function h is uniquely determined a.e. and is called the Radon-Nikodym derivative of ν with respect to μ.

When μ is the Lebesgue measure, this tells us that there is a 1-1 correspondence between integrable functions $h \in L^1$ and absolutely continuous measures $d\nu = f dx \in \mathcal{M}$ (f is then called the density of $d\nu$). So identifying L^1 with its image through the correspondence, we set $L^1 \subset \mathcal{M}$.

A measure $\nu \in \mathcal{M}$ is called singular if there exists a negligible set N, $|N| = 0$, such that $\nu(N^c) = 0$. By the Lebesgue decompsition theorem, any measure μ defined in $\mathbb{R}^n$ may be uniquely expressed as a sum $\mu = \mu_1 + \nu$ with an absolutely continuous measure μ_1 and a singular measure ν.

The indefinite integral of $f \in L^1(\mathbb{R})$ is defined as $F(x) = \int_0^x f$; hence $F(x + h) - F(x - h) = \int_{x-h}^{x+h} f = \int_{I_x} f = F(I_x)$ for $I_x = (x - h, x + h)$, an interval (or one dimensional sphere) of center x and radius h. In $\mathbb{R}^n$, we define $F(I_x) = \int_{I_x} f$, where $I_x = \{y : |x - y| < h\}$ is the sphere with center x and radius h. The Lebesgue theorem on differentiation of the integral (or fundamental theorem of calculus) asserts that if $f \in L_{loc}$ then

$$\lim_{I_x \to x} \frac{F(I_x)}{|I_x|} = \lim_{I_x \to x} \frac{1}{|I_x|} \int_{I_x} f(y)dy = f(x) \quad \text{a.e.} \tag{1.4}$$

(and the same result holds taking cubes instead of spheres). In Chapter 5 we shall give an independent proof of this theorem, but we will use it earlier (as a result in integration theory).

In this connection, the Lebesgue set of a function $f \in L_{loc}$, denoted by $\mathcal{L}_f$, is the set of points $x \in \mathbb{R}^n$ such that

$$r^{-n}\int_{|y|<r} |f(x, y) - f(x)|dy \to 0 \text{ as } r \to 0 \tag{1.5}$$

We claim that $\mathscr{L}_f$ includes almost all points in $\mathbb{R}^n$. In order to see this, remark that (1.4) can be rewritten as

$$r^{-n}\int_{|y|<r} (f(x - y) - f(x))dy \to 0 \text{ when } r \to 0 \tag{1.4a}$$

for almost every $x \in \mathbb{R}^n$. Notice that $g(x) = |f(x) - \rho| \in L_{loc}$ for any constant ρ, whenever $f \in L_{loc}$. So that by (1.4a), the set $\mathscr{F}_\rho$ of all $x \in \mathbb{R}^n$ such that

$$\lim_{r\to 0} r^{-n}\int_{|y|<r} (|f(x - y) - \rho| - |f(x) - \rho|)dy \neq 0$$

has measure zero. Thus, it also has measure zero the set $\mathscr{F} = \bigcup_\rho \mathscr{F}_\rho$, where the union is taken over all ρ rational. Now let $x \in \mathbb{R}^n - \mathscr{F}$ and let us see that, for such x, (1.5) holds. We fix x, $\varepsilon > 0$ and take ρ rational such that $|f(x) - \rho| < \varepsilon$. Then

$$r^{-n}\int_{|y|<r} |f(x - y) - f(x)|dy \leq r^{-n}\int_{|y|<r} |f(x - y) - \rho|dy$$

$$+ r^{-n}\int_{|y|<r} |f(x) - \rho|dy$$

and the last term of the sum is equal to a constant times $|f(x) - \rho| < \varepsilon$, while the first one tends to $|f(x) - \rho|$ when $r \to 0$ since $x \in \mathbb{R}^n - \mathscr{F}$. Therefore $r^{-n}\int_{|y|<r} |f(x - y) - f(x)|dy < C_\varepsilon$ for all $\varepsilon > 0$ if r is sufficiently small and (1.5) holds a.e. in $\mathbb{R}^n$. ▽

In the set $\mathcal{M}$ of all finite measures we define the norm given by the total mass, i.e.

$$\|\mu\| = \inf\{\mu_1(\mathbb{R}^n) + \mu_2(\mathbb{R}^n) : \mu = \mu_1 = \mu_2,\ \mu_1 \geq 0,\ \mu_2 \geq 0\}$$
$$= \int_{\mathbb{R}^n} |d\mu|$$

$\mathcal{M}$ is then a Banach space. Observe that even though $\tau_h\mu(A) = \mu(\tau_{-h}A)$ implies $\|\tau_h\mu\| = \|\mu\|$ for all $h \in \mathbb{R}^n$, in general $\|\tau_h\mu - \mu\|$ does not tend to zero when $|h| \to 0$ (e.g., for $\mu = \delta$). In fact, we have that $\|\tau_h\mu - \mu\| \to 0$ as $|h| \to 0$ if and only if μ is absolutely continuous (see Section 3).

We shall frequently use polar coordinates in $\mathbb{R}^n$ and provide the details now.

2. POLAR COORDINATES IN $\mathbb{R}^n$

Let $\Sigma = \Sigma_n$ denote the unit sphere in $\mathbb{R}^n$, ω_n its surface area and Ω_n its volume.

For $x = (x_1, \ldots, x_n) \in \mathbb{R}^n$, $n > 1$, we consider the transformation given by

$$x_1(r, \varphi_1, \ldots, \varphi_{n-1}) = r \cos \varphi_1$$
$$x_2(r, \varphi_1, \ldots, \varphi_{n-1}) = r \sin \varphi_1 \cos \varphi_2$$
$$x_k(r, \varphi_1, \ldots, \varphi_{n-1}) = r \sin \varphi_1 \sin \varphi_2 \ldots \sin \varphi_{k-1} \cos \varphi_k, \quad 2 \leq k \leq n - 2$$
$$x_n(r, \varphi_1, \ldots, \varphi_{n-1}) = r \sin \varphi_1 \sin \varphi_2 \ldots \sin \varphi_{n-1}$$

where $0 \leq \varphi_k \leq \pi$, $k = 1, \ldots, n - 2$, $0 \leq \varphi_{n-1} \leq 2\pi$, $r = |x|$ and $x' = (\varphi_1, \ldots, \varphi_{n-1}) \in \Sigma$.

The Jacobian associated with the above transformation is equal to $r^{n-1}(\sin \varphi_1)^{n-2}(\sin \varphi_2)^{n-3} \ldots (\sin \varphi_{n-2})$, thus if f is integrable in $\mathbb{R}^n$,

$$\int_{\mathbb{R}^n} f(x)dx = \int\int\cdots\int\int r^{n-1}(\sin\varphi_1)^{n-2}(\sin\varphi_2)^{n-3}\cdots\sin\varphi_{n-2}f(rx')\, dr d\varphi_1 \cdots d\varphi_{n-1}$$

We write

$$\int_0^{\pi}\int_0^{\pi}\cdots\int_0^{2\pi}(\sin\varphi_1)^{n-2}\cdots\sin\varphi_{n-2}d\varphi_1\cdots d\varphi_{n-1} = \int_{\Sigma} dx'$$

and thus

$$\int_{\mathbb{R}^n} f(x)dx = \int_{\Sigma}\int_0^{\infty} r^{n-1}f(rx')drdx' \tag{2.1}$$

Here dx' is called the surface area element on Σ.

Lemma 2.1. For $n > 1$, $\omega_n = 2\pi^{n/2}(\Gamma(n/2))^{-1}$ and $\Omega_n = \pi^{n/2}(\Gamma(1+n/2))^{-1}$.

Proof. By Fubini's theorem,

$$I = \int_{\mathbb{R}^n} \exp(-|x|^2)dx = \prod_{k=1}^{n}\int_{-\infty}^{\infty}\exp(-x_k^2)dx_k$$

$$= \left(\int_{-\infty}^{\infty}\exp(-t^2)dt\right)^n = \pi^{n/2}$$

since $(\int_{-\infty}^{\infty}\exp(-t^2)dt)^2 = \int\int_{\mathbb{R}^2}\exp(-|x|^2)dx = \int_0^{2\pi}\int_0^{\infty} re^{-r^2}drd\varphi = \pi$.

On the other hand, by (2.1),

$$I = \int_\Sigma \int_0^\infty \exp(-r^2) r^{n-1} dr dx'$$

$$= \int_\Sigma dx' \int_0^\infty \exp(-r^2) r^{n-1} dr$$

$$= \omega_n \int_0^\infty \exp u \cdot u^{n/2-1} du = 2^{-1} \omega_n \Gamma(n/2)$$

As

$$\Omega_n = \int_{|x| \leq 1} dx = \int_\Sigma \int_0^1 r^{n-1} dr dx' = \omega_n \cdot (1/n)$$

the thesis follows, since $u\Gamma(u) = \Gamma(u + 1)$. ∇

Observe that for any $x \in \mathbb{R}^n$ and $a > 0$ the area of the sphere $S_a(x)$ of center x and radius a is equal to $a^{n-1}\omega_n$ and its volume is equal to $a^n \Omega_n = a^n \omega_n / n$.

Lemma 2.2. The function $f(x) = |x|^\lambda$ is integrable in any neighborhood of the origin if and only if $\lambda > -n$ and is integrable in the complement of any neighborhood of the origin if and only if $\lambda < -n$.

Proof. For a fixed $\lambda \neq -n$ and $0 < a < b < \infty$, we have

$$\int_{a<|x|<b} |x|^\lambda dx = \int_\Sigma \int_a^b r^{n-1+\lambda} dr$$

$$= \omega_n (n + \lambda)^{-1} (b^{n+\lambda} - a^{n+\lambda})$$

and the thesis follows. ∇

Furthermore, $\int_\Sigma f(x' \cdot y') dx'$ is independent of $y' \in \Sigma$ if $x' \cdot y'$ stands for the scalar product of x' and y' and f is any function of one variable.

3. THE SPACES C AND L AND THEIR DUALS

Given a set A, $C(A)$ stands for the class of continuous functions on A, $C_b(A)$ for those that are continuous and bounded, $C^m(A)$ for those that are continuous and have continuous derivatives up to the order m and $C^\infty(A)$ for those that have continuous derivatives of all orders. Likewise, $C_0(A)$, etc., stand for the subsets of the above where the elements are functions with compact support. $C_\infty(\mathbb{R}^n) = \{f \in C(\mathbb{R}^n) : \lim_{|x|\to\infty} f(x) = 0\}$ is the space of continuous functions in $\mathbb{R}^n$ that vanish at infinity, and $\mathcal{S} = \mathcal{S}(\mathbb{R}^n) = \{f \in C^\infty(\mathbb{R}^n) : \sup_x |x^\alpha(\partial^\beta f)(x)| < \infty$ for all $\alpha, \beta\}$ is the Schwartz space. Thus $C_0^\infty \subset \mathcal{S} \subset C_\infty$.

$L^p = L^p(\mathbb{R}^n)$, $1 \leq p \leq \infty$, is the space of (equivalence classes of) measurable complex valued functions such that $\|f\|_p = (\int |f(x)|^p dx)^{1/p}$ is finite. L^p is a Banach space with the norm $\|\cdot\|_p$ for $1 \leq p < \infty$. Thus the imbedding $L^1 \subset \mathcal{M}$ is continuous in the respective norms.

$L^\infty = L^\infty(\mathbb{R}^n)$ is the space of (classes of equivalence of) measurable (essentially) bounded functions, normed by $\|f\|_\infty = (\text{ess})\sup_x |f(x)|$. The space $C_b, C_0, C_0^\infty, C_\infty$ can also be normed by $\|\cdot\|_\infty$ and so are Banach subspaces of L^∞. C_0 <u>is dense in L^p for all</u> $1 \leq p < \infty$ <u>and in</u> $C_\infty \subset L^\infty$ <u>(but not in</u> L^∞).

A function f is called <u>simple</u> if $f(x) = \Sigma_{k=1}^m c_k \chi_{A_k}(x)$, where the A_k's are measurable sets. If their measures are finite, f is called <u>elementary</u> and if the A_k's are compact intervals, f is called a <u>step function</u>. The classes of elementary functions and of step functions are dense in all L^p, $1 \leq p < \infty$ (but not in L^∞, since every elementary function vanishes outside a set of finite measure and cannot approximate $f(x) \equiv 1.$).

The following is a fundamental property of L^p functions.

Proposition 3.1. Every function $f \in L^p(\mathbb{R}^n)$, $1 \leq p < \infty$, is continuous in the L^p norm, i.e. $\|\tau_h f - f\|_p \to 0$ as $|h| \to 0$.

Proof. As $\|\tau_h f - f\|_p^p = \int |f(x - h) - fx|^p dx$, the thesis immediately holds for $f \in C_0$. For a general $f \in L^p$, $1 \leq p < \infty$, and $\varepsilon > 0$, there is a $g \in C_0$ such that $\|f - g\|_p < \varepsilon$, and so

$$\|\tau_h f - f\|_p < \|\tau_h f - \tau_h g\|_p + \|\tau_h g - g\|_p + \|g - f\|_p < 2\varepsilon + \|\tau_h g - g\|_p$$

and the proposition holds. ∇

Similarly, if $f \in L^p$, $1 \leq p < \infty$, then $\|\tau_h f + f\|_p$ tends to $2^{1/p}\|f\|_p$ as $|h| \to \infty$, since $\int |f(x + h) + f(x)|^p dx = \int |f(x + h)|^p dx + \int |f(x)|^p dx$ for f of compact support, $|h|$ large.

Given a sequence $\{f_k\}$ of L^p functions, for $1 \leq p \leq \infty$, we say that

(1) $f_k \to f$ pointwise a.e.
    if $\lim_{k\to\infty} f_k(x)$ exists a.e. and is equal to $f(x)$

(2) $f_k \to f$ in the norm of L^p
    if $\|f_k - f\|_p \to 0$ as $k \to \infty$.

If $p = \infty$, convergence in the norm of L^∞ coincides with uniform convergence.

Pointwise convergence a.e. does not imply convergence in the norm or vice versa, for functions in L^p, $p < \infty$, but convergence in the norm implies the existence of a subsequence of the original sequence that converges pointwise a.e. . Also if $f_k(x) \uparrow f(x)$ (i.e., $0 \leq f_1(x) \leq f_2(x) \leq \ldots$ and $f(x) = \lim f_k(x)$) a.e., then $\|f_k - f\|_p \to 0$ by the theorem of Beppo Levi, and if $f_k(x) \to f(x)$ a.e. and there is a majorant in L^p for the sequence (i.e., $|f_k(x)| < F(x)$ a.e. with $F \in L^p$) then $\|f_k - f\|_p \to 0$ by the Lebesgue dominated convergence theorem.

Given p, $1 \leq p \leq \infty$, p' will always stand for its conjugate index such that $1/p + 1/p' = 1$ (we use the convention $1/\infty = 0$).

There exist $f, g \in L^1$ for which $fg \notin L^1$, but if $f \in L^1$ and $g \in L^\infty$, then $fg \in L^1$ always (in particular, if $g(x) = e^{ix}$). More generally, Hölder's inequality asserts that, if $f \in L^p$ and $g \in L^{p'}$, $1 \leq p \leq \infty$, then $fg \in L^1$ and

$$\left| \int f(x)g(x)dx \right| \leq \|f\|_p \|g\|_{p'}$$

We write $I_f(g) = \langle f, g\rangle = \int f(x)g(x)dx$ and $I_\mu(\varphi) = \langle \mu, \varphi\rangle = \int \varphi(x)d\mu$, so that if $f \in L^p$ and $g \in L^p$, $|\langle f, g\rangle| \leq \|f\|_p \|g\|_{p'}$, and if $\mu \in \mathcal{M}$, $\varphi \in L^\infty$, $|\langle \mu, \varphi\rangle| \leq \|\varphi\|_\infty \|\mu\|$.

The dual spaces (i. e., the spaces of the bounded linear functionals) of the Banach spaces L^p are characterized by the F. Riesz representation theorem, which asserts that $I \in (L^p)'$ for any p, $1 \leq p < \infty$ if and only if there exists a $g \in L^{p'}$ such that $I(f) = \langle f, g\rangle$ for all $f \in L^p$. In this sense we write $(L^p)' = L^{p'}$ for $1 \leq p < \infty$. Note that as $(p')' = p$, then $(L^{p'})' = L^p$ for $1 < p < \infty$, but while $(L^1)' = L^\infty$, $(L^\infty)' \neq L^1$. The F. Riesz representation theorem also asserts that $I \in (C_\infty)'$ if and only if there exists a $\mu \in \mathcal{M}$ such that $I(f) = \langle \mu, f\rangle$ for all $f \in C_\infty$.

We shall repeatedly use the fact that if f is a measurable function such that $\sup \{ |\langle f, g\rangle| : g \text{ simple}, \|g\|_p = 1\} < \infty$, then $f \in L^{p'}$, for $1 < p \leq \infty$, and $\|f\|_{p'} = \sup |\langle f, g\rangle|$. Similarly, $\|\mu\| = \sup |\langle \mu, \varphi\rangle|$ and, if $\mu \geq 0$, then $\|\mu\| = \langle \mu, 1\rangle$.

Definition 3.1. A sequence of functions $\{f_k\} \subset L^p$, $1 < p < \infty$, converges weakly to f if $\langle f_k, g\rangle \to \langle f, g\rangle$ for all g $L^{p'}$. Similarly, $\{\mu_k\} \subset \mathcal{M}$ converges weakly-* to μ if $\langle \mu_k, \varphi\rangle \to \langle \mu, \varphi\rangle$ for all $\varphi \in C_\infty$.

The Helly-Bray theorem asserts that if a given sequence in $\mathcal{M}$ is uniformly bounded (in the norm) then there exists a measure μ and a subsequence such that the subsequence converge weakly-* to μ. (Equivalently, the unit sphere of $\mathcal{M}$ is weakly-* compact, a version of the Alaoglu-Bourbaki theorem; see Section 4).

An essential tool in the theory of L^p spaces is the following

Proposition 3.2 (The Minkowski integral inequality). Let $(\mathcal{X}, \mu)$, $(\mathcal{Y}, \nu)$ be two measure spaces of positive (σ-finite) measure and $F(x, y)$ a function defined in $\mathcal{X} \times \mathcal{Y}$ and $\mu \otimes \nu$-measurable. If $F(\cdot, y) \in L^p(\mathcal{X}, \mu)$, $1 \leq p \leq \infty$, for a.e. y fixed and, furthermore, $\int_{\mathcal{Y}} \|F(\cdot, y)\|_{p,\mu} d\nu(y) = A < \infty$, then $\int_{\mathcal{Y}} F(x, y) d\nu(y)$ converges for a.e. x and

$$\left\| \int_{\mathcal{Y}} F(x, y) d\nu(y) \right\|_{p,\mu} \leq \int_{\mathcal{Y}} \|F(\cdot, y)\|_{p,\mu} d\nu(y) \tag{3.1}$$

Proof. The conclusion is obvious if $p = \infty$. If $p < \infty$, let p' be its conjugate index, so that $L^{p'}$ is the dual of L^p. Let $f(x) = \int_{\mathcal{Y}} |F(x, y)| d\nu(y)$. Since it is a μ-measurable function (prove it), we need only show that $\sup_g |\langle f, g\rangle|$ is finite when taken over all g simple with $\|g\|_{p',\mu} = 1$. Since $|\langle f, g\rangle| \leq \int |f(x)g(x)| d\mu(x) \leq \int_{\mathcal{X}} (\int_{\mathcal{Y}} |F(x, y)| d\nu(y)) |g(x)| d\mu(x)$, Fubini's theorem and Hölder's inequality yield

$$\begin{aligned} |\langle f, g\rangle| &\leq \int_{\mathcal{Y}} \int_{\mathcal{X}} |F(x, y)| \, |g(y)| d\mu(x) d\nu(y) \\ &\leq \int_{\mathcal{Y}} \|F(\cdot, y)\|_{p,\mu} \|g\|_{p',\mu} d\nu(y) = A \end{aligned}$$

so that $f \in L^p$ and (3.1) is satisfied. ∇

Another important inequality bears the name of Young. Let $t = \varphi(s)$, $s \geq 0$, be nonnegative, increasing, equal to 0 at the origin and tending to infinity with s. Let $s = \psi(t)$, $t \geq 0$, be the inverse function. It is geometrically easy to see that for $a, b \geq 0$ we have the Young inequality:

$$ab \leq \int_0^a \varphi(s)ds + \int_0^b \psi(t)dt \tag{3.2}$$

The special case $\varphi(s) = s^{p-1} (p>1)$, $\psi(t) = t^{1/(p-1)}$, gives

$$ab \leq (a^p / p) + (b^{p'} / p')$$

an inequality which easily leads to Hölder's inequality. If $\varphi(s) = \log(s + 1)$, then $\psi(t) = e^t - 1$ and we have

$$ab \leq a \log(a + 1) + e^b \tag{3.3}$$

4. HILBERT AND BANACH SPACES

We assume the reader is familiar with the notions of linear normed and Banach spaces. Recall that in a normed space a linear operator is continuous if and only if it is bounded.

We will make frequent use of the following result on extension of operators.

Proposition 4.1. Let E and F be two Banach spaces and D a dense subspace of F. If $T_0 : D \to F$ is a bounded linear operator, such that $\|T_0 x\|_F \leq C \|x\|_E$ for all $x \in D$, then there is a unique bounded linear operator $T : E \to F$ such that (i) $T = T_0$ on D, (ii) $\|Tx\|_F \leq C \|x\|_E$ for all $x \in E$. T is called the extension of T_0.

Proof. Since D is dense in E, for each $x \in E$ there is a sequence $\{x_k\} \subset D$ such that $\|x_k - x\|_E \to 0$ as $k \to \infty$. The sequence $\{T_0 x_k\}$ is a Cauchy sequence in F since $\|T_0 x_k - T_0 x_j\|_E = \|T_0(x_k - x_j)\|_F \leq C\|x_k - x_j\|_E \to 0$ as $k, j \to \infty$. F is complete, hence $\{T_0 x_k\}$ tends to a limit $y \in F$. Let us define $Tx = y$. Then $Tx = T_0 x$ if $x \in D$, and $\|Tx\|_F \leq \|Tx - Tx_k\|_F + \|Tx_k\|_F < \varepsilon + C\|x_k\|_E$. This proves part (ii). ∇

Furthermore, let $\{T_k\}$ be a sequence of operators such that $\|T_k\| < C$ for all k. If $\{T_k x\}$ converge in a dense subspace D, then $\{T_k x\}$ converges for all $x \in E$.

If $T : E \to F$ is a bounded operator, its norm as an operator from E to F is given by

$$\|T\| = \|T\|_{E,F} = \inf\{C : \|Tx\|_F \leq C\|x\|_E, \forall x \in E\}$$

A Hilbert space H is a Banach space with a scalar product, such that $\|x\| = (x, x)^{1/2}$, where (x, y) denotes the scalar product of x and y. Two vectors x and y are orthogonal if $(x, y) = 0$, and we write $x \perp y$ in such case. The orthogonal complement of a set A is $A^\perp = \{x \in H : (x, a) = 0 \text{ for } a \in A\}$. The following important results will be often used in the text.

Proposition 4.2. The orthogonal complement of a closed proper subspace $S \subsetneq H$ contains a nonzero vector.

Corollary 4.3. For any subset $A \subset H$, the set $A \oplus A^\perp$ is dense in H.

An operator $T : H_1 \to H_2$ between Hilbert spaces is an isometry if $(Tx, Ty) = (x, y)$ for all $x, y \in H_1$. An isometry is called a unitary operator if it is surjective.

Every unitary operator U has an inverse operator $U^{-1} : H_2 \rightarrow H_1$ that coincides with the adjoint U^* of U defined by $(Ux, y) = (x, U^* y)$. U^{-1} is also unitary.

It was already remarked that the Lebesgue spaces L^p, $1 \leq p \leq \infty$, are Banach spaces. But only L^2 is a Hilbert space, where the scalar product is given by $(f, g) = \int f\bar{g} = \langle f, \bar{g} \rangle$ for all $f, g \in L^2$.

A (complex) linear algebra is a (complex) linear space $\mathscr{A}$ in which a product is defined such that, for all $x, y, z \in \mathscr{A}$, $a, b \in \mathbb{C}$,

$$x(yz) = (xy)z$$

$$(ax)y = x(ay) = a(xy)$$

$$x(ay + bz) = a(xy) + b(xz)$$

If a linear algebra $\mathscr{A}$ is equipped with a norm under which it is a Banach space, $\mathscr{A}$ is a Banach algebra if

$$\|xy\| \leq \|x\| \, \|y\|$$

for all $x, y \in \mathscr{A}$. If furthermore $\mathscr{A}$ has an identity for multiplication, $e = ex = xe$ for all $x \in \mathscr{A}$, then $\|e\| = 1$.

The dual of a Banach space E is also a Banach space E' with the norm $\|f\|_{E'} = \sup_{x \in E} |f(x)|$, so we can consider in E' the topology given by the norm. E' induces a weak topology on E such that

$$x_n \rightarrow x \text{ weakly} \quad \text{if} \quad f(x_n) \rightarrow f(x) \text{ for all } f \in E'$$

Since $E \subset (E')'$ but $E \neq (E')'$ in general, there is a weaker topology induced on E' by E, that is called the weak-* topology:

$$f_n \rightarrow f \text{ weakly-*} \quad \text{if} \quad f_n(x) \rightarrow f(x) \text{ for all } x \in E$$

The Alaoglu-Bourbaki theorem asserts that the unit sphere of E' is compact in the weak-* topology. (It cannot be compact in the norm topology unless E is finite dimensional. Compare this result with the Helly-Bray theorem in Section 3.).

Generally, let E be a linear space (topological or not) and F the set of the linear functionals on E (not necessarily continuous). We assume that F is also a linear space. A topology can be defined on E as follows: a (generalized) sequence x_α in E converges to x if $f(x_\alpha) \to f(x)$ for every $f \in F$. We call this topology the weak topology (E, F). Endowed with this weak topology (E, F), E becomes a topological linear space, whose dual is precisely F. If E is a Banach space, let us take $F = E'$. Then the weak-* topology on E' is the weak topology (F, E) on F. In general E is not the dual of F (in the norm topology) but E is the dual of F with respect to the weak-* topology on F.

A set K is convex if $x, y \in K$ imply $\lambda x + (1 - \lambda)y \in K$ for all $0 \leq \lambda \leq 1$. An element $z \in K$ is called extreme if $z = \lambda x + (1 - \lambda)y$ for $x, y \in K$, $0 \leq \lambda \leq 1$, implies $z = x = y$. This is what happens to the vertices of a polygon. A polygon is determined by its vertices and the same happens for a convex in a topological linear space. More precisely, if F is a topological linear space and $K \subset F$ is a convex and compact subset, then every element of K is the limit of convex combinations $(\lambda_1 e_1 + \cdots + \lambda_k e_k$, $0 \leq \lambda_j \leq 1$, $j = 1, \ldots, k$, $\lambda_1 + \cdots + \lambda_k = 1)$ of extreme points $e_1, \ldots, e_k$ of K. In other words, K is contained in the closure of the convex hull of its extreme points. This is the Krein-Milman theorem (for a proof see [1], p. 179).

A frequent use of the Krein-Milman theorem is as follows. Let E be a Banach space, E' its dual space and $F = E'_w$, the space E' endowed with the weak-* topology. By the Alaoglu-

Bourbaki theorem, every set $K \subset E'$ that is bounded (in the norm topology of E') and closed (in the weak-* topology) is compact in F. If K is also convex, K is contained in the closure (in F) of the convex hull of its extreme points.

* What follows, up to the end of Section 4, will not be used until the last section of Chapter 6. For proofs see for instance [2].

Again, let E be a Banach space and F its dual. If the elements of F are real linear functionals we have the following corollaries (stemming from the Hahn-Banach theorem) which will be invoked at the end of the book:

Proposition 4.4. Let F_w denote the topological linear space F under the weak-* topology. If K is a convex set of F_w and $g \in F_w$ then $g \in$ closure of K if and only if there exists $z \in E$ such that $f(z) < 1$ for all $f \in K$ and $g(z) > 1$.

Proposition 4.5 (The polar theorem). If $K \subset F_w$ is convex, $K^0 = \{x \in E : f(x) \leq 1 \text{ for all } f \in K\}$, $K^{00} = \{g \in F_w : g(x) \leq 1, \text{ for all } x \in K^0\}$, then K^{00} coincides with the closure of K.

In particular, if K is a cone (i.e., K is a convex and if $f \in K$ then $\lambda f \in K$ for all $\lambda > 0$) and if $x \in E$ is such that $f(x) \leq 1$ for all $f \in K$, then $\lambda f(x) \leq 1$ for all $\lambda > 0$ and all $f \in K$, so $f(x) \leq 0$ for all $f \in K$. Therefore, we have

Corollary 4.6. If $K \subset F_w$ is a cone and $g \in F_w$, then g belongs to closure of K if and only if for all $x \in E$, $f(x) \geq 0$ for all $f \in K$ implies $g(x) \geq 0$.

5. THE THREE LINES THEOREM

We only assume the basic notions of complex variable theory, such as the maximum principle for analytic functions which asserts that if $f(z)$ is analytic in a bounded domain D of the complex plane, then $|f(z)|$ cannot attain its maximum at an interior point of D unless $f(z)$ reduces to a constant.

In Chapter 4 we shall need an extension of this principle to an unbounded domain.

Now let $D = \{z \in \mathbb{C} : z = x + iy, x, y \in \mathbb{R}, 0 \leq x \leq 1\}$ be a strip in the complex plane delimited by the lines $\Delta_0 = \{z \in \mathbb{C} : z = iy, y \in \mathbb{R}\}$ and $\Delta_1 = \{z \in \mathbb{C} : z = 1 + iy, y \in \mathbb{R}\}$.

Proposition 5.1 (Phragmén-Lindelöf maximum principle). Let $f(z)$ be a function that is analytic in the interior D^0 of D, and continuous and bounded in all of D. If $|f(z)| \leq M$ for all $z \in \Delta_0 \cup \Delta_1$, then $|f(z)| \leq M$ for all $z \in D^0$. Furthermore if $|f(z_0)| = M$ for some $z_0 \in D^0$ then f reduces to a constant.

Proof. If $f(z) = f(x + iy) \to 0$ as $y \to \infty$ uniformly in $0 \leq x \leq 1$, the proof is easy. In this case, there is a large positive N such that $|f(x + iy)| \leq M$ for $|y| \geq N$. On the other hand, inside the rectangle $\{(x, y) : 0 \leq x \leq 1, |y| \leq N\}$ the estimate $|f(x + iy)| \leq M$ follows from the maximum principle for a bounded region. In the general case, define, for each k,

$$f_k(z) = f(z)\exp(z^2/k) = f(z)\exp((x^2 - y^2)/k)\exp(2ixy/k)$$

Since $|f_k(z)| = |f(z)|\exp((x^2 - y^2)/k)$, $\lim_{y \to +\infty} f_k(x + iy) = 0$ uniformly in $0 \leq x \leq 1$ and if $z \in \Delta_0 \cup \Delta_1$ (or $x = 0$, $x = 1$), then $|f_k(z)| \leq M \exp(1/k)$. By the above case, $|f_k(z)| \leq M \exp(1/k)$ for all $z \in D^0$. Letting $k \to \infty$ proves the proposition. ∇

Proposition 5.2 (The three lines theorem). Assume that $f(z)$ is analytic in D^0 and bounded and continuous in D. If $|f(z)| \leq M_0$ for all $z \in \Delta_0$ and $|f(z)| \leq M_1$ for all $z \in \Delta_1$, then $|f(z)| \leq$

$\leq M_0^{1-\theta} M_1^{\theta}$ for all $z \in \Delta_\theta = \{z : z = \theta + iy,\ y \in \mathbb{R}\}$ and $0 < \theta < 1$. In particular, $|f(\theta)| \leq M_0^{1-\theta} M_1^{\theta}$.

Proof. Let $g(z) = f(z)e^{az}$ for a real, such that $|g(z)| = |f(z)|e^{ax}$, $x = \text{Re } z$. Then $|g(iy)| < M_0$ and $|g(1+iy)| < M_1 e^a$, for all $y \in \mathbb{R}$. By Proposition 5.1, $|g(\theta + iy)| < \max\{M_0, M_1 e^a\}$. Choosing a such that $M_0 = M_1 e^a$, or $e^a = M_0/M_1$, we obtain

$$|g(\theta + iy)| = e^{a\theta} |f(\theta + iy)| \leq M_0$$

or

$$|f(\theta + iy)| < e^{-a\theta} M_0 = (M_1/M_0)^{\theta} M_0 = M_0^{1-\theta} M_1^{\theta} \qquad \nabla$$

REFERENCES

1. H. L. Royden, Real Analysis, MacMillan, New York, 1964.
2. K. Yosida, Functional Analysis, Springer-Verlag, New York-Heidelberg-Berlin, 1974.

Chapter 1

CONVOLUTION UNITS AND THE GROUP ALGEBRA

1. CONVOLUTION OF FUNCTIONS

In L^1, the space of integrable functions, in addition to the linear operations there is a product operation which endows it with a structure richer than the Banach space structure, making it a Banach algebra. This operation is called convolution.

Definition 1.1. Given two (Lebesgue) measurable functions f and g defined in $\mathbb{R}^n$, their convolution $f * g$ is well defined, and $f * g = h$ if, for fixed x, $f(x - y)g(y)$ is an integrable function of y, a.e. in x, and

$$h(x) = \int_{\mathbb{R}^n} f(x - y)g(y)dy \tag{1.1}$$

The convolution operation is of great significance in analysis. Some of the reasons are:

(a) $L^1 = L^1(\mathbb{R}^n, dx)$ is not an algebra with respect to the ordinary (pointwise) product, i.e., $f, g \in L^1$ does not generally imply that $f \cdot g \in L^1$(e.g., $f(x) = g(x) = x^{-1/2}$ for $x \in (0, 1)$ and zero otherwise). Nevertheless, the product becomes integrable if one of the two functions is translated as in (1.1). Then $f, g \in L^1$ implies $f * g \in L^1$, as we shall see later, and the convolution product makes L^1 into an algebra. This fact is basic in harmonic analysis.

(b) If just one of the two functions f, g is differentiable (up to a certain order), the convolution $f * g$ will also be differentiable (up to the same order). More generally, the convolution is an operation that "ameliorates" the functions, in the sense that the product gets the "good" properties of each factor. From this fact stems the method of regularization of functions which consists in approximating general functions by smooth ones.

(c) The operation of convolution interacts with other important analytical operations in a specific way. For instance, an important fact of harmonic analysis that will be studied in the next chapter is that the Fourier transform turns convolution products into ordinary products and viceversa.

(d) In probability theory, the law of the sum of random variables is expressed in terms of convolutions.

Before dealing formally with the properties of convolutions let us recall some facts that gave rise to formula (1.1).

With each sequence $f = \{f_n\}_{n \in \mathbb{Z}}$ with finite support (i.e., $f_n = 0$ but for a finite number of the n) can be associated in a 1-1 way a polynomial $F(z) = \Sigma f_n z^n$ which is a function of the complex variable z; in writing, $f \sim F$. If $f \sim F$, $g \sim G$ and $h \sim H$, then $H(z) = F(z) + G(z)$, for all z, if and only if $h_n = f_n + g_n$, for each n, and, in this case, we write $h = f + g$. Similarly, $H(z) = F(z) \cdot G(z)$, for all z, if and only if $h_n = \Sigma_k f_{n-k} g_k$, for each n, and, in this case, we write $h = f * g$. Furthermore, if $f = \{f_n\}$, $g = \{g_n\}$ and $h = \{h_n\}$ are sequences without restrictions on their supports, we still write $h = f * g$ whenever the series $h_n = \Sigma_k f_{n-k} g_k$ converges absolutely for each n. So, changing the sequences f, g to functions of real variable, $f = f(x)$, $g = g(x)$, $x \in (-\infty, \infty)$, the definition of $h = f * g$ becomes (1.1) (for the case $\mathbb{R}^n = \mathbb{R}^1$). But given two general sequences or functions of infinite supports, the series or integral defining $f * g$ can be divergent, so we must find conditions on f and g that will imply convergence of (1.1). Such criteria are given in Theorems 1.1 and 1.2 and in the exercises that follow.

From now on, unless stated otherwise our functions are defined on $\mathbb{R}^n$, $n \geq 1$ and L^1 is $L^1(\mathbb{R}^n, dx)$, where dx is the n-dimensional Lebesgue measure. For each $y \in \mathbb{R}^n$, let τ_y denote the translation operator such that for every function f,

$$\tau_y f(x) = f(x - y) \tag{1.2}$$

We know (see Chapter 0, (1.1)) that the Lebesgue measure dx is characterized by translation invariance: for every $f \in L^1$ and every $y \in \mathbb{R}^n$ we have

$$\int_{\mathbb{R}^n} \tau_y f(x) = \int_{\mathbb{R}^n} f(x)dx \tag{1.3}$$

and, in particular, for every $f \in L^p$,

$$\int_{\mathbb{R}^n} |\tau_y f(x)|^p dx = \int_{\mathbb{R}^n} |f(x)|^p dx$$

or

$$\|\tau_y f\|_p = \|f\|_p \tag{1.3a}$$

As was already pointed out, the product $f(x) \cdot g(x)$ of two integrable functions f and g can be nonintegrable but, thanks to the invariance property expressed in (1.3), $\tau_y f(x) \cdot g(x)$ will be integrable for a.e. y. More precisely:

Theorem 1.1. If $f, g \in L^1$, then $f(x - y)g(y)$ is integrable as a function of y, for almost all $x \in \mathbb{R}^n$, i.e., the convolution $h = f * g$ exists as defined in (1.1) and, furthermore, $h \in L^1$ and

$$\|h\|_1 \leq \|f\|_1 \|g\|_1 \tag{1.4}$$

Proof. Since

$$\left|\int h(x)dx\right| \leq \int\int |f(x - y)|\,|g(y)|dydx$$

the assertion follows from Fubini's theorem (see Exercise 1.1). ∇

This is a particular case of

Theorem 1.2 (Young's inequality). If $f \in L^p$, $1 \leq p \leq \infty$, $g \in L^1$ then $h = f * g$ is well defined (i.e., the integral in (1.1) converges absolutely a.e. in x) and belongs to L^p. Moreover,

$$\|f * g\|_p \leq \|f\|_p \|g\|_1 \tag{1.4a}$$

Proof. Assume first $f \geq 0$, $g \geq 0$ and let $d\mu = g(y)dy$, $F(x, y) = f(x - y)$. Since f is a measurable function in $\mathbb{R}^n$, F will be a measurable function in $\mathbb{R}^n \times \mathbb{R}^n$ (see Exercise 1.1). By (1.3a) we have that

$$\int \|F(.\,, y)\|_p d\mu(y) = \int \left(\int (f(x - y))^p dx\right)^{1/p} g(y)dy$$

$$= \|f\|_p \int g(y)dy = \|f\|_p \|g\|_1 < \infty \tag{1.5}$$

By Minkowski's integral inequality (see Chapter 0, Proposition 3.2), (1.5) implies that

$$h(x) = \int F(x, y)d\mu(y) = \int f(x - y)g(y)dy$$

is finite for a.e. x, so that $h = f * g$ is well defined and belongs to L^p, since

$$\|h\|_p = \left(\int \left(\int f(x - y)g(y)dy\right)^p dx\right)^{1/p}$$

$$\leq \int \left(\int (f(x - y))^p dx\right)^{1/p} g(y)dy = \|f\|_p \cdot \|g\|_1$$

If f and g are not positive then $f = f^+ - f^-$ and $g = g^+ - g^-$, for $f^+ = \max(f, 0)$, $f^- = -\min(f, 0)$ and g^+, g^- likewise. Since f^+, f^-, g^+, g^- are positive integrable functions and

$$f * g = (f^+ * g^+) - (f^+ * g^-) - (f^- * g^+) + (f^- * g^-),$$

$f * g$ is still well defined and belongs to L^p. Moreover, since $h \leq h_1 = |f| * |g|$, we get

$$\|h\|_p \leq \|h_1\|_p \leq \| |f| \|_p \| |g| \|_1 = \|f\|_p \|g\|_1$$

∇

Exercise 1.1. Show that $F(x, y) = f(x - y)$ is a measurable function in $\mathbb{R}^n \times \mathbb{R}^n$ for every measurable function f in $\mathbb{R}^n$. (Hint: Show it for characteristic functions of intervals and then apply the standard limit procedure.)

Remark 1.1. In the proof of Theorem 1.2 the invariance property (1.3) of the Lebesgue measure is crucial. Therefore it is essential that the integral in (1.1) be taken with respect to the ordinary Lebesgue measure dx and extended to the whole space

$\mathbb{R}^n$. The theory we are developing does not hold for other measures in $\mathbb{R}^n$. More generally, in any locally compact topological group there is a measure invariant under group translations, called the Haar measure of the group (Lebesgue measure is the Haar measure of $\mathbb{R}^n$) and convolutions of functions defined on the group can be introduced.

Corollary 1.3. The convolution operation is defined for every $f, g \in L^1$ and satisfies the following properties:

(a) it is commutative: $f * g = g * f$;

(b) it is associative: $f * (g * k) = (f * g) * k = f * g * k$;

(c) it is linear: $(af + bg) * k = a(f * k) + b(g * k)$, $a, b \in \mathbb{C}$;

(d) it is continuous: $\|f * g\|_1 \leq \|f\|_1 \|g\|_1$.

Proof of (a). Changing variables

$$\int f(x - y)g(y)dy = \int f(y)g(x - y)dy$$

since the integrals are taken on the whole $\mathbb{R}^n$ and in the proof of Theorems 1.1, 1.2, the change in the order of the variables of $F(x, y)$ is justified. ∇

Exercise 1.2. Prove (b) - (c) of Corollary 1.3.

Corollary 1.3 asserts that L^1 is a Banach algebra with respect to the operations of sum, multiplication by scalars and convolution. In Section 2 we shall return to the consideration of L^1 as a Banach algebra, but now we proceed with the study of convolution.

Remark 1.2. Theorem 1.2 is a special case of the generalized Young inequality (see Chapter 4, Theorem 1.6) which asserts that if $f \in L^p$, $g \in L^q$, $1 \leq p, q \leq \infty$, $1/p + 1/q \geq 1$, then $h = f * g \in L^r$, $1/r = 1/p + 1/q - 1$, and $\|f * g\|_r \leq \|f\|_p \|g\|_q$.

Note that Theorem 1.2 states in particular that the convolution operator K defined by $K : f \to Kf = f * k$ for a fixed $k \in L^1$ (k is called the kernel of the operator K) is continuous as an operator from L^p to L^p, $1 \leq p \leq \infty$, with norm $\|K\|_p \leq \|k\|_1$.

Exercise 1.3. If $f \in L^p$, $g \in L^{p'}$, $1/p + 1/p' = 1$, $1 \leq p \leq \infty$, then $f * g$ is not only in L^∞ but is also uniformly continuous. Prove it.

Exercise 1.4. Prove that if $\operatorname{supp} f \subset A$, $\operatorname{supp} g \subset B$, then $\operatorname{supp} f * g \subset A + B$.

Exercies 1.5. Prove that if $f \in L^p$, $g \in L^{p'}$, $1/p + 1/p' = 1$, $1 < p < \infty$, then $f * g \in C_\infty$. (Hint: Prove it first for functions in C_0, using Exercises 1.3 and 1.4.)

Exercise 1.6. Prove that the convolution operator K defined in L^1 by $Kf = f * k$, $k \in L^1$ fixed, commutes with translations, i.e., $K\tau_y = \tau_y K$ for every $y \in \mathbb{R}^n$.

Exercise 1.7. Let C_b^m be the class of C^m functions that are bounded and with bounded (and continuous) derivatives up to the order m. Prove that, for a fixed $k \in L^1$, the corresponding convolution operator K transforms $f \in C_b^m$ into $Kf \in C_b^m$. (Hint: Use Lebesgue dominated convergence theorem to apply the C_b^m condition under the integral sign).

The last exercise shows that convolving a function can make it smoother. This is true as a general principle of regularization that we shall state precisely. For that purpose, fix a $\phi \in L^1$ with $\int \phi(x)dx = 1$ and let, for each $\varepsilon > 0$,

$$\phi_\varepsilon(x) = \varepsilon^{-n}\phi\left(\frac{x}{\varepsilon}\right) \tag{1.6}$$

It is immediate that the family $\{\phi_\varepsilon\}_{\varepsilon>0}$ satisfies

$$\int \phi_\varepsilon(x)dx = \int \phi(x)dx = 1, \quad \forall \varepsilon > 0 \tag{1.7}$$

As an example of such a family of functions, let us consider $\phi(x) = \chi_{(-1,+1)}(x)$, for $x \in \mathbb{R}^1$. Then for each $\varepsilon > 0$, $\phi_\varepsilon(x) = 1/\varepsilon$ for $|x| < \varepsilon$ and zero elsewhere, and $\lim_{\varepsilon\to 0} \phi_\varepsilon(x) = \delta(x)$ with $\delta(x) = 0$ for $x \neq 0$ and $\delta(0) = \infty$, the Dirac function. (For a more precise statement, see the remark following Proposition 3.2.)

In view of the following result, such a family of functions is called a <u>convolution unit</u> (or an <u>approximation of the identity</u>).

<u>Theorem 1.4.</u> Let $\phi \in L^1$ and assume that $\int\phi = 1$. Let ϕ_ε be as in (1.6). If $f \in L^p$, $1 \leq p < \infty$ (respec. $f \in C_\infty \subset L^\infty$), then $f * \phi_\varepsilon$ tends to f in L^p (respec. in L^∞) for $\varepsilon \to 0$.

<u>Proof.</u>

$$\begin{aligned} f * \phi_\varepsilon(x) - f(x) &= \int f(x-y)\phi_\varepsilon(y)dy - f(x)\int \phi_\varepsilon(y)dy \\ &= \int (f(x-y) - f(x))\phi_\varepsilon(y)dy \\ &= \int (f(x-\varepsilon y) - f(x))\phi(y)dy \\ &= \int (\tau_{\varepsilon y}f(x) - f(x))\phi(y)dy \end{aligned}$$

By Minkowski's integral inequality,

$$\|f * \phi_\varepsilon - f\|_p \leq \int \|\tau_{\varepsilon y}f - f\|_p |\phi(y)|dy = \int_{|y|\leq R} + \int_{|y|>R}$$

As for $f \in L^p$, $\|\tau_{\varepsilon y}f - f\|_p \leq 2\|f\|_p$, and as $\phi \in L^1$, we can choose R so that the last integral be arbitrarily small. For that fixed R and a given $\delta > 0$ we choose ε_0 so that $\|\tau_{\varepsilon y}f - f\|_p < \delta$

for $\varepsilon < \varepsilon_0$ and for all $|y| \leq R$ (see Chapter 0, Prop. 3.1). Therefore the first integral tends to zero with ε and the result is proved.∇

Exercise 1.8. Let $\phi \in L^1$ with $\int \phi = 0$. Prove that $\|f * \phi_\varepsilon\|_p \to 0$ as $\varepsilon \to 0$ for all $f \in L^p$, $1 \leq p < \infty$, or $f \in C_\infty \subset L^\infty$.

Exercise 1.9. Show that

$$\rho(x) = \begin{cases} e^{1/(|x|^2 - 1)} & \text{for } |x| < 1 \\ 0 & \text{for } |x| \geq 1 \end{cases}$$

belongs to C_0^∞, the class of infinitely differentiable functions of compact support.

The last exercise shows that the class C_0^∞ is not empty. In fact, it is dense in L^p, $1 \leq p < \infty$, and in C_∞, as indicated in the following result.

Corollary 1.5 (Regularization of functions). If $f \in L^p$, $1 \leq p < \infty$ or if $f \in C_\infty \subset L^\infty$, then there exist $\{g_\varepsilon\}_{\varepsilon>0} \subset C_0^\infty$ such that g_ε tends to f in L^p or in L^∞, respectively, when $\varepsilon \to 0$.

Proof. For $f \in L^p$, $1 \leq p \leq \infty$, with compact support, we take $g_\varepsilon = f * \rho_\varepsilon$, for a fixed $\rho \in C_0^\infty$. Then $f * \rho_\varepsilon \in C_0^\infty$ (see Exercises 1.4 and 1.7). For general $f \in L^p$, $1 \leq p < \infty$ or $f \in C_\infty$ and $\delta > 0$, there exists a function g with compact support such that $\|f - g\|_p < \delta/2$, $1 \leq p \leq \infty$. Taking $g_\varepsilon = g * \rho_\varepsilon$, with ρ as before, $\|f - g_\varepsilon\|_p \leq \|f - g\|_p + \|g - g_\varepsilon\|_p < \delta$ for δ arbitrary and ε sufficiently small. ∇

Exercise 1.10. Let $f \in L^1$ and assume that $\int f\phi = 0$ for all $\phi \in C_0^\infty$. Prove that $f = 0$ a.e. (Hint: Apply the hypothesis to $\phi(x) = \rho_\varepsilon(x - y)$, $\rho \in C_0^\infty$).

Exercise 1.11. For a fixed $k \in L^1$, let K be the corresponding convolution operator. Prove that the norm $\|K\|^{(1)}$ of K as an operator acting on L^1 is equal to $\|k\|_1$. (Hint: Show that for every $\varepsilon > 0$, $\|K\phi_\varepsilon\|_1 > \|k\|_1 - \varepsilon$ for ϕ_ε as in (1.6).)

2. POINTWISE CONVERGENCE

The approximation of functions given in Theorem 1.4 in the L^p norm can be also obtained in the pointwise convergence sense under additional conditions.

For instance, we have the following pointwise version for the second result in Theorem 1.4:

Exercise 2.1. Let $\phi \in L^1$ be of compact support and $\int \phi = 1$.

If f is uniformly continuous (in particular, if $f \in C_\infty$) then $(f * \phi_\varepsilon)(x)$ tends uniformly to $f(x)$, for all x, as ε tends to zero. (Hint: If for instance $\phi(x) = 0$ for $|x| > 1$ then $(f * \phi_\varepsilon)(x) - f(x) = \int_{|y|<\varepsilon} (f(x - y) - f(x))\phi_\varepsilon(y)dy$).

A deeper result is given by Theorem 2.1 that follows, where, in addition to $\phi \in L^1$, $\int \phi = 1$, it is assumed that if

$$\psi(x) = \sup_{|y| \geq |x|} |\phi(y)| \tag{2.1}$$

then

$$\psi \in L^1 \tag{2.1a}$$

Before stating the theorem observe that from (2.1), (2.1a) it is clear that

(i) ψ is radial: $|x| = |y|$ implies $\psi(x) = \psi(y)$;

(ii) for $|x| = r$, the function $\psi_0(r) = \psi(x)$ is decreasing in $r > 0$;

(iii) $|x|^n\psi(x) \to 0$ when $|x|$ tends to zero or to infinity and, in particular, there exists a constant $A > 0$ such that $|x|^n\psi(x) \leq A$ for all $0 < |x| < \infty$;

(iv) for every $0 < \eta < \infty$ and $1 \leq p \leq \infty$, $\|\chi_\eta \psi_\varepsilon\|_{p'} \to 0$ as $\varepsilon \to 0$, where χ_η is the characteristic function of the set $\{x : |x| > \eta\}$ and $\psi_\varepsilon(x) = \varepsilon^{-n}\psi(x/\varepsilon)$.

In fact, to show (iii), note that by (ii)

$$\int_{r/2<|x|<r} \psi(x)dx = \int_\Sigma dx' \int_{r/2}^{r} \psi_0(\rho)\rho^{n-1}d\rho \geq \Omega_n \psi_0(r)r^n(1 - 2^{-n}) \tag{2.2}$$

and since by (2.1a) we assume that ψ is integrable, the first integral in (2.2) tends to zero as r tends to zero or to infinity. To prove (iv) observe that

$$\|\chi_\eta \psi_\varepsilon\|_{p'}^{p'} = \int_{|x|>\eta} (\psi_\varepsilon(x))^{p'} dx$$

$$= \int_{|x|>\eta} \psi_\varepsilon(x).(\psi_\varepsilon(x))^{p'/p} dx$$

$$\leq (\varepsilon^{-n}\psi_0(\tfrac{\eta}{\varepsilon}))^{p'/p} \int_{|x|>\eta} \psi_\varepsilon(x)dx$$

and by (iii), $(\varepsilon/\eta)^{-n}\psi_0(\eta/\varepsilon)$ tends to zero as $\varepsilon \to 0$ for fixed η.

<u>Theorem 2.1.</u> Let $\phi \in L^1$ with $\int \phi = 1$ and let $\psi(x) = \sup_{|y|\geq|x|} |\phi(y)|$, $\phi_\varepsilon(x) = \varepsilon^{-n}\phi(x/\varepsilon)$, $\varepsilon > 0$.

If $\psi \in L^1$ then, for $f \in L^p$, $1 \leq p \leq \infty$,

$$\lim_{\varepsilon \to 0} f * \phi_\varepsilon(x) = f(x)$$

whenever $x \in \mathscr{L}_f$, the Lebesgue set of f (in particular, almost everywhere).

(Recall the properties of the Lebesgue set $\mathscr{L}_f$ as given in Chapter 0, Section 1.)

Proof. Let us fix $x \in \mathcal{L}_f$ and $\delta > 0$. We can find $\eta > 0$ such that, for all $0 < r < \eta$,

$$r^{-n}\int_{|y|\leq r} |f(x - y) - f(x)|dy < \delta \tag{2.3}$$

Consider, as in the proof of Theorem 1.4, the difference

$$f * \phi_\varepsilon(x) - f(x) = \int_{\mathbb{R}^n} (f(x - y) - f(x))\phi_\varepsilon(y)dy$$

$$= \int_{|y|\leq\eta} + \int_{|y|>\eta} = I_1 + I_2 \tag{2.4}$$

Changing variables to polar coordinates $y \to (y', s)$, (2.3) becomes

$$r^{-n}\int_0^r \int_\Sigma |f(x - sy') - f(x)|s^{n-1}ds\, dy' < \delta$$

and calling

$$g(s) = \int_\Sigma |f(x - sy') - f(x)|dy' \tag{2.5}$$

$$G(r) = \int_0^r g(s)s^{n-1}ds, \quad \Delta_x(r) = r^{-n}G(r) \tag{2.5a}$$

we get $\Delta_x(r) < \delta$ for $0 < r < \eta$. Thus,

$$|I_1| \leq \int_{|y|\leq\eta} |f(x - y) - f(x)|\varepsilon^{-n}\psi(y/\varepsilon)dy$$

$$= \int_0^\eta r^{n-1}g(r)\varepsilon^{-n}\psi_0(r/\varepsilon)dr$$

(continued)

(integrating by parts)

$$= G(r)\varepsilon^{-n}\psi_0(r/\varepsilon)\Big|_0^\eta - \int_0^\eta G(r)d(\varepsilon^{-n}\psi_0(r/\varepsilon))$$

$$= \Delta_x(r)(r/\varepsilon)^n\psi_0(r/\varepsilon)\Big|_0^\eta - \int_0^{\eta/\varepsilon} G(\varepsilon s)\varepsilon^{-n}d\psi_0(s)$$

$$\leq \Delta_x(\eta). A - \int_0^{\eta/\varepsilon} \Delta_x(\varepsilon s)s^n d\psi_0(s)$$

$$< \delta(A - \int_0^\infty s^n d\psi_0(s)) = \delta(A + B)$$

where A is the constant in (iii) and

$$B = -\int_0^\infty s^n d\psi_0(s) = +n\int_0^\infty s^{n-1}\psi_0(s)ds = (n/\omega_n)\int_{\mathbb{R}^n} \psi(x)dx > 0$$

and, since $\psi \in L^1$, $B < \infty$.

To estimate I_2, we let $\psi_\varepsilon(y) = \varepsilon^{-n}\psi(y/\varepsilon)$ and then

$$|I_2| \leq \int_{|y|>\eta} |f(x-y)|\psi_\varepsilon(y)dy + |f(x)|\int_{|y|>\eta} \psi_\varepsilon(y)dy \quad (2.6)$$

Since $\int_{|y|>\eta} \psi_\varepsilon(y)dy = \int_{|y|>\eta/\varepsilon} \psi(y)dy \to 0$ when $\varepsilon \to 0$, being $\psi \in L^1$, the second term on the right of (2.6) tends to zero with ε. The same is true for the first term: applying Hölder's inequality to the functions $|f(x - y)|$ and $\chi_\eta(y)\psi_\varepsilon(y)$, where χ_η is the characteristic function of the set $\{y \in \mathbb{R}^n: |y| > \eta\}$, we get

$$\int_{|y|>\eta} |f(x-y)|\psi_\varepsilon(y)dy \leq \|f\|_p \cdot \|\chi_\eta\psi_\varepsilon\|_{p'}$$

for $1/p + 1/p' = 1$. The last term tends to zero by property (iv). Collecting the estimates for I_1 and I_2, we see that $(f * \phi_\varepsilon)(x) - f(x) \to 0$ as $\varepsilon \to 0$, as desired. ∇

Remark 2.1. Under the slightly more general hypothesis $\phi, \psi \in L^1$, the proof yields

$$\lim_{\varepsilon \to 0} f * \phi_\varepsilon(x) = f(x)\left(\int \phi(t)dt\right)$$

whenever $x \in \mathscr{L}_f$.

3. CONVOLUTION OF FINITE MEASURES

The operation of convolution can be extended to include finite Borel measures. Measures $\mu \in \mathcal{M}$ can be identified with bounded linear functionals in C_∞ since $\mathcal{M}$ is the dual space of C_∞ (cfr. Chapter 0). In particular, a measure μ is determined by its values $I_\mu(\phi) = \int \phi d\mu$ for all $\phi \in C_\infty$.

Definition 3.1. The convolution of two measures $\mu_1, \mu_2 \in \mathcal{M}$ is the measure $\mu = \mu_1 * \mu_2$ defined by

$$I_\mu(\phi) = \int_{\mathbb{R}^n} \phi(x)d\mu(x) = \iint_{\mathbb{R}^n \times \mathbb{R}^n} \phi(x + y)d\mu_1(x)d\mu_2(y) \qquad (3.1)$$

for all $\phi \in C_\infty$.

Remark 3.1. The functional I_μ defined by (3.1) is clearly linear and, from Lemma 3.1(b) below, is bounded in C_∞. So I_μ determines a measure $\mu \in \mathcal{M}$. From the definition, $\mu_1 * \mu_2 = \mu_2 * \mu_1$.

Proposition 3.1. Let $\mu_1, \mu_2 \in \mathcal{M}$ and $\mu = \mu_1 * \mu_2$.

(a) The definition of convolution of two finite Borel measures is consistent, i.e., if μ_1 and μ_2 are absolutely continuous, $d\mu_i = f_i dx$, $f_i \in L^1$, $i = 1, 2$, then μ is absolutely continuous and $d\mu = (f_1 * f_2)dx$.

(b) The functional I_μ corresponding to μ is bounded, and

$$\|\mu\| \leq \|\mu_1\| \cdot \|\mu_2\| \tag{3.2}$$

(c) For every B Borel set,

$$\mu(B) = \int_{\mathbb{R}^n} \mu_1(B - y)d\mu_2(y) \tag{3.3}$$

<u>Proof.</u> (a) Let $d\mu_1 = f_1 dx$, $d\mu_2 = f_2 dx$, $f = f_1 * f_2$, $\phi \in C_\infty$. Then

$$\int_{\mathbb{R}^n} \phi(x)f(x)dx = \int_{\mathbb{R}^n} \phi(x)\left(\int_{\mathbb{R}^n} f_1(x - y)f_2(y)dy\right)dx$$

$$= \int_{\mathbb{R}^n}\int_{\mathbb{R}^n} \phi(x)f_1(x - y)f_2(y)dy\, dx \quad \text{(Fubini)}$$

$$= \int_{\mathbb{R}^n} f_2(y)\left(\int_{\mathbb{R}^n} f_1(x)\phi(x + y)dx\right)dy$$

$$= \iint_{\mathbb{R}^n \times \mathbb{R}^n} \phi(x + y)f_1(x)f_2(y)dx\, dy$$

$$= \iint_{\mathbb{R}^n \times \mathbb{R}^n} \phi(x + y)d\mu_1(x)d\mu_2(y)$$

(b) I_μ is a linear functional on C_∞. Furthermore, by (3.1),

$$|I_\mu(\phi)| = \left|\iint \phi(x + y)d\mu_1(x)d\mu_2(y)\right| \leq \|\phi\|_\infty \|\mu_1\| \, \|\mu_2\|$$

Thus,

$$\|\mu\| = \sup\{I_\mu(\phi) : \|\phi\|_\infty \leq 1\} \leq \|\mu_1\| \, \|\mu_2\|$$

(c) Let $\nu(B) = \int \mu_1(B - y)d\mu_2(y)$. Since $x + y \in B$ if and only if $x \in B - y$, then $\chi_B(x + y) = \chi_{B-y}(x)$ (χ_A is the characteristic function of the set A). Then,

$$\begin{aligned} \nu(B) &= \int(\int \chi_{B-y}(x)d\mu_1(x))d\mu_2(y) \\ &= \int(\int \chi_B(x + y)d\mu_1(x))d\mu_2(y) \\ &= \int\int \chi_B(x + y)d\mu_1(x)d\mu_2(y) \qquad (3.4) \end{aligned}$$

From (3.4) it follows that ν is a measure (this is immediate for μ_1 and μ_2 positive measures, and is obtained in the general case by linear combinations). By (3.4), the defining relation (3.1) is satisfied for $\phi = \chi_B$. Hence it holds for ϕ any simple function and, by the standard monotone limit process, it holds for any continuous ϕ of compact support. Thus $I_\nu(\phi) = I_\mu(\phi)$ for all $\phi \in C_0$ and hence also for all $\phi \in C_\infty$, as we wanted to prove. ∇

Proposition 3.2. If $f \in L^p$, $1 \leq p \leq \infty$ and $\mu \in \mathcal{M}$, then $h = f * \mu$ is a function such that

(a) $h(x) = \int_{\mathbb{R}^n} f(x - y)d\mu(y), \qquad x \in \mathbb{R}^n,$ (3.5)

(b) the integral in (a) exists in the Lebesgue sense a.e.,

(c) $h \in L^p$ and $\|h\|_p \leq \|f\|_p \cdot \|\mu\|$, (3.6)

(d) if $\mu = \delta_a$ is the Dirac measure at a, then $h(x) = f(x - a)$. In particular, if $\mu = \delta_0 = \delta$, then $\mu * f = f$. (3.7)

Proof. (a) follows from Definition 3.1, (d) follows from (a), (b) follows from the proof of (c). As for (c), since

$$|h(x)| \leq \int |f(x-y)|\,|d\mu(y)|$$

by Minkowski's integral inequality,

$$\|h\|_p = \left(\int |h(x)|^p dx\right)^{1/p}$$

$$\leq \int\left(\int |f(x-y)|^p dx\right)^{1/p} |d\mu(y)|$$

$$= \|f\|_p \int |d\mu(y)| = \|f\|_p \|\mu\| \qquad \nabla$$

We conclude this section with some remarks on last property (d) which says that $\delta * f = f$ for every $f \in L^1$.

As was said in Section 1, every kernel $k \in L^1$ defines, by convolution, an operator $K : f \to k * f$, acting in L^p, $1 \leq p \leq \infty$. The converse is of course not true: take I, the identity operator, $If = f$ for all $f \in L^1$. Is there a function $\delta \in L^1$ such that $If = \delta * f$ for all $f \in L^1$?

If there were such function δ it would mean that $f(x) = \delta * f(x) = \int \delta(x-y)f(y)dy$. Since the value of f at x does not depend on the values f takes at other points, $\delta(x-y) = 0$ for almost all $y \neq x$. Otherwise, changing f at $y \neq x$ would change the value of $f(x)$. So $\delta(x) = 0$ a.e., but $\delta * f = 0 \neq If$ if f is not identically zero.

For many years it was assumed that the so-called Dirac function δ, defined by $\delta(x) = 0$ if $x \neq 0$, $\delta(0) = \infty$, had the properties $\int_{|x|<\varepsilon} \delta(x)dx = 1$ for all $\varepsilon > 0$ and $\delta * f = f$ for all $f \in L^1$. It is evident that these assumptions are self-contradictory. A correct treatment deals with δ <u>as a measure</u>, rather than a function, namely, the Dirac measure concentrated at the origin.

On the other hand, the theory of singular kernels consists essentially in giving integral representation of the identity operator I in terms of functions. Since it is impossible to assign <u>one</u> kernel to this operator, one assigns to it a family of kernels, such that the corresponding convolution operators tend (in some sense) to I. This is precisely the idea of the convolution unit, introduced in Theorem 1.4, consisting in a family of functions $\{\phi_\varepsilon\}_{\varepsilon>0}$, generated by a fixed $\phi \in L^1$ with $\int \phi = 1$, such that the corresponding $\{K_\varepsilon\}_{\varepsilon>0}$, $K_\varepsilon : f \to f * \phi_\varepsilon$, tend to I in the sense that for every $f \in L^1$, $K_\varepsilon f \to f$ either in norm (Theorem 1.4) or pointwise (Theorem 2.1). (For further information on the subject, see [1]).

For fixed $k \in L^1$, the operator $K : f \to k * f$ acts, in particular, on the Hilbert space L^2, and has an adjoint operator K^*:

$$(Kf, g) = (f, K^*g) \quad \text{for all} \quad f, g \in L^2 \tag{3.8}$$

where $(f, g) = \int f(x)\overline{g(x)}dx$.

Proposition 3.3. The adjoint K^* of the convolution operator K on L^2 defined by $Kf = f * k$ for a fixed $k \in L^1$, is given by $K^*f = f * k^*$ where $k^*(x) = \overline{k(-x)}$, $x \in \mathbb{R}^n$.

More generally, $K^* : f \to f * k^*$ is the adjoint of $K : f \to f * k$, considered as operators on $L^{p'}$ and L^p, respectively, for $1 < p < \infty$.

Proof. For all $f \in L^p$, $g \in L^{p'}$, $1 < p < \infty$, $1/p + 1/p' = 1$,

$$\begin{aligned}(Kf, g) &= \int(\int k(x - y)f(y)dy)\overline{g(x)}dx \\ &= \int f(y)(\int k(x - y)\overline{g(x)}dx)dy \\ &= \int f(y)(\overline{\int \overline{k(-(y - x))}g(x)dx})dy \\ &= \int f(y)(\overline{\int k^*(y - x)g(x)dx})dy \\ &= (f, K^*g)\end{aligned}$$

and the assumption follows. ∇

Consistently with the notation introduced in Proposition 3.3, we denote in what follows by h^* the function

$$h^*(x) = \overline{h(-x)} \tag{3.9}$$

The following properties are immediate:

$$\text{if} \quad h_1 = h^* \quad \text{then} \quad h = h_1^*, \quad \text{so} \quad (h^*)^* = h \tag{3.10}$$

$$(\lambda h)^* = \overline{\lambda} h^*, \quad (h_1 + h_2)^* = h_1^* + h_2^* \tag{3.11}$$

$$(h_1 * h_2)^* = h_1^* * h_2^* \tag{3.12}$$

$$\|h^*\|_p = \|h\|_p, \; 1 \le p \le \infty, \quad \text{in particular} \quad \|k^*\|_1 = \|k\|_1 \tag{3.13}$$

Also, if $k \in L^1$, $f \in L^p$, $g \in L^{p'}$, then

$$(k * f, g^*) = (k * g, f^*) \tag{3.14}$$

(same proof as in Proposition 3.3).

For fixed $k \in L^1$, $\|K\|^{(p)}$ is the norm of the corresponding convolution operator K as acting on L^p. In Section 1 (Exercise 1.11) we determined that $\|K\|^{(1)} = \|k\|_1$. Now observe that, since $(L^p)' = L^{p'}$, for $1 \le p \le \infty$,

$$\begin{aligned} \|K\|^{(p)} &= \sup\{ \|Kf\|_p : \|f\|_p \le 1\} \\ &= \sup\{ |(Kf, g)| : \|f\|_p \le 1, \; \|g\|_{p'} \le 1\} \end{aligned}$$

From (3.13) and (3.14) we obtain that $\|K\|^{(p')} = \|K\|^{(p)}$ for $1/p + 1/p' = 1$, $1 < p < \infty$, if $K : f \to k * f$ for $k \in L^1$.

4. THE GROUP ALGEBRA OF $\mathbb{R}^n$ AND ITS CHARACTERS

A (commutative) Banach algebra is a Banach space $\mathcal{B}$ with a product operation $(x, y) \to x.y$ satisfying the conditions:

(i) $\mathcal{B}$ is a (commutative) algebra with respect to the linear operations and the product;

(ii) $\|x.y\| \leq \|x\| \|y\|$ for every $x, y \in \mathcal{B}$;

(iii) when the algebra $\mathcal{B}$ has a multiplicative unit e $(e.x = x.e = x$ for all $x \in \mathcal{B})$, $\|e\| = 1$.

If $\mathcal{B}$ has a unit, it is called unital. If $\mathcal{B}$ is not unital, we associate canonically to $\mathcal{B}$ an unital $\mathcal{A}$ as follows. Let $\mathcal{A} = \mathcal{B} + \{\lambda e\}$ be the set of all symbols $\xi = \{x, \lambda\} = x + \lambda e$, where $x \in \mathcal{B}$, $\lambda \in \mathbb{C}$ and e is a fixed symbol, and define in $\mathcal{A}$ the linear and product operations and the norm by

$$\begin{aligned}
c\xi &= c\{x, \lambda\} = c(x + \lambda e) = cx + (c\lambda)e = \{cx, c\lambda\} \\
\xi_1 + \xi_2 &= \{x_1, \lambda_1\} + \{x_2, \lambda_2\} = (x_1 + \lambda_1 e) + (x_2 + \lambda_2 e) \\
&= (x_1 + x_2) + (\lambda_1 + \lambda_2)e = \{x_1 + x_2, \lambda_1 + \lambda_2\} \\
\xi_1 \cdot \xi_2 &= \{x_1, \lambda_1\} \cdot \{x_2, \lambda_2\} = (x_1 + \lambda_1 e) \cdot (x_2 + \lambda_2 e) \\
&= (x_1 x_2 + \lambda_1 x_2 + \lambda_2 x_1) + \lambda_1 \lambda_2 e \\
&= \{x_1 x_2 + \lambda_1 x_2 + \lambda_2 x_1, \lambda_1 \lambda_2\} \\
\|\xi\| &= \|\{x, \lambda\}\| = \|x + \lambda e\| = \|x\| + |\lambda|
\end{aligned} \tag{4.1}$$

where $\xi_j = x_j + \lambda_j e$, $j = 1, 2$.

It is easy to see that $\mathcal{A}$ is a unital Banach algebra with e as the unit, and that $\mathcal{B}$ may be identified with the subalgebra of $\mathcal{A}$ consisting of the elements $x \equiv x + 0e$. If $\mathcal{B}$ has a unit, then $\mathcal{B} \cong \mathcal{A}$, but $\mathcal{B} \neq \mathcal{A}$ otherwise.

Exercise 4.0. Prove that $\mathscr{A}$ is a complete normed (thus, Banach) space.

Definition 4.1. A nonunital Banach algebra $\mathscr{B}$ is said to have an approximative unit if there exists a (generalized) sequence $\{u_j\} \subset \mathscr{B}$ such that $\lim_j u_j x = x$ for all $x \in \mathscr{B}$.

From Corollary 1.3 it follows that $L^1 = L^1(\mathbb{R}^n, dx)$ is a commutative Banach algebra with respect to the operations $(f, g) \to af + bg$ and $(f, g) \to f * g$ and the norm $\|f\| = \|f\|_1$. From the remarks in Section 3 it is clear that L^1 is not unital. The associated algebra $\mathscr{A} = L^1 + \{\lambda e\}$ is called the group algebra of $\mathbb{R}^n$. The elements of L^1 will be denoted by $f, g, h, \ldots$ and those of $\mathscr{A}$ by $\Phi = f + \lambda e$, $\psi = g + \rho e, \ldots$ where $f, g, \ldots \in L^1$, $\lambda, \rho, \ldots \in \mathbb{C}$.

From Theorem 1.4 it follows that L^1 has an approximative unit, because if we fix $\phi \in L^1$, with $\int \phi = 1$, and let $u_j(x) = \phi_{1/j}(x)$, $j = 1, 2, \ldots$, then $\|u_j * f - f\|_1 \to 0$ for all $f \in L^1$.

If $\mathscr{A}$ is a unital Banach algebra and $x \in \mathscr{A}$, $\|x\| = a < 1$, then there exists $z \in \mathscr{A}$ such that

$$(e - x)z = e \tag{4.2}$$

In fact, since $\|x^2\| = \|x.x\| \leq \|x\|\,\|x\| = \|x\|^2$, we have $\|x^n\| \leq \|x\|^n = a^n$, thus $\Sigma \|x^n\| \leq \Sigma a^n = 1/(1 - a) < \infty$, and there is an element z such that $z = \Sigma x^n = \lim_N \Sigma_{n=0}^N x^n$. Since $(e - x)\cdot(e + x + \ldots + x^N) = e - x^{N+1} \to e - 0 = e$, we obtain (4.2) by letting N tend to infinity.

Definition 4.2. A continuous linear functional $I(x)$, defined in a Banach algebra $\mathscr{B}$, is called multiplicative if $I(x.y) = I(x)I(y)$ for all $x, y \in \mathscr{B}$.

Definition 4. 2a. If $\mathscr{A}$ is a unital Banach algebra and I is a nonzero continuous linear multiplicative functional in $\mathscr{A}$, then I is called a character of $\mathscr{A}$.

If I is a character of $\mathscr{A}$, then

$$I(e) = 1 \tag{4.3}$$

By definition, there exists $z \in \mathscr{A}$ such that $I(z) \neq 0$. Then $I(z) = I(e.z) = I(e)I(z)$ and hence $I(e) = 1$.

Let $\mathscr{A}$ be the unital algebra associated with the algebra $\mathscr{B}$. For each multiplicative functional I in $\mathscr{B}$ there is a unique character I_1 of $\mathscr{A}$ whose restriction to $\mathscr{B}$ is I.

In fact, it is enough to write $I_1(x+\lambda e) = I(x) + \lambda$. Thus there is a 1 - 1 correspondence between a subset of characters of $\mathscr{A}$ and the continuous linear multiplicative functionals in $\mathscr{B}$.

If I is a character in $\mathscr{A}$, then

$$\|I\| = 1 \tag{4.4}$$

In fact, if it were $\|I\| > 1$, there would be an $x \in \mathscr{A}$ with $\|x\| < 1$ but $I(x) = 1$ (why?) and, by (4. 2), there would exist $z \in \mathscr{A}$ such that $(e - x)z = e$. Hence, by (4. 3),

$$1 = I(e) = I(e - x)I(z) = (I(e) - I(x))I(z) = (1 - 1)I(z) = 0$$

and $1 = 0$ is a contradiction! Thus $\|I\| \leq 1$ and, by (4. 3), $\|I\| \geq 1$ (since $\|e\| = 1$), hence $\|I\| = 1$. ∇

We are going to determine the characters of the group algebra $\mathscr{A}$ of $\mathbb{R}^n$. With this purpose in mind we introduce the following

Definition 4. 3. A character of $\mathbb{R}^n$ is a continuous function $h : \mathbb{R}^n \to \mathbb{C}$ such that $h \neq 0$, and for all $t, s \in \mathbb{R}^n$,

$$h(t + s) = h(t)h(s) \tag{4.5}$$

and

$$|h(t)| \leq 1 \tag{4.6}$$

For every $x \in \mathbb{R}^n$ set

$$h_x(t) = e^{ix.t} \tag{4.7}$$

Lemma 4.1. For every $x \in \mathbb{R}^n$, h_x is a character of $\mathbb{R}^n$ and, conversely, every character of $\mathbb{R}^n$ is of the form $h = h_x$ for some $x \in \mathbb{R}^n$.

Proof. It is immediate that $h_x(t) = e^{ix.t}$ is a character of $\mathbb{R}^n$ and we have only to prove that every character is of this form.

(i) Consider first the case $n = 1$ and let $h : \mathbb{R} \to \mathbb{C}$ be a character of $\mathbb{R}$. Taking $s = 0$ in (4.5), we obtain $h(0) = 1$. Since there exists $\delta > 0$ such that $\int_0^\delta h(t)dt = a \neq 0$, we get $ah(s) = h(s)\int_0^\delta h(t)dt = \int_0^\delta h(s)h(t)dt = \int_0^\delta h(s + t)dt = \int_s^{s+\delta} h(t)dt.$

Since h is a continuous function, the last integral is differentiable and h has a continuous derivative. Differentiating (4.5) with respect to s and taking then $s = 0$,

$$h'(t) = Ch(t), \qquad C = h'(0) \tag{4.8}$$

By (4.6) and the fact that $h(0) = 1$, we have $h(t) = e^{ix.t}$ for some $x \in \mathbb{R}$.

(ii) Let h be a character of $\mathbb{R}^n$ and, for $t \in \mathbb{R}$, define $h_1(t) = h(t, 0, \ldots, 0)$. Then h_1 will be a character of $\mathbb{R}$ and, by (i), $h_1(t) = e^{ix_1 t}$. Letting $h_2(t) = h(0, t, 0, \ldots, 0), \ldots, h_n(t) = h(0, \ldots, 0, t)$, we find that $h_k(t) = e^{ix_k t}$ and therefore, for

$t \in \mathbb{R}^n$, $t = (t_1, \ldots, t_n)$,

$$h(t) = h(t_1, \ldots, t_n) = h(t_1, \ldots, 0) \ldots h(0, \ldots, t_n)$$

$$= h_1(t_1), \ldots, h_n(t_n) = e^{ix_1 t_1} \ldots e^{ix_n t_n} = e^{ix.t}$$

for $x = (x_1, \ldots, x_n) \in \mathbb{R}^n$, since x_k is a fixed real number for every $k = 1, \ldots, n$. ∇

For every $x \in \mathbb{R}^n$, let I_x be the functional defined in L^1 by

$$I_x(f) = \int_{\mathbb{R}^n} f(t)h_x(t)dt = \int_{\mathbb{R}^n} f(t)e^{ix.t}dt \tag{4.9}$$

<u>Lemma 4.2.</u> For every $x \in \mathbb{R}^n$, I_x defined as in (4.9) is a continuous linear multiplicative functional in L^1.

<u>Proof.</u> Since the functional I_x is defined by means of an integral, it is obviously linear. Furthermore, $|\exp(ix.t)| = 1$ for all $x, t \in \mathbb{R}^n$, so

$$|I_x(f)| \leq \|f\|_1$$

and I_x is continuous. To see that it is also multiplicative, write, for $f, g \in L^1$,

$$I_x(f * g) = \int(\int f(t - s)g(s)ds)e^{ix.t}dt \tag{4.10}$$

$F(t, s) = f(t - s)g(s)\exp(ix.t)$ is, for each fixed x, an integrable function in $\mathbb{R}^n \times \mathbb{R}^n$ (see Exercise 4.1) and we apply Fubini's theorem to (4.10) to obtain

$$\begin{aligned} I_x(f * g) &= \int g(s)\left(\int f(t)e^{ix.(t+s)}dt\right)ds \\ &= \int g(s)\left(\int f(t)e^{ix.t}dt\right)e^{ix.s}ds \\ &= \left(\int f(t)e^{ix.t}dt\right)\left(\int g(s)e^{ix.s}ds\right) \\ &= I_x(f)I_x(g) \end{aligned} \tag{4.11}$$

∇

<u>Exercise 4.1.</u> Show that if $f, g \in L^1(\mathbb{R}^n)$ and $F(t, s) = f(t - s)g(s)\cdot \exp(ix.t)$, then $F \in L^1(\mathbb{R}^n \times \mathbb{R}^n)$.

As we said before, such I_x correspond to characters of the group algebra $\mathcal{A}$ of $\mathbb{R}^n$ given by

$$I_x(f + \lambda e) = I_x(f) + \lambda \tag{4.9a}$$

These are almost all the characters of $\mathcal{A}$. To be precise let

$$I_\infty(f) = 0 \quad \text{for all} \quad f \in L^1 \quad \text{and} \quad I_\infty(e) = 1 \tag{4.9b}$$

<u>Exercise 4.2.</u> Show that I_∞ is a character of $\mathcal{A}$.

<u>Proposition 4.3.</u> Every character of $\mathcal{A} = L^1 + \{\lambda e\}$ is of the form I_x (4.9a) or I_∞ (4.9b). Moreover, if $x \neq y$ then $I_x(f) \neq I_y(f)$ for some $f \in L^1$, so that the characters of $\mathcal{A}$ are in 1 - 1 correspondence with the elements of $\overline{\mathbb{R}^n} = \mathbb{R}^n \cup \{\infty\}$.

<u>Proof.</u>

(1) Let I be a multiplicative functional in L^1 and let us see that $I = I_x$ for some $x \in \mathbb{R}^n$. By (4.4), $\|I\| = 1$, so that by the Riesz representation theorem (see Chapter 0, Section 3) there

is a $p \in L^\infty$ such that $\|p\|_\infty = 1$ and $I(f) = \int f(t)p(t)dt$ for all $f \in L^1$. From $I(f * g) = I(f)I(g)$ and Fubini's theorem we get

$$\int(\int f(t - s)g(s)ds)p(t)dt = \int\int p(t + s)f(t)g(s)dtds$$

$$= (\int p(t)f(t)dt)(\int p(s)g(s)ds)$$

Taking Q and D, two cubes in $\mathbb{R}^n$, centered at t_0 and at s_0 respectively, and of measures $|Q|$ and $|D|$, let $f = \chi_Q$ and $g = \chi_D$ be their characteristic functions. Then,

$$\int_{Q\times D}\int p(t + s)dtds = \int\int p(t + s)f(t)g(s)dtds$$

$$= (\int_Q p(t)dt)(\int_D p(s)ds)$$

and

$$\frac{1}{|Q\times D|}\int_{Q\times D}\int p(t + s)dtds = (\frac{1}{|Q|}\int_Q p(t)ds)(\frac{1}{|D|}\int_D p(s)ds) \tag{4.12}$$

Letting $|Q| \to 0$ and $|D| \to 0$ in (4.12) we obtain from the classical Lebesgue theorem on the differentiation of the integral (for an independent proof of this theorem see Chapter 4, Theorem 1.1) that

$$p(t_0 + s_0) = p(t_0)p(s_0) \tag{4.13}$$

for almost all $t_0, s_0 \in \mathbb{R}^n$.

Letting $Q_0 = t_0 + Q$, for Q a fixed cube, (4.13) implies that

$$\int_{Q_0} p(s)ds = \int_Q p(t_0 + s)ds = p(t_0)\int_Q p(s)ds = Cp(t_0)$$

$$p(t_0) = \frac{1}{C}\int_{Q_0} p(s)ds \tag{4.14}$$

for almost all $t_0 \in \mathbb{R}^n$. Since the integral in (4.14) varies continuously with t_0, p is a continuous function a.e. and we may choose p continuous. Then (4.13) holds for <u>all</u> t_0, s_0 and, since $|p(t)| \leq 1$, p is a character of $\mathbb{R}^n$. By Lemma 4.1, $p(t) = \exp(ix.t)$ for some $x \in \mathbb{R}^n$ and $I = I_x$ as claimed.

(2) If $I_x(f) = \int f(t)\exp(ix.t)dt = I_y(f) = \int f(t)\exp(iy.t)dt$ for all $f \in L^1$, then $\exp(ix.t) = \exp(iy.t)$ for almost all t and, by continuity, for all t. This implies that $x = y$. Hence, if $x \neq y$ then $I_x(f) \neq I_y(f)$ for some $f \in L^1$. ∇

5. REMARKS ON THE PERIODIC CASE

The results of the preceding sections for the groups $\mathbb{R}$ and $\mathbb{R}^n$, still hold--but for some slight modifications and in a more elementary way--for the group $\mathbb{T}$ of the unit circle.

The correspondence that assigns to every point on the real line, $t \in \mathbb{R}$, a point on the unit circle, $\exp(it) \in \mathbb{T}$, enables us to identify $\mathbb{T}$ with $\mathbb{R}(\bmod 2\pi)$, and the functions defined on $\mathbb{T}$ with the <u>periodic</u> functions defined in $\mathbb{R}$. The points of $\mathbb{T}$ will be denoted <u>indistinctly</u> by $\exp(it)$ or by t. Thus $\mathbb{T}$ is a group under the sum $t + s$, and the Lebesgue measure is translation invariant with respect to it:

$$\int_{\mathbb{T}} f(t + s)dt = \int_{\mathbb{T}} f(t)dt$$

Therefore, the convolution of two functions defined on $\mathbb{T}$, may be defined as in Section 1, and the results of Theorems 1.1-1.5 hold. We may also consider the algebra $L^1 = L^1(\mathbb{T})$ and the corresponding group algebra $\mathscr{A} = \mathscr{A}(\mathbb{T})$. These are Banach algebras with involution and $L^1(\mathbb{T})$ has approximative unit.

Every continuous function $h(t)$ defined on $\mathbb{T}$ must be periodic, and since $h_x(t) = \exp(ix.t)$ is not periodic unless x is an integer, the characters h of $\mathbb{T}$ will be of the form $h = h_m$, where

$$h_m(t) = \exp(imt), \quad m \in \mathbb{Z} \tag{5.1}$$

and the characters I of $\mathscr{A}(\mathbb{T})$ will now be of the form $I_m(f + \lambda e) = I_m(f) + \lambda$ or $I_\infty(f + \lambda e) = \lambda$, where

$$I_m(f) = \frac{1}{2\pi} \int_{\mathbb{T}} f(t)e^{imt}dt, \quad f \in L^1(\mathbb{T}) \tag{5.2}$$

Thus the characters $\{h_m\}$ of $\mathbb{T}$ are in 1 - 1 correspondence with the multiplicative functionals $\{I_m\}$ in $L^1(\mathbb{T})$ and with the integers $m = 0, \pm 1, \pm 2, \ldots,$ while in the case of $\mathbb{R}$ there are as many characters I_x of $\mathscr{A}(\mathbb{R})$ as real numbers $x \in \overline{\mathbb{R}}$.

Similar results hold for the discrete line, i.e., for the group $\mathbb{Z} = \{0, \pm 1, \pm 2, \ldots\}$. Here the Lebesgue measure will be the discrete measure with unit mass at each point $m \in \mathbb{Z}$, so $L^1(\mathbb{Z}) = \ell^1$, and every $f \in L^1(\mathbb{Z})$ is a sequence $f = \{f_m\}$ with $\int f(m)dm = \Sigma_{-\infty}^{+\infty} f_m$.

Thus, if f and $g \in L^1(\mathbb{Z})$, the convolution $h = f * g$ is given by $h = \{h_m\}$ with $h_m = \Sigma_k f_{m-k}g_k$ (cfr., the motivation given at the beginning of Section 1). Now $L^1(\mathbb{Z})$ does have a unit e given by $e(0) = 1$, $e(m) = 0$ if $m \neq 0$, so that $L^1(\mathbb{Z}) = \mathscr{A}(\mathbb{Z})$. As in the case of $\mathbb{R}$, for each character $h = \{h_m\}$ of the group $\mathbb{Z}$ there is an $x \in \mathbb{R}$ such that $h = h_x$, i.e.,

$$h_m = \exp(imx), \quad m \in \mathbb{Z} \tag{5.3}$$

but since $\exp(im(x + 2\pi)) = \exp(imx)$ for all $m \in \mathbb{Z}$, the characters of $\mathbb{Z}$ are in a 1 - 1 correspondence with the points on the circle $x \in [0, 2\pi) \sim \mathbb{T}$. Similarly, the characters of $\mathscr{A}(\mathbb{Z})$ are of the form $I = I_x$, $x \in \mathbb{T}$, where

$$I_x(f) = \sum_{m=-\infty}^{\infty} f_m e^{imx} \tag{5.4}$$

for $f \in L^1(\mathbb{Z})$. Observe that now $I_\infty(f) \equiv 0$ is not a character, since $e \in L^1(\mathbb{Z})$.

Therefore while the set of characters of $\mathscr{A}(\mathbb{R})$ can be identified with $\overline{\mathbb{R}}$, that of $\mathscr{A}(\mathbb{T})$ can be identified with $\overline{\mathbb{Z}}$ and that of $\mathscr{A}(\mathbb{Z})$ with $\mathbb{T}$. These facts that can be expressed as: the dual group of $\mathbb{R}$ is $\mathbb{R}$, while the dual group of $\mathbb{T}$ is $\mathbb{Z}$ and the dual group of $\mathbb{Z}$ is $\mathbb{T}$.

In the case of $\mathbb{T}$, as the total Lebesgue measure of the circle is finite, the characters $h_m(t) = \exp(imt)$ are elements of $L^2(\mathbb{T})$ and they form a complete orthonormal set in the Hilbert space $L^2(\mathbb{T})$. Every function $f \in L^2(\mathbb{T})$ admits a development in Fourier series

$$f(t) \sim \sum_{m=-\infty}^{\infty} c_m e^{imt} \tag{5.5}$$

where

$$c_m = \frac{1}{2\pi} \int_{\mathbb{T}} f(t) e^{-imt} dt = I_{-m}(f) \tag{5.6}$$

and the series in (5.5) converges to f in $L^2(\mathbb{T})$.

The sequence of the coefficients $\{c_m\} = \{I_{-m}\}_{m \in \mathbb{Z}}$ is called the Fourier transform of f and is also denoted as $\{\hat{f}_m\}$. Formula (5.5) expresses the function f by means of its Fourier transform (in the L^2 sense).

More generally, for every $f \in L^1(\mathbb{T})$, its Fourier transform may be defined as the sequence $\{\hat{f}_m\}$ where $\hat{f}_m = I_{-m}(f)$, $m \in \mathbb{Z}$, and it is associated to the <u>Fourier series</u> $\Sigma_m \hat{f}_m \exp(imt) = \Sigma_m I_{-m}(f)\exp(imt)$.

This situation gives rise to the classical inversion problem of how to obtain $f \in L^1(\mathbb{T})$ from its Fourier transform $\{\hat{f}_m\}$ or from its associated Fourier series.

As in the Lebesgue integration theory functions are defined a.e., a function f defined in the interval $[0, 2\pi]$ (or in any finite interval $[a, b]$) can be modified so that $f(0) = f(2\pi)$. Then f will be periodic and can be treated as a function defined on $\mathbb{T}$. Therefore the preceding considerations on Fourier series still hold for functions defined on a finite interval of $\mathbb{R}$. This is not true for functions on $\mathbb{R}$, since in order to become periodic, such functions would need to be modified in a set of positive measure.

In the case of $\mathbb{R}$ there are as many characters as real numbers, so if $f \in L^1(\mathbb{R})$, by analogy to the case of $\mathbb{T}$, its Fourier transform may be defined as the function

$$\hat{f}(x) = I_{-x}(f) = \int_{\mathbb{R}} f(t)e^{-ix.t}dt$$

Note that in the case of $\mathbb{T}$ the Fourier transform of f is a function $\hat{f}$ defined in $\mathbb{Z}$, the dual group of $\mathbb{T}$, and in the case of $\mathbb{R}$ the Fourier transform of f is a function $\hat{f}$ defined in $\mathbb{R}$, the dual group of $\mathbb{R}$.

The Fourier transform in $\mathbb{R}$ is associated to an integral rather than a series and the inversion problem is: given $\hat{f}$ determine f or, more generally, in what sense does the Fourier integral

$$\int \hat{f}(x)e^{ix.t}dx \tag{5.7}$$

determine $f \in L^1(\mathbb{R})$?

The theory in $\mathbb{R}$ or in $\mathbb{R}^n$ is more difficult than in $\mathbb{T}$, for two reasons. First of all, the characters $h_x(t) = \exp(ix.t)$ are neither integrable nor square integrable functions on $\mathbb{R}$ or $\mathbb{R}^n$, and therefore do not provide an orthonormal basis of L^2. Secondly, we are forced to deal with (Fourier) integrals instead of (Fourier) series because in these cases the dual groups are $\mathbb{R}$ and $\mathbb{R}^n$.

In Chapter 2 we study the properties of the Fourier transform for functions in L^1 and in Chapter 3, the inversion problem and the L^2 theory. In what follows we shall develop Fourier theory for $\mathbb{R}$ and $\mathbb{R}^n$, indicating only briefly the corresponding results for the periodic case of $\mathbb{T}$, that may be obtained by the analogy between $\mathbb{T}$ and $\mathbb{R}$.

REFERENCES

1. J. C. Merlo, Nociones elementales sobre núcleos singulares, Cursos y seminarios de matemática, Universidad de Buenos Aires fasc. 7, Buenos Aires, 1960.

Chapter 2

FOURIER TRANSFORMS OF INTEGRABLE FUNCTIONS AND FINITE MEASURES

1. FOURIER TRANSFORMS IN $L^1(\mathbb{R}^n)$

Definition 1.1. The Fourier transform of $f \in L^1(\mathbb{R}^n)$ is denoted by $\hat{f}$ and defined by

$$\hat{f}(x) = \int_{\mathbb{R}^n} f(t)e^{-2\pi i x.t}dt \tag{1.1}$$

for all $x \in \mathbb{R}^n$. The transformation $f \to \hat{f}$ will be denoted by $\mathscr{F}$.

Remark 1.1. The factor 2π appearing in expression (1.1) allows a symmetric formulation of the inversion theorem (see Chapter 3). The Fourier transform may be defined in an equivalent way without it, as in (4.6) of Chapter 1.

Let us observe that, since $|\exp(-i2\pi x.t)| = 1$, the simplest class of functions for which it is possible to define the Fourier transform as in (1.1) is indeed $L^1(\mathbb{R}^n)$. In the case of $f \in L^1$, $\hat{f}(x)$ is defined for all $x \in \mathbb{R}^n$.

In what follows, $(\ )^{\wedge}$ stands for $\mathscr{F}(\ )$.

Proposition 1.1 (Basic properties). Let $f \in L^1$, then

(a) the mapping $f \to \hat{f}$ is linear;

(b) $\hat{f}$ is a bounded function and $\|\hat{f}\|_\infty \leq \|f\|_1$;

(c) $\hat{f}$ is uniformly continuous;

(d) if $f \geq 0$ then $\|\hat{f}\|_\infty = \|f\|_1 = \hat{f}(0)$.

Proof.

(a) follows from the linearity of the defining integral in (1.1).

(b) $|\hat{f}(x)| \leq \int |f(t)e^{-2\pi ix.t}dt| = \int |f(t)|dt = \|f\|_1$ so $\|\hat{f}\|_\infty \leq \|f\|_1$.

(c) Let $h \in \mathbb{R}^n$, then

$$\hat{f}(x+h) - \hat{f}(x) = \int f(t)(e^{-2\pi ih.t} - 1)e^{-2\pi ix.t}dt$$

and

$$|\hat{f}(x+h) - \hat{f}(x)| \leq \int |f(t)||e^{-2\pi ih.t} - 1|dt$$

$$\leq \int_{|t|\leq R} |f(t)||e^{-2\pi ih.t} - 1|dt + 2\int_{|t|>R} |f(t)|dt$$

As $f \in L^1$, R can be chosen so large that the last integral is arbitrarily small. Fixing such R, h is chosen so that $|\exp(-2\pi ih.t) - 1| < \delta$, δ arbitrary, for $|t| \leq R$. Then the difference $|\hat{f}(x+h) - \hat{f}(x)|$ can be made arbitrarily small independently of x.

(d) By (b), $\|f\|_1 = \int f(t)dt = \hat{f}(0) \leq \|\hat{f}\|_\infty \leq \|f\|_1$. ∇

Remark 1.2. Let $\mathcal{F}$ be the Fourier transform operator given by $\mathcal{F} : f \to \hat{f}$. The preceding lemma states that $\mathcal{F}$ is a bounded linear operator from L^1 to L^∞, with norm $\|\mathcal{F}\| \leq 1$. In particular, if a sequence $\{f_n\} \subset L^1$ converges to f in L^1, then the corresponding transformed sequence $\{\hat{f}_n\}$ is uniformly convergent to $\hat{f}$.

Example 1.1. Let $n = 1$, $f(t) = \chi_{(a,b)}(t)$. Then

$$\hat{f}(x) = \int_{\mathbb{R}^1} \chi_{(a,b)}(t) e^{-2\pi i x.t} dt = \int_a^b e^{-2\pi i x.t} dt$$

$$= \frac{e^{-i2\pi bx} - e^{-i2\pi ax}}{2\pi i x}$$

Note that in this case

(i) $\hat{f}$ is not integrable in $\mathbb{R}^1$,
(ii) $\hat{f}(x) \to 0$ when $|x| \to \infty$. More precisely, $|\hat{f}(x)| \leq (\pi|x|)^{-1}$.

Example 1.2. Let $n = 1$, $f(t) = \psi_{a,c}(t)$ defined by

$$\psi_{a,c}(t) = \begin{cases} 1 \text{ if } -a \leq t \leq a \\ 0 \text{ if } t \geq a + c \text{ or if } t \leq -a - c \\ (a - t + c)/c \text{ if } a < t < a + c \\ (a + t + c)/c \text{ if } -a - c < t < -a \end{cases} \tag{1.2}$$

Then

$$\hat{f}(x) = 2\int_0^\infty \psi_{a,c}(t) \cos 2\pi x.t \, dt = -\frac{1}{\pi x}\int_0^\infty \psi'_{a,c}(t) \sin 2\pi x.t \, dt$$

Since $\psi'_{a,c}(t)$ for $t > 0$ is zero outside the interval $(a, a + c)$ where it takes the value $-\frac{1}{c}$, we have

$$\hat{f}(x) = \frac{1}{\pi c x}\int_a^{a+c} \sin 2\pi x.t \, dt = -\frac{1}{\pi c x^2}(\cos 2\pi a.t - \cos 2\pi(a + c)t)$$

Thus $|\hat{f}(x)| \leq 2(\pi c x^2)^{-1}$ and $\hat{f}$, being continuous and bounded, is integrable in $\mathbb{R}^1$ and tends to zero at infinity (more rapidly than the characteristic function of Example 1.1.).

Example 1.3. Suppose that $n > 1$ and $f \in L^1(\mathbb{R}^n)$ is of the form

$$f(t) = f_1(t_1).f_2(t_2)...f_n(t_n) \tag{1.3}$$

for $t = (t_1, t_2, ..., t_n) \in \mathbb{R}^n$. Then, Fubini's theorem implies that

$$\hat{f}(x) = \hat{f}_1(x_1).\hat{f}_2(x_2)...\hat{f}_n(x_n) \tag{1.3a}$$

for $x = (x_1, x_2, ..., x_n) \in \mathbb{R}^n$.

This result is valid, in particular, for characteristic functions of n-dimensional intervals, which are products of characteristic functions of intervals on the line. Thus Example 1.1 shows that $\hat{\chi}_I(x)$ tends to zero whenever $|x| \to \infty$ if $\chi_I = \chi_{I_1} \cdot \chi_{I_2} \cdots \chi_{I_n}$, where the I_j, $j = 1, ..., n$ are intervals on the line.

Theorem 1.2 (Riemann-Lebesgue lemma). If $f \in L^1$ then $\hat{f}(x) \to 0$ when $|x| \to \infty$. In particular, $\mathscr{F} : L^1 \to C_\infty \subset L^\infty$ is a continuous mapping.

Proof. In Example 1.3 we saw that the thesis is valid for characteristic functions of intervals in $\mathbb{R}^n$ and, by linearity, it holds also for step functions. The class of step functions is dense in L^1, i.e., given $f \in L^1$ and $\varepsilon > 0$ there exists a step function ϕ such $\|f - \phi\|_1 < \varepsilon/2$. If we write $\hat{f} = \mathscr{F}(f - \phi) + \mathscr{F}(\phi)$, then

$$|\hat{f}(x)| \leq \|f - \phi\|_1 + |\hat{\phi}(x)| < \varepsilon$$

for $|x|$ sufficiently large, since $|\hat{\phi}(x)| \to 0$ when $|x| \to \infty$.

Remark 1.3. By Theorem 1.2, $\mathcal{F}: L^1 \to C_\infty$ continuously. This gives a necessary condition (to belong to C_∞) for a function to be the Fourier transform of an L^1 function. This condition is not sufficient, as we see in the following example, and there is no simple way to characterize the image of L^1 under the operator $\mathcal{F}$.

Example 1.4. For simplicity, let be $n = 1$ and let f be an odd function, so that $\hat{f}(x) = -i\int_{-\infty}^{\infty} f(t)\sin 2\pi xt\, dt$. An application of Fubini's theorem, shows that $|\int_1^b (\hat{f}(x)/x)dx| \leq A$, where A is independent of $b > 1$. This last assertion follows from the well-known estimate $|\int_\alpha^\beta (\sin x/x)dx| \leq B$ independently of α, β, since

$$\int_1^b \frac{\hat{f}(x)}{x}\, dx = -i \int_{-\infty}^{\infty} f(t)\Big(\int_1^b \frac{\sin 2\pi xt}{x}\, dx\Big)dt$$

$$= -i \int_{-\infty}^{\infty} f(t)dt. \int_{2\pi t}^{2\pi bt} \frac{\sin y}{y}\, dy$$

and therefore

$$\Big|\int_1^b \frac{\hat{f}(x)}{x}dx\Big| \leq B\|f\|_1 \quad \text{independently of } b$$

To give an example of a C_∞ function that is not the Fourier transform of an integrable function it is enough to look for a $g \in C_\infty$, odd and such that $\int_1^b (g(x)/x)dx$ is not bounded for $b \to \infty$. This is clearly the case if $g(x) = (\log x)^{-1}$ for large x.

Theorem 1.3 (Multiplication formula). Given $f, g \in L^1$, it holds that

$$\int \hat{f}(x)g(x)dx = \int f(x)\hat{g}(x)dx \tag{1.4}$$

Proof. Both terms of the equation are well defined since $f, g \in L^1$ imply $\hat{f}, \hat{g} \in L^\infty$ and $\hat{f}g, f\hat{g} \in L^1$. Then

$$\int \hat{f}(x)g(x)dx = \int\left(\int f(t)e^{-2\pi i x.t}dt\right)g(x)dx$$

Since $F(x, t) = \exp(-i2\pi x.t)f(t)g(x)$ is a measurable function in $\mathbb{R}^n \times \mathbb{R}^n$ and

$$\iint_{\mathbb{R}^n\times\mathbb{R}^n} |F(x, t)|dx\,dt = \int_{\mathbb{R}^n} |f(t)|dt.\int_{\mathbb{R}^n} |g(x)|dx$$

$$= \|f\|_1 \|g\|_1$$

$F(x, t)$ is also integrable in $\mathbb{R}^n \times \mathbb{R}^n$ and we can apply Fubini's theorem to obtain

$$\int \hat{f}(x)g(x)dx = \int f(x)\left(\int e^{-2\pi i x.t}g(x)dx\right)dt = \int f(t)\hat{g}(t)dt \qquad \nabla$$

An essential feature of harmonic analysis is that the Fourier transform of the convolution product of two functions is the (pointwise) product of their Fourier transforms.

Theorem 1.4. Given $f, g \in L^1$ then

$$(f * g)\hat{}\,(x) = \hat{f}(x).\hat{g}(x) \tag{1.5}$$

Proof. This was already proved as Lemma 4.2 of Chapter 1, by direct computation and Fubini's theorem. ∇

Other important operations are related to the Fourier transform in a simple way. For instance, we have the two following symmetric relations between $\mathcal{F}$ and τ_h, the translation operator, for $f \in L^1$:

(i) $(\tau_h f)\hat{}(x) = e^{-2\pi i h.t}\hat{f}(x)$

$$(\tau_h f)\hat{}(x) = \hat{f}_h(x) = \int f_h(t)e^{-2\pi i x.t}dt$$

$$= \int f(t)e^{-2\pi i(x+h).t}dt$$

$$= \int f(t)e^{-2\pi i x.t}e^{-2\pi i h.t}dt$$

$$= e^{-2\pi i h.t}\hat{f}(x)$$

(ii) $\tau_h f(x) = (e^{2\pi i h.t}f(t))\hat{}(x)$

$$(e^{2\pi i h.t}f(t))\hat{}(x) = \int f(t)e^{2\pi i h.t}e^{-2\pi i x.t}dt$$

$$= \int f(t)e^{-2\pi i(x-h).t}dt$$

$$= \hat{f}(x - h) = \tau_h\hat{f}(x)$$

Before looking for consequences of these symmetric relations, that appear when the Fourier transform acts upon translations, let us see how it acts upon dilations.

Let $a > 0$ and δ_a be the dilation operator defined by $\delta_a f(t) = f(at)$.

(iii) $(\delta_a f)\hat{}(x) = a^{-n}\hat{f}(\delta_{a^{-1}}x)$

$$(\delta_a f)\hat{}(x) = \int f(at)e^{-2\pi i x.t}dt$$

$$= \int f(t)e^{-2\pi i x.(t/a)}a^{-n}dt$$

$$= a^{-n}\hat{f}(x/a)$$

Exercise 1.1. If A is a linear transformation from $\mathbb{R}^n$ onto $\mathbb{R}^n$, given by an $n \times n$ matrix (a_{jk}), show that $(f(At))\hat{}(x) = |\det(a_{jk})|^{-1}f(A^{\#}x)$, where $A^{\#}$ is the contragradient (inverse of the adjoint) of A. In particular, this implies that $(f(0t))^\wedge(x) = \hat{f}(0x)$ for every 0 orthogonal linear transformation from $\mathbb{R}^n$ onto $\mathbb{R}^n$.

Remark 1.4. Exercise 1.1 shows that the Fourier transform of an odd function is an odd function, and that of a radial function is radial.

Going back to properties (i) and (ii) and taking $h = (0, \ldots, h_k, \ldots, 0)$, we get

$$\text{(iv)} \quad \frac{\hat{f}(x + h) - \hat{f}(x)}{|h|} = \frac{\tau_{-h}\hat{f}(x) - \hat{f}(x)}{|h|}$$

$$= \mathscr{F}(|h|^{-1}(e^{-2\pi i h.t} f(t) - f(t)))$$

$$= \mathscr{F}((e^{-2\pi i h.t} - 1)|h|^{-1} f(t))$$

and

$$\text{(v)} \quad \mathscr{F}\left(\frac{f(t + h) - f(t)}{|h|}\right) = \mathscr{F}(|h|^{-1}(\tau_{-h}f(t) - f(t)))$$

$$= (e^{2\pi i h.x} - 1)|h|^{-1}\hat{f}(x)$$

If $f \in L^1(\mathbb{R}^1)$ and we let $|h| \to 0$ in (iv) and (v), we obtain formally

(vi) $\hat{f}'(x) = (-2\pi i t f(t))^\wedge(x)$

(vii) $(f')^\wedge(x) = 2\pi i x \hat{f}(x)$

which again are symmetric results.

We establish now these results rigorously, stating them for $n \geq 1$.

Proposition 1.5. Let $f \in L^1(\mathbb{R}^n)$ and $t_k f(t) \in L^1(\mathbb{R}^n)$, where t_k is the k-th coordinate of $t \in \mathbb{R}^n$. Then $\hat{f}$ is differentiable with respect to x_k (k-th coordinate of $x \in \mathbb{R}^n$) and

$$\frac{\partial f}{\partial x_k} = (-2\pi i t_k f(t))^\wedge(x) \tag{1.6}$$

Proof. Let $h = (0, \ldots, h_k, \ldots, 0)$ with $h_k \neq 0$. By (iv) it is

$$\frac{\hat{f}(x + h) - \hat{f}(x)}{h_k} = \int \frac{e^{-2\pi i h.t} - 1}{h_k} f(t) e^{-2\pi i x.t} dt$$

where $h.t = h_k t_k$. Pointwise $h_k^{-1}(\exp(-2\pi i h.t) - 1) \to 2\pi i t_k$ when $h_k \to 0$. The thesis follows by an application of Lebesgue dominated convergence theorem, since

$$|f(t) e^{-2\pi i x.t} \frac{e^{-2\pi i h.t} - 1}{h_k}| = |f(t)| |\frac{e^{-2\pi i h.t} - 1}{h_k}| \leq C |f(t)| |t_k|$$

and $t_k f(t) \in L^1$. ∇

This result says that to apply Fourier transform after multiplying by the k-th coordinate is equivalent (up to a constant) to take partial derivative of the Fourier transform with respect to the k-th variable.

The Fourier transforms of such derivatives can also be obtained (up to a constant) by multiplying the Fourier transform of the function by the corresponding coordinate function. Although there are many versions of this result, we shall give here one of the simplest, in terms of differentiability in the L^p sense. For this purpose let us consider

Definition 1.2. A function $f \in L^p(\mathbb{R}^n)$ is differentiable in the L^p sense with respect to the k-th variable if there exists a function $g \in L^p(\mathbb{R}^n)$ such that, for $h = (0, \ldots, h_k, \ldots, 0)$,

$$\int |\frac{f(t + h) - f(t)}{|h|} - g(t)|^p dt \to 0 \text{ when } |h| \to 0$$

The function g is called the partial derivative of f in the L^p sense with respect to the k-th variable.

Remark 1.5. The concept of differentiability in the L^p sense is related to that of continuity in L^p (see Chapter 0, Section 3).

To better understand it, observe that

(1) In the case $n = 1$, f has a derivative in the L^p sense if and only if f coincides a.e., with a function that is absolutely continuous locally and has a derivative that belongs to L^p (see [1], page 9);

(2) In general, $f \in L^p$ has a partial derivative in the L^p sense if and only if, considered as a tempered distribution, its corresponding partial derivative (in the sense of distributions) coincides with an L^p function.

Proposition 1.6. Let $f \in L^1(\mathbb{R}^n)$ and $g \in L^1(\mathbb{R}^n)$ be the partial derivative of f in the L^1 sense with respect to the k-th variable. Then

$$\hat{g}(x) = 2\pi i x_k \hat{f}(x) \tag{1.7}$$

Proof. By (v), for $h = (0, \ldots, h_k, \ldots, 0)$,

$$\mathscr{F}\left(\frac{f(t+h) - f(t)}{|h|} - g(t)\right) = \frac{e^{-2\pi i x.h} - 1}{|h|}\hat{f}(x) - \hat{g}(x)$$

By Lemma 1.1(b),

$$\left|\mathscr{F}\left(\frac{f(t+h) - f(t)}{|h|} - g(t)\right)\right| \le \left\|\frac{\tau_{-h}f - f}{|h|} - g\right\|_1$$

and the last term tends to zero when $|h| \to 0$ by definition of g. Thus we obtain the pointwise limit

$$\hat{g}(x) = \lim_{h_k \to 0} (\exp(-2\pi i x_k h_k) - 1) h_k^{-1} \hat{f}(x)$$

$$= -2\pi i x_k \hat{f}(x)$$

∇

The results of Propositions 1.5 and 1.6 extend immediately to higher derivatives. We have then

(viii) $(P(\partial)f)^{\wedge}(x) = P(-2\pi i x)\hat{f}(x)$

and

(ix) $P(\partial)\hat{f}(x) = (P(2\pi i t)f(t))^{\wedge}(x)$

where $P(x)$ is a polynomial in the n variables $x_1, \ldots, x_n$ and $P(\partial)$ is the associated differential polynomial obtained by replacing x^α by ∂^α in $P(x)$.

Remark 1.6. If $f \in L^1$ and g is its partial derivative of order m in the L^1 sense with respect to the k-th variable, $\hat{g}(x) = (-2\pi i x_k)^m \hat{f}(x)$. As $g \in L^1$, by the Riemann-Lebesgue lemma (Theorem 1.2), $(-2\pi i x_k)^m \hat{f}(x) \to 0$ when $|x| \to \infty$. If f has all the partial derivatives in the L^1 sense in every variable up to the order m, it follows that

$$(1 + |x|)^m \hat{f}(x) \to 0 \quad \text{when} \quad |x| \to \infty \tag{1.8}$$

i.e., $\hat{f}(x)$ tends to zero at infinity faster than $(1 + |x|)^{-m}$. Functions that tend to zero at infinity faster than $(1 + |x|)^{-m}$ for all m are called rapidly decreasing functions. So, by (1.8), if f has all the partial derivatives in L^1 of all orders, then $\hat{f}$ is a rapidly decreasing function. This fact illustrates a general principle in Fourier theory: the smoother f is, the smaller $\hat{f}$ is at infinity.

Exercise 1.2. Prove that the set $\mathscr{F}(L^1) = \{\hat{f} : f \in L^1\}$ has the following properties:

(a) $g, h \in \mathscr{F}(L^1)$ implies $g.h \in \mathscr{F}(L^1)$ and $\overline{g} \in \mathscr{F}(L^1)$;

(b) given $x_1 \neq x_2 \in \mathbb{R}^n$ there exists $g \in \mathscr{F}(L^1)$ with $g(x_1) \neq g(x_2)$.

Exercise 1.3. Prove that given $h \in C_\infty$ and $\varepsilon > 0$ there exists $f \in L^1$ with $\|h - \hat{f}\|_\infty < \varepsilon$. Thus $\mathscr{F}(L^1)$ is a dense self-adjoint subalgebra of C_∞. (Hint: use Exercise 1.2 and apply the Stone-Weierstrass theorem to $C(\overline{\mathbb{R}^n}) = C_\infty + \{\lambda.1\}$.)

Exercise 1.4. Given $h \in C_\infty$ and $\varepsilon > 0$, prove that there exists $\phi \in C_0$ such that $\|h - \hat{\phi}\|_\infty < \varepsilon$.

2. FOURIER TRANSFORM OF FINITE MEASURES

Many of the above considerations extend to finite Borel measures $\mu \in \mathscr{M}(\mathbb{R}^n)$.

Definition 2.1. The Fourier transform of $\mu \in \mathscr{M}$ is denoted by $\hat{\mu}$ and is defined by

$$\hat{\mu}(x) = \int_{\mathbb{R}^n} e^{-2\pi i x.t} d\mu(t) \tag{2.1}$$

Proposition 2.1 (Analog to Proposition 1.1). Let $\mu \in \mathscr{M}$, $\hat{\mu}$ its Fourier transform, then

(a) $\mu \to \hat{\mu}$ is a linear mapping.

(b) $\hat{\mu}$ is a bounded function and $\|\hat{\mu}\|_\infty \leq \|\mu\|$.

(c) $\hat{\mu}$ is a uniformly continuous function.

(d) If $\mu \geq 0$ then $\|\mu\| = \hat{\mu}(0) = \|\hat{\mu}\|_\infty$.

Proof.

(a) Follows from the linearity of the integral in (1.1).

(b) $|\hat{\mu}(x)| = |\int e^{-2\pi i x.t} d\mu(t)| \leq \int |d\mu(t)| = \|\mu\|$.

(c) The same proof as in Proposition 1.1 (c).

(d) If $\mu \geq 0$ then $\|\mu\| = \int d\mu(t) = \hat{\mu}(0) \leq \|\hat{\mu}\|_\infty$ and, by (b), this is less than or equal to $\|\mu\|$. ∇

Remark 2.1. The Riemann-Lebesgue lemma (Theorem 1.2) is not valid for finite measures. In fact, if $\mu = \delta$, the Dirac measure, then $\hat{\delta} \equiv 1$, and $\hat{\delta}$ never tends to zero.

The multiplication formula and the convolution theorem still hold for finite measures.

Proposition 2.2. Given $\mu, \nu \in \mathcal{M}$,

$$\int \hat{\mu}(x) d\nu(x) = \int \hat{\nu}(x) d\mu(x) \tag{2.2}$$

Proof. Apply Fubini's theorem to obtain

$$\begin{aligned}\int \hat{\mu}(x) dx &= \int (\int e^{-2\pi i x.t} d\mu(t)) d\nu(x) \\ &= \int (\int e^{-2\pi i x.t} d\nu(x)) d\mu(t) \\ &= \int \hat{\nu}(t) d\mu(t)\end{aligned}$$
∇

Since $\mathcal{M} = (C_\infty)'$ and $L^\infty = (L^1)'$, we may consider, for $\mu \in \mathcal{M}$, $h \in L^\infty$, the functionals I_μ, I_h given by $I_\mu(\phi) = \int \phi d\mu$, $\phi \in C_\infty$, and $I_h(f) = \int fh\, dx$, $f \in L^1$, and the weak-* topologies defined on $\mathcal{M}$ and L^∞.

Proposition 2.3. The linear mapping from $\mathscr{M}$ to L^∞ given by $\mu \to \hat{\mu}$ is continuous with respect to the weak-* topologies, i.e., if $I_{\mu_\alpha}(\phi) \to I_\mu(\phi)$, $\forall \phi \in C_\infty$ then $I_{\hat{\mu}_\alpha}(f) \to I_{\hat{\mu}}(f)$, $\forall f \in L^1$.

Moreover,

$$I_{\hat{\mu}}(f) = I_\mu(\hat{f}) \tag{2.3}$$

Proof. Formula (2.3) corresponds to (2.2) for $\nu \in \mathscr{M}$ given by $d\nu = f\,dx$ for $f \in L^1$. Since $f \in L^1$ implies $\hat{f} \in C_\infty$, $I_{\mu_\alpha}(\hat{f}) \to I_\mu(\hat{f})$, if $\{\mu_\alpha\}$ converges to μ in the weak-* topology, and by (2.2)

$$I_{\hat{\mu}_\alpha}(f) = I_{\mu_\alpha}(\hat{f}) \to I_\mu(\hat{f}) = I_{\hat{\mu}}(f) \qquad \nabla$$

Proposition 2.4. Given $\mu, \nu \in \mathscr{M}$, let $\mu * \nu$ be their convolution. Then

$$(\mu * \nu)\hat{} = \hat{\mu}.\hat{\nu} \tag{2.4}$$

Proof. Let $\sigma = \mu * \nu \in \mathscr{M}$. Its Fourier transform $(\mu * \nu)\hat{} = h \in L^\infty$ defines a continuous linear functional I_h on L^1. By the multiplication formula (2.3), for any $f \in L^1$,

$$I_h(f) = I_{\hat{\sigma}}(f) = I_\sigma(\hat{f}) \tag{2.5}$$

But $\hat{f} \in C_\infty$, and by definition of convolution of two measures,

$$\begin{aligned}
I_\sigma(\hat{f}) &= \int\int \hat{f}(x+y)d\mu(x)d\nu(y) \\
&= \int\int(\int f(t)e^{-2\pi i(x+y).t}dt)d\mu(x)d\nu(y) \\
&= \int f(t)(\int e^{-2\pi ix.t}d\mu(x))(\int e^{-2\pi iy.t}d\nu(y))dt \\
&= \int f(t)\hat{\mu}(t)\hat{\nu}(t)dt \\
&= I_g(f) \qquad (2.6)
\end{aligned}$$

where $g = \hat{\mu}.\hat{\nu} \in L^\infty$. Thus by (2.5) and (2.6), $I_h(f) = I_g(f)$ for all $f \in L^1$ and so $h = g$, which is the thesis. ∇

For each fixed t the function $h(x) = e^{-2\pi ix.t}$ satisfies the identity $h(x - y) = h(x)\overline{h(y)}$, and h has the property that

$$\sum_{j=1}^{m} \sum_{k=1}^{m} h(x_j - x_k)\lambda_j\overline{\lambda}_k \geq 0 \tag{2.7}$$

for every $x_1, \ldots, x_m, \lambda_1, \ldots, \lambda_m$ with $x_j \in \mathbb{R}^n, \lambda_j \in \mathbb{C}$, $j = 1, \ldots, m$. In fact, the left hand of (2.7) is equal to $\Sigma_j \Sigma_k h(x_j)\overline{h(x_k)}\lambda_j\overline{\lambda}_k =$ $(\Sigma_j h(x_j)\lambda_j)(\overline{\Sigma_k h(x_k)\lambda_k}) = |\Sigma_j h(x_j)\lambda_j|^2$ which is always nonnegative.

Definition 2.2. A function $h(x)$, defined for all $x \in \mathbb{R}^n$, is positive definite (p.d.) in the classical sense if it has property (2.7) for every $x_1, \ldots, x_m, \lambda_1, \ldots, \lambda_m$, $x_j \in \mathbb{R}^n$, $\lambda_j \in \mathbb{C}$, $j = 1, \ldots, m$.

Thus, $e^{-2\pi ix.t}$ is p.d. for every fixed $t \in \mathbb{R}^n$. Furthermore,

Proposition 2.5. If $\mu \in \mathcal{M}$, $\mu \geq 0$, then $h = \hat{\mu}$ is a p.d. function.

Proof. If $h(x) = \hat{\mu}(x) = \int e^{-2\pi ix.t} d\mu(t)$, then

$$\Sigma_j \Sigma_k h(x_j - x_k)\lambda_j\overline{\lambda}_k = \int (\Sigma_j \Sigma_k e^{-2\pi i(x_j - x_k).t}\lambda_j\overline{\lambda}_k) d\mu(t)$$

$$= \int |\Sigma_j e^{-2\pi ix_j.t}\lambda_j|^2 d\mu(t) \geq 0$$

since the integrand is nonnegative for every t and $\mu \geq 0$. ∇

In Section 4 we shall prove the Bochner theorem, the converse to Proposition 2.5, which characterizes the Fourier transforms of positive finite measures.

We conclude this section with some basic properties of the p.d. functions that provide a motivation for Section 3.

Exercise 2.1. Prove that if $h(x)$ is a p.d. function then

(a) $h(0) \geq 0$, $h(-x) = \overline{h(x)}$ and $|h(x)| \leq h(0)$ for all x. (Hint: apply (2.7) with $m = 1$ and $m = 2$, $x_1 = x$, $x_2 = 0$.)

(b) $h(0)(1 + 2\lambda^2) + 2\lambda|h(x) - h(y)| - 2\lambda^2 \operatorname{Re} h(x - y) \geq 0$,

hence the discriminant of this polynomial in λ is negative or zero and then

$$|h(x) - h(y)|^2 \leq 2h(0)(h(0) - \operatorname{Re} h(x - y))$$
$$= 2h(0)\operatorname{Re}(h(0) - h(x - y))$$

(Hint: apply (2.7) with $m = 3$, $x_1 = x$, $x_2 = 0$, $x_3 = y$, $\lambda_1 = 1$, $\lambda_2 = \lambda|h(x) - h(y)|/(h(x) - h(y))$, $\lambda_3 = -\lambda_2$ and use (a).)

(c) If h is p.d. and continuous at $x = 0$ then h is uniformly continuous in $\mathbb{R}^n$.

By the exercises above, if h is p.d. and continuous at the origin, then h is continuous and bounded, so that $I_h(f) = \int h(x)f(x)dx$ defines a functional on $f \in L^1$.

Lemma 2.6. Every continuous function h that is p.d. in the classical sense has the property

$$I_h(f * f^*) \geq 0 \quad \text{for all} \quad f \in L^1 \tag{2.8}$$

Remark 2.2.

$$I_h(f * f^*) = \int h(x)(f * f^*)(x)dx = \int h(x)\left(\int f(x - y)\overline{f(-y)}dy\right)dx \tag{2.8a}$$

and, by Fubini's theorem, (2.8) can be rewritten as

$$I_h(f * f^*) = \iint h(x - y)f(x)\overline{f(y)}dx\, dy \geq 0 \tag{2.8b}$$

for all $f \in L^1$.

Proof. Given h continuous and as in Definition 2.2, let us see that the expression (2.8b) is positive for all $f \in L^1$. Since the step functions are dense in L^1 it will be sufficient to prove that (2.8b) is positive for $f = \Sigma_j \lambda_j \chi_{A_j}$, where the sum is finite and the A_j's are disjoint n-dimensional intervals, that can be taken arbitrarily small in measure. Then we have that

$$f(x)\overline{f(y)} = \Sigma_j \Sigma_k \lambda_j \overline{\lambda}_k \chi_{A_j}(x)\chi_{A_k}(y)$$

so that (2.8b) becomes

$$\Sigma_j \Sigma_k \lambda_j \overline{\lambda}_k \iint_{A_j \times A_k} h(x - y)dx\, dy \tag{2.9}$$

Given $\delta > 0$ we choose the A_j's so small that the continuous function $H(x, y) = h(x - y)$ has oscillation less than δ in each $A_j \times A_k$. Then, if for each j we fix a point $x_j \in A_j$,

$$|h(x - y) - h(x_j - x_k)| < \delta \quad \text{for all } (x, y) \in A_j \times A_k$$

Therefore the integral in (2.8b) can be approximated as close as wanted by a finite double sum of the form

$$\begin{aligned} &\Sigma_j \Sigma_k \lambda_j \overline{\lambda}_k \iint_{A_j \times A_k} h(x_j - x_k)dx\, dy \\ &\quad = \Sigma_j \Sigma_k \lambda_j \overline{\lambda}_k h(x_j - x_k)|A_j||A_k| \\ &\quad = \Sigma_j \Sigma_k h(x_j - x_k)\lambda'_j \overline{\lambda'_k} \end{aligned}$$

which is positive by hypothesis. ∇

Definition 2.3. A function $h \in L^\infty$ is positive definite (p.d.) in the integral sense if it satisfies (2.8) (or, equivalently, (2.8b)).

Thus the preceding lemma can be restated as: for a continuous function h, if h is p.d. in the classical sense then h is p.d. in the integral sense.

The converse implication also holds, as we shall prove using Fourier transforms. Nevertheless, this fact can be proved directly.

Exercise 2.2. Prove that every continuous function h that is p.d. in the integral sense is also p.d. in the classical sense. (Hint: given $x_1, \ldots, x_m \in \mathbb{R}^n$, $\lambda_1, \ldots, \lambda_m \in \mathbb{C}$, take $\delta > 0$ so small that the intervals $I_j = \{x : |x - x_j| < \delta\}$, $j = 1, \ldots, m$, do not overlap, and $|h(x - y) - h(x_j - x_k)| < \varepsilon$ for $x \in I_j$, $y \in I_k$; then apply (2.8) with $f(x) = \lambda_j$ if $x \in I_j$ and zero otherwise.)

Remark 2.3. For $h \in L^\infty$, I_h is a continuous linear functional on the Banach algebra L^1. The last definition can be formulated for continuous linear functionals on general Banach algebras, as shown in the next section, where the notion above will be studied in more detail.

3. POSITIVE FUNCTIONALS ON L^1 AND $\mathscr{A}$

In this section we use the notations and concepts of Section 4 of Chapter 1.

For every $\Phi = f + \lambda e \in \mathscr{A}$, the group algebra of $\mathbb{R}^n$, let $\Phi^* = f^* + \overline{\lambda} e$, where $f^*(x) = \overline{f(-x)}$, as in (3.9) of Chapter 1. We have thus defined an operation $\Phi \to \Phi^*$ in $\mathscr{A}$ with the following properties, analog to (3.10) - (3.13) of Chapter 1:

$$(\Phi^*)^* = \Phi, (a\Phi + b\Psi)^* = \overline{a}\Phi^* + \overline{b}\Psi^*, (\Phi * \Psi)^* = \Phi^* * \Psi^* \tag{3.1}$$

$$e^* = e, \|\Phi^*\| = \|\Phi\| \tag{3.2}$$

A Banach algebra with involution is a Banach algebra in which an operation satisfying properties (3.1) and (3.2) is defined. The group algebra $\mathcal{A}$ of $\mathbb{R}^n$ is a Banach algebra with involution. We shall work with the algebra L^1 and its associate unital group algebra $\mathcal{A}$, but the following notions apply to the general case.

Definition 3.1. An element $\Phi \in \mathcal{A}$ is called hermitian if $\Phi = \Phi^*$ and positive if $\Phi = \Psi * \Psi^*$ for some $\Psi \in \mathcal{A}$.

Clearly $e = e^* = e * e^*$, so e is both hermitian and positive. Every $\Phi \in \mathcal{A}$ has a unique representation of the form

$$\Phi = \Psi_1 + i\Psi_2 \text{ with } \Psi_1, \Psi_2 \text{ hermitian} \tag{3.3}$$

since it suffices to take $\Psi_1 = (\Phi + \Phi^*)/2$ and $\Psi_2 = (\Phi - \Phi^*)/2i$. Then $\Phi^* = \Psi_1 - i\Psi_2$.

Lemma 3.1. If $\Phi \in \mathcal{A}$ is hermitian and $\|\Phi\| < 1$, then $e - \Phi$ is positive, i.e., $e - \Phi = \Psi * \Psi^*$, and moreover we may take $\Psi = \Psi^*$, so $e - \Phi = \Psi * \Psi$.

Proof. If λ is real number such that $|\lambda| < 1$, the series $1 - \frac{1}{2}\lambda - \frac{1}{2!}\frac{1}{2}\frac{1}{2}\lambda^2 - \ldots$ is absolutely convergent and its sum s satisfies $s^2 = 1 - \lambda$. The proof of this fact uses only properties which hold in any Banach algebra with involution, hence the series

$$e - \frac{1}{2}\Phi - \frac{1}{2!}\frac{1}{2}\frac{1}{2}\Phi^2 - \ldots$$

where $\Phi^2 = \Phi * \Phi$, $\Phi^3 = \Phi^2 * \Phi$, etc., is absolutely convergent to some element $\Psi \in \mathcal{A}$ and

$$\Psi^2 = \Psi * \Psi = e - \Phi$$

Since $\Phi = \Phi^*$ we also have $\Psi = \Psi^*$ and the thesis holds. ∇

The "polarization identity"

$$f * g^* = \frac{1}{4}((f + g) * (f + g)^* - (f - g) * (f - g)^*$$
$$- i(f + ig) * (f + ig)^* + i(f - ig) * (f - ig)^*) \qquad (3.4)$$

shows that every element of the form $f * g^*$ is a linear combination of positive elements. Hence in the unital algebra $\mathscr{A}$ every element $\Phi = \Phi * e = \Phi * e^*$ is a linear combination of positive elements. In the nonunital algebra L^1 (or in any Banach algebra with involution and approximative unit) we have:

Lemma 3.2. The linear combinations of elements of the form $f * f^*$, $f \in L^1$, form a dense set in L^1.

Proof. By (3.3) and (3.4) it is enough to prove that every hermitian element $h = h^* \in L^1$ is the limit of elements of the form $f * g^*$.

But $h = \lim h * u_j = \lim h^* * u_j$, where $\{u_j\}$ is an approximative unit. ∇

Definition 3.2. A linear functional I in $\mathscr{A}$ is called real if $I(\Phi)$ is real whenever Φ is hermitian.

Every linear functional I in $\mathscr{A}$ has the representation

$$I = I_1 + iI_2, \; I_1 \text{ and } I_2 \text{ real} \qquad (3.5)$$

since it is enough to take $I_1(\Phi) = (I(\Phi) + \overline{I(\Phi^*)})/2$, $I_2(\Phi) = (I(\Phi) - \overline{I(\Phi^*)})/2$ for every $\Phi \in \mathscr{A}$.

From (3.3) it follows that I is real if and only if

$$I(\Phi^*) = \overline{I(\Phi)} \tag{3.6}$$

From (4.6) of Chapter 1 it follows that every character I of is real.

Definition 3.3. A linear functional I in $\mathscr{A}$ (or in L^1) is called positive if $I(\Phi * \Phi^*) \geq 0$ for all $\Phi \in \mathscr{A}$ (respectively, $I(f * f^*) \geq 0$ for all $f \in L^1$).

Every character I in $\mathscr{A}$ is positive since it is real, i.e., $I(\Phi^*) = \overline{I(\Phi)}$, and so

$$I(\Phi * \Phi^*) = I(\Phi)I(\Phi^*) = I(\Phi)\overline{I(\Phi)} = |I(\Phi)|^2 \geq 0$$

If I is positive then

$$<\Phi, \Psi> = I(\Phi * \Psi^*) \tag{3.7}$$

is an (eventually degenerate) scalar product.

Lemma 3.3. If I is a positive functional in $\mathscr{A}$ then

(a) I is real;
(b) $I(\Phi * \Psi^*) = \overline{I(\Psi * \Phi^*)}$;
(c) $|I(\Phi * \Psi^*)|^2 \leq I(\Phi * \Phi^*)I(\Psi * \Psi^*)$;
(d) $|I(\Phi)|^2 \leq I(e)I(\Phi * \Phi^*)$;
(e) I is continuous and $\|I\| = I(e)$.

Remark 3.1. Properties (b) and (c) hold for I positive in L^1.

Proof. For every $\lambda \in \mathbb{C}$ it is $I((\Phi + \lambda\Psi) * (\Phi + \lambda\Psi)^*) \geq 0$, thus

$$I(\Phi * \Phi^*) + \lambda I(\Psi * \Phi^*) + \bar{\lambda} I(\Phi * \Psi^*) + |\lambda|^2 I(\Psi * \Psi^*) \geq 0$$

hence $\lambda I(\Psi * \Phi^*) + \bar{\lambda} I(\Phi * \Psi^*)$ is real, so $I(\Psi * \Phi^*) = \overline{I(\Phi * \Psi^*)}$. This proves (b) and, letting $\Psi = e$, $I(\Phi) = \overline{I(\Phi^*)}$, which proves (a). Since $I(\Phi * \Psi^*) = <\Phi, \Psi>$ has the properties of a scalar product, (c) holds by Schwarz inequality. Letting $\Psi = e$ we get (d) from (c). If $\Phi = \Phi^*$ and $\|\Phi\| < 1$ then, by (a), $I(\Phi)$ is real and, by Lemma 3.1, $e - \Phi$ is positive, so $I(e - \Phi) \geq 0$ or $I(\Phi) \leq I(e)$. Similarly, $I(\Phi) \geq - I(e)$, so $|I(\Phi)| \leq I(e)$ under the conditions on Φ. If Φ is arbitrary but $\|\Phi\| < 1$ then $\Phi^* * \Phi$ is hermitian and $\|\Phi^* * \Phi\| < 1$, so $|I(\Phi^* * \Phi)| \leq I(e)$ and, by (d), $|I(\Phi)|^2 \leq (I(e))^2$ and $|I(\Phi)| \leq I(e)$. Hence $\|I\| \leq I(e)$ and, since $\|e\| = 1$, $I(e) = I(e)\|e\|$, we get $\|I\| \geq I(e)$, and $\|I\| = I(e)$. ∇

Proposition 3.4. If I is a positive continuous linear functional in L^1 then there exists a (continuous) positive functional I_1 in $\mathscr{A}$ such that $I_1(f) = I(f)$ for all $f \in L^1$ and $I_1(e) = \|I_1\| = \|I\|$, so that $\|I_1\|_{\mathscr{A}} = \|I_1\|_{L^1}$. Thus, there is a 1-1 correspondence between the positive continuous functionals I in L^1 and the positive functionals I_1 in $\mathscr{A}$, such that $\|I_1\|_{\mathscr{A}} = \|I\|_{L^1}$.

Proof. Since 3.3(c) holds in L^1, if $\|e_j\| \leq 1$ then

$$|I(e_j * f)|^2 \leq I(e_j * e_j^*) I(f^* * f) \leq \|I\| I(f * f^*)$$

Taking $\{e_j\}$ an approximative unit, we get

$$|I(f)|^2 = \lim |I(e_j * f)|^2 \leq \|I\| I(f * f^*) \tag{3.8}$$

Now for each $f + \lambda e \in \mathscr{A}$ we define $I_1(f + \lambda e) = I(f) + \lambda \|I\|$. Then by (3.8) we get

$$I_1((f + \lambda e) * (f + \lambda e)^*) = I(f * f^*) + \lambda I(f^*) + \bar{\lambda} I(f) + \lambda\bar{\lambda} \|I\|$$

$$\geq \|I\|^{-1}(\lambda\bar{\lambda}\|I\|^2 + \lambda\|I\| I(f^*) + \bar{\lambda}\|I\| I(f) + |I(f)|^2)$$

$$= \|I\|^{-1}(\lambda\|I\| + I(f))(\bar{\lambda}\|I\| + \overline{I(f)}) \geq 0$$

Thus I_1 is positive in $\mathscr{A}$, $I_1 = I$ in L^1 and, by 3.3(e), $\|I_1\| = I_1(e) = \|I\|$. ∇

We are going to determine the precise relationship between positive functionals and characters in $\mathscr{A}$.

$\mathscr{A}$ is a Banach space and has a dual space which will be denoted by $\mathscr{A}'$. Let $\Sigma = \{I \in \mathscr{A}' : \|I\| = 1\}$ be the unit sphere of $\mathscr{A}'$, let $\mathscr{P} = \{I$ positive functional in $\mathscr{A}\}$, and let $\mathscr{P}_1 = \{I \in \mathscr{P} : I(e) = 1\} = \{I \in \mathscr{A}'$ positive: $\|I\| = 1\}$.

Lemma 3.5. $\mathscr{P}_1 \subset \Sigma$ and $\mathscr{P}_1$ is convex and compact in the weak-* topology.

Proof. By definition $\mathscr{P}_1 \subset \Sigma$, and since Σ is compact in the weak-* topology (by the theorem of Alaoglu-Bourbaki), it is enough to prove that $\mathscr{P}_1$ is closed in this topology. Let $\{I_\alpha\} \subset \mathscr{P}_1$, $I \in \Sigma$, $\lim I_\alpha(\Phi) = I(\Phi)$ for every $\Phi \in \mathscr{A}$. We claim that $I \in \mathscr{P}_1$: this follows immediately from $I(\Phi * \Phi^*) = \lim I_\alpha(\Phi * \Phi^*) \geq 0$ and $I(e) = \lim I_\alpha(e) = 1$. ∇

As it was already remarked, every character of $\mathscr{A}$ belongs to $\mathscr{P}_1$. Moreover,

Proposition 3.6. The characters of $\mathscr{A}$ are precisely the extreme points of the convex set $\mathscr{P}_1$.

Proof.

(1) Let $I \in \mathscr{P}_1$ be an extreme point; we claim that I is multiplicative, i.e., $I(\Phi * \Psi) = I(\Phi)I(\Psi)$ for every $\Phi, \Psi \in \mathscr{A}$. We may assume that Ψ is positive and that $\|\Psi\| < 1$. Thus,

let $\Psi = \Psi_1 * \Psi_1^*$, so that by Lemma 3.1, $e - \Psi = \Psi_2 * \Psi_2^*$, and $I(\Psi) \geq 0$, $I(e - \Psi) \geq 0$ and $I(\Psi) < 1$ since $\|\Psi\| < 1$. Consider the case $I(\Psi) > 0$ so that $0 < I(\Psi) < 1$, $0 < I(e - \Psi) < 1$. Let us fix Ψ and define I_1, I_2 by

$$I_1(\Phi) = I(\Phi * \Psi)/I(\Psi), \quad I_2(\Phi) = I(\Phi * (e - \Psi))/I(e - \Psi) \tag{3.9}$$

Since $I_1(\Phi * \Phi^*) = I(\Phi * \Phi^* * \Psi_1 * \Psi_1^*)/I(\Psi) \geq 0$, $I_1(e) = I(\Psi)/I(\Psi) = 1$, we have that $I_1 \in \mathscr{P}_1$ and similarly $I_2 \in \mathscr{P}_1$. By (3.9),

$$I = \lambda I_1 + (1 - \lambda)I_2$$

where $0 < \lambda = I(\Psi) < 1$ and, since I is extreme, it must be $I_1 = I$, which amounts to $I(\Phi * \Psi) = I(\Phi)I(\Psi)$.

If $I(\Psi) = 0$, by 3.3(c) it is

$$\begin{aligned} |I(\Phi * \Psi)|^2 &= |I(\Phi * \Psi_1 * \Psi_1^*)|^2 \\ &\leq I(\Phi * \Psi_1 * \Phi^* * \Psi_1^*)I(\Psi_1^* * \Psi_1) = 0 \end{aligned}$$

since $I(\Psi_1^* * \Psi_1) = I(\Psi) = 0$, and so, $I(\Phi * \Psi) = 0 = I(\Phi)I(\Psi)$.

(2) Let $I \in \mathscr{P}_1$ be multiplicative and let us prove that I is extreme i.e., if $I = (I_1 + I_2)/2$ with $I_1, I_2 \in \mathscr{P}_1$ then $I = I_1 = I_2$. Since $I(e) = I_1(e)$ it will be enough to prove that $I_i(\Phi) = 0$, $i = 1, 2$, whenever $I(\Phi) = 0$. But if $I(\Phi) = 0$, then $0 = 2I(\Phi^*)I(\Phi) = 2I(\Phi * \Phi^*) = I_1(\Phi * \Phi^*) + I_2(\Phi * \Phi^*)$, so $I_i(\Phi * \Phi^*) = 0$ and $0 \leq I_i(\Phi) = I_i(e * \Phi) \leq I_i(e * e)I_i(\Phi * \Phi^*) = 0$ for $i = 1, 2$. ∇

<u>Corollary 3.7.</u> Every element of $\mathscr{P}_1$ is the weak-* limit of convex combinations of characters of $\mathscr{A}$.

<u>Proof.</u> Follows from a direct application of the Krein-Milman theorem (see Chapter 0, Section 4). ∇

4. THE BOCHNER THEOREM

To each function $h \in L^{\infty}$ corresponds the functional $I_h \in (L^1)'$. We write $h \in L^{\infty}_{p.d.}$ if h is (bounded and) p.d. in the integral sense (Definition 2.3.). Let $\mathscr{P}^*$ denote the set of all positive functionals $I \in \mathscr{P}$, acting on $\mathscr{A} = L^1 + \{\lambda e\}$, such that $I(e) = \|I\|$, the norm of I as functional in L^1 (cfr. Proposition 3.4).

Lemma 4.1. $h \longleftrightarrow I_h$ is a 1-1 correspondence between $L^{\infty}_{p.d.}$ and $\mathscr{P}^*$.

Proof. If $h \in L^{\infty}_{p.d.}$ then, by definition, I_h is a positive functional in L^1, and by Proposition 3.4, we may write $I_h \in \mathscr{P}^*$. Conversely, if $I \in \mathscr{P}^*$ then I is a positive linear functional in L^1 and $I = I_h$ for some $h \in L^{\infty}$. By (2.8), such $h \in L^{\infty}_{p.d.}$ ∇

Theorem 4.2 (The Bochner theorem). The following conditions on a function h are equivalent:

(a) $h(x) = \hat{\mu}(x)$ for all $x \in \mathbb{R}^n$, where μ is a positive finite Borel measure;

(b) h is a continuous function p.d. in the classical sense;

(c) h is a (bounded) continuous function p.d. in the integral sense.

Proof. That (a) implies (b) was already proved as Proposition 2.5, and that (b) implies (c) as Lemma 2.6. It remains to prove that (c) implies (a).

Let h be as in (c), i.e., $h \in L^{\infty}_{p.d.}$ and is continuous. By Lemma 4.1 the corresponding $I_h \in \mathscr{P}^*$. Since the equality $I_h(f) = I_{\hat{\mu}}(f)$ for all $f \in L^1$ would imply $h(x) = \hat{\mu}(x)$ a.e., and therefore $h(x) = \hat{\mu}(x)$ for all x (h and $\hat{\mu}$ are continuous), it will suffice to prove that every $I \in \mathscr{P}^*$ coincides on L^1 with a functional of the form I_μ for some $\mu \geq 0$. Moreover, we may clearly assume (without loss) that $I(e) = 1$, i.e., that $I \in \mathcal{P}_1^* = \mathcal{P}_1 \cap \mathcal{P}^*$.

Let $\mathcal{M}_1^+ = \{\mu \in \mathcal{M} : \mu \geq 0, \|\mu\| \leq 1\}$ and let $F(\mu) = I_{\hat{\mu}}$. By the implication (a) $\Longrightarrow$ (c) and Lemma 4.1, F maps $\mathcal{M}_1^+$ into $\mathcal{P}^*$. If we endow $\mathcal{M}_1^+$ with the weak $(\mathcal{M}_1^+, C_\infty)$-topology and $\mathcal{P}^*$ with the weak $(\mathcal{P}^*, L^1)$-topology (see Chapter 0, Section 4, for the notation), then, by Proposition 2.3, F is a continuous map. Since by the Helly-Bray theorem $\mathcal{M}_1^+$ is convex and compact in that topology (Chapter 0, Section 3), so is $F(\mathcal{M}_1^+)$ (in its topology).

Observe that all the characters of $\mathbb{R}^n$ are given by the Fourier transform of a measure $\mu \in \mathcal{M}_1^+$. In fact, if $\mu = \mu_z$ is the Dirac measure at z, then $\hat{\mu}_z(x) = \int \exp(-2\pi i x . t) d\mu_z(t) =$ $= \exp(-2\pi i x . z) = h_{-2\pi z}(x)$, the character corresponding to $-2\pi z$, while $h_\infty(x) \equiv 0 = \hat{\mu}(x)$ for all x, if $\mu = 0$. Proposition 4.3 of Chapter 1 states that every character of $\mathcal{A}$ is of the form I_h, h character of $\mathbb{R}^n$, and therefore coincides on L^1 with some functional $I_{\hat{\mu}} = F(\mu) \in F(\mathcal{M}_1^+)$. Since $F(\mathcal{M}_1^+)$ is convex and closed in the weak $(\mathcal{P}^*, L^1)$-topology, the same is true for every element of $\mathcal{P}^* \subset \mathcal{P}$ which is a weak $(\mathcal{P}, \mathcal{A})$-limit of convex combinations of characters I_h of $\mathcal{A}$ (the $(\mathcal{P}, \mathcal{A})$-convergence implies the $(\mathcal{P}, L^1)$-convergence).

But by Corollary 3.7, every $I \in \mathcal{P}_1^* \subset \mathcal{P}_1$ is a weak $(\mathcal{P}, \mathcal{A})$-limit of convex combinations of characters I_h. Hence, every $I \in \mathcal{P}_1^*$ coincides on L^1 with some $I_{\hat{\mu}}$ and the thesis follows. ▽

<u>Corollary 4.3.</u> Every function $h \in L^\infty$ that is p.d. in the integral sense is equal a.e. to a continuous function that is p.d. in the classical sense.

Similar results hold for the discrete line $\mathbb{Z}$. We have already remarked in Section 5 of Chapter 1, that $\mathbb{Z}$ is the dual group of $\mathbb{T}$ and that $L^1(\mathbb{Z}) = \ell^1$ being a unital Banach algebra, $\mathscr{A}(\mathbb{Z}) = \ell^1$. The elements of $L^1(\mathbb{Z})$ are not functions but sequences, thus p.d. sequences are to be considered.

Definition 4.1. A numerical sequence $\{a_k\}$ is positive definite (p.d.) if for any finite sequence $\{\lambda_k\}$,

$$\sum_{j=1}^{m} \sum_{k=1}^{m} a_{j-k} \lambda_j \bar{\lambda}_k \geq 0 \tag{4.1}$$

The following theorem characterizes the Fourier coefficients of positive measures defined in $\mathbb{T}$.

Theorem 4.4 (The Herglotz-Bochner theorem). A numerical sequence $\{a_k\}$ is p.d. if and only if there exists a positive measure μ in $\mathbb{T}$ such that $a_k = \hat{\mu}(k) = \int_{\mathbb{T}} e^{-2\pi ikt} d\mu(t)$ for every $k \in \mathbb{Z}$.

Exercise 4.1. Develop in detail the theory of Sections 2, 3 and 4 for the case $\mathbb{Z}$ and deduce the Herglotz-Bochner theorem. (Note that the analog of the passage from $L^1(\mathbb{R}^n)$ to $\mathscr{A}(\mathbb{R}^n)$ in Section 3 is unnecessary in this case). (For further information see [2], Chapter 1).

Remark 4.1. The groups $\mathbb{R}^n$, $\mathbb{Z}$ and $\mathbb{T}$ are examples of locally compact topological groups. As on general locally compact topological groups there exists an invariant measure, the Haar measure, that replaces the Lebesgue measure, the whole contents of the preceding Sections 2, 3 and 4 as well of the following Section 5 admit an extension to the abstract case. (For further information see [2], Chapter 7 and [3], Chapter 4. On a more advanced level, see [4] and [5].)

Remark 4.2. In this chapter we have considered the class of finite measures $\mathcal{M}(\mathbb{R}^n)$ as an extension of $L^1(\mathbb{R}^n)$. The theory extends also for nonfinite measures, by defining convolution, Fourier transform and positive definiteness in terms of (tempered) distributions, ((tempered) measures are a particular case of these). In such a context a generalization of Theorem 4.3, the Bochner-Schwartz theorem, can be proved, characterizing the Fourier transforms of positive tempered measures as positive definite distributions.

5. AN APPLICATION TO CONVERGENCE THEOREMS

We shall derive some convergence theorems as a consequence of the Bochner theorem. They are essential in applications of the Fourier transform to probability theory.

In $\mathcal{M} = (C_\infty)'$ the weak-* convergence is defined by $\mu_k \to \mu$ weakly-* if $I_{\mu_k}(\phi) \to I_\mu(\phi)$ for all $\phi \in C_\infty$. By the Banach-Steinhaus theorem (see [6], p. 171), if $\{\mu_k\}$ converges weakly-* then it is uniformly bounded, i.e., $\|\mu_k\| \le C$ for all k, and $\|\mu\| \le \lim \|\mu_k\| \le C$. However it is not necessarily true that $\|\mu\| = \lim \|\mu_k\|$.

Exercise 5.1. Show that $\{\mu_k = \delta_k\}$ converges weakly-* but $\|\lim \mu_k\| \ne \lim \|\mu_k\|$.

For positive measures $\mu \in \mathcal{M}^+$, $\|\mu\| = I_\mu(1)$, where 1 stands for the function constantly equal to one. If $\mu, \mu_k \in \mathcal{M}_1^+$, $k = 1, 2, \ldots$, and $I_\mu(1) = \lim I_{\mu_k}(1)$, then the measures are uniformly tight in the following sense:

Lemma 5.1 (Uniform tightness). If $\mu, \mu_k \in \mathcal{M}^+$, $k = 1, 2, \ldots, \mu_k \to \mu$ weakly-* and $I_{\mu_k}(1) \to I_\mu(1)$, then for every $\varepsilon > 0$ there exist a compact set K and a function $\Psi \in C_0$, $0 \le \Psi \le 1$, such that

$$\mu(K^c) < \varepsilon, \quad \mu_k(K^c) < \varepsilon \tag{5.1}$$

where K^c is the complement of K, and

$$I_\mu(1 - \Psi) < \varepsilon, \quad I_{\mu_k}(1 - \Psi) < \varepsilon \tag{5.2}$$

for $k = 1, 2, \ldots$.

<u>Proof</u>. Since $\mu \in \mathcal{M}^+$, for every $\varepsilon > 0$ there exists a compact set $K_\varepsilon \subset \mathbb{R}^n$ such that $\mu(K_\varepsilon^c) < \varepsilon/2$. Let $K \supset K_\varepsilon$ be a larger compact so that again $\mu(K^c) < \varepsilon/2$, and take $\Psi \in C_0$ with supp $\Psi \subset K$ and $\Psi = 1$ on K_ε, $0 \leq \Psi \leq 1$, so that $\chi_{K_\varepsilon} \leq \Psi \leq \chi_K$ and $I_\mu(1 - \Psi) < \varepsilon$.

If $\mu_k \to \mu$ weakly-* then $\mu_k(K^c) < \varepsilon$ and $I_{\mu_k}(1 - \Psi) < \varepsilon$ for all k sufficiently large, and hence for all k, since

$$\mu_k(K^c) = I_{\mu_k}(1) - \int_K d\mu_k \leq I_{\mu_k}(1) - \int \Psi d\mu_k$$

$$\leq I_\mu(1) - \int \Psi d\mu + \varepsilon/2$$

$$\leq I_\mu(1) - \int_{K_\varepsilon} d\mu + \varepsilon/2$$

$$= \mu(K_\varepsilon^c) + \varepsilon/2 < \varepsilon$$

for k large. ∇

The weak-* topology of $\mathcal{M}^+$ is given by the I_μ's considered as functionals on C_∞. But I_μ can also be considered as a continuous functional on C_b, the set of all continuous bounded functions. For every $\phi \in C_b$ the integral $I_\mu(\phi)$ exists and $|I_\mu(\phi)| \leq \|\phi\|_\infty I_\mu(1)$. Since $C_\infty \subset C_b$, the convergence $I_{\mu_k}(\phi) \to I_\mu(\phi)$ for every $\phi \in C_b$ clearly implies the weak-* convergence. The converse holds under an additional condition.

Proposition 5.2. If $\mu, \mu_k \in \mathcal{M}^+$, $k = 1, 2, \ldots$, then $I_{\mu_k}(\phi) \to I_\mu(\phi)$ for all $\phi \in C_b$ if and only if $\mu_k \to \mu$ weakly-* and $I_{\mu_k}(1) \to I_\mu(1)$.

Proof. The "only if" implication is immediate since $1 \in C_b$ and $C_\infty \subset C_b$. Conversely, for a given $\varepsilon > 0$, let K and Ψ be as in Lemma 5.1. If $\phi \in C_b$ then $\phi . \Psi \in C_0 \subset C_\infty$ and $I_{\mu_k}(\phi . \Psi) \to I_\mu(\phi . \Psi)$. Thus,

$$|I_{\mu_k}(\phi) - I_\mu(\phi)| \leq |I_{\mu_k}(\phi . \Psi) - I_\mu(\phi . \Psi)| + |I_{\mu_k}(\phi(1 - \Psi))|$$
$$+ |I_\mu(\phi(1 - \Psi))| < \varepsilon + 2\|\phi\|_\infty \varepsilon$$

for k sufficiently large, and the thesis follows. ∇

The next propositions enable us to deal with the convergence of a sequence of positive finite measures as well as with the sequence of their Fourier transforms.

Theorem 5.3. Let $\mu, \mu_k \in \mathcal{M}^+$, $k = 1, 2, \ldots$, and let $\{F_k = \hat{\mu}_k\}$ be the sequenee of the Fourier transforms. If $\mu_k \to \mu$ weakly-*, and if $I_{\mu_k}(1) \to I_\mu(1)$, then $F_k(x) \to F(x) = \hat{\mu}(x)$ for all $x \in \mathbb{R}^n$ and F is continuous.

Proof. By Proposition 5.2, $I_{\mu_k}(\phi) \to I_\mu(\phi)$ for all $\phi \in C_b$. In particular, for $\phi_x(t) = e^{-2\pi i x . t}$, $x \in \mathbb{R}^n$, it is

$$I_{\mu_k}(\phi_x) = \hat{\mu}_k(x) = F_k(x) \to I_\mu(\phi_x) = \hat{\mu}(x) = F(x)$$

F is continuous since it is the Fourier transform of $\mu \in \mathcal{M}$. ∇

Remark 5.1. Under the hypothesis of Theorem 5.3 more can be said on the sequence $\{F_k\}$. In fact, since $|\exp(-2\pi i h . t) - 1| \in C_b$, Proposition 5.2 asserts

$$|F_k(x+h) - F_k(x)| \leq \int |e^{-2\pi i h.t} - 1| d\mu_k(t)$$
$$< \int |e^{-2\pi i h.t} - 1| d\mu(t) + \varepsilon/2$$

for every $\varepsilon > 0$ and $k \geq N = N(\varepsilon)$. Now $|\exp(-2\pi i h.t) - 1| < \varepsilon/2\|\mu\|$ for $|h|$ sufficiently small, independently of x.

Therefore, $\{F_k\}$ is an uniformly equicontinuous family. An application of the Ascoli theorem shows that $F_k(x) \to F(x)$ uniformly on compacts. (For this and the concept of equicontinuity, see [6], pp. 153-155.)

Theorem 5.4. Assume that the sequence $\{\mu_k\} \subset \mathcal{M}^+$ be uniformly bounded and let $\{F_k = \hat{\mu}_k\}$ be the sequence of their Fourier transforms. If $F_k(x) \to F(x)$ for all $x \in \mathbb{R}^n$ and F is continuous at the origin, there exists a $\mu \in \mathcal{M}^+$ such that (a) $\mu_k \to \mu$ weakly-*, and (b) $\hat{\mu} = F$.

Proof. For each k, $F_k = \hat{\mu}_k$ is a continuous function that, by Proposition 2.5, is p.d. (in the classical sense). Therefore the limit function F is also p.d. (in the classical sense). Being continuous at the origin, F is also uniformly continuous in $\mathbb{R}^n$ (see Exercise 2.3) and, by the Bochner theorem (Theorem 4.3), there exists a $\mu \in \mathcal{M}^+$ such that $\hat{\mu} = F$, so (b) is satisfied.

To prove (a) is to show $I_{\mu_k}(\phi) \to I_\mu(\phi)$ for all $\phi \in C_\infty$. By Exercise 1.4, given $\phi \in C_\infty$ and $\varepsilon > 0$, there exists $\theta \in C_0$ with $\|\phi - \hat{\theta}\|_\infty < \varepsilon$. Moreover, by Proposition 2.2,

$$I_{\mu_k}(\hat{\theta}) = \int \hat{\theta} d\mu_k = \int \hat{\mu}_k \theta \, dx$$

where $\hat{\mu}_k . \theta \in L^1$ and $\hat{\mu}_k(x).\theta(x) \to \hat{\mu}(x).\theta(x)$ for all $x \in \mathbb{R}^n$. By dominated convergence, then,

$$I_{\mu_k}(\hat{\theta}) \to \int \hat{\mu}\theta \, dx = \int \hat{\theta} \, d\mu = I_\mu(\hat{\theta})$$

Therefore, as $\mu_k, \mu \in \mathcal{M}^+$,

$$|I_{\mu_k}(\phi) - I_\mu(\phi)| \leq I_{\mu_k}(|\phi - \hat{\theta}|) + |I_{\mu_k}(\hat{\theta}) - I_\mu(\hat{\theta})| + I_\mu(|\hat{\theta} - \phi|)$$

$$\leq \|\phi - \hat{\theta}\|_\infty (\|\mu_k\| + \|\mu\|) + |I_{\mu_k}(\hat{\theta}) - I_\mu(\hat{\theta})|,$$

and (a) follows from the uniform boundedness of $\{\mu_k\}$. ∇

A measure $\mu \in \mathcal{M}^+$ is called a probability measure if $\|\mu\| = I_\mu(1) = 1$. The following corollary is an immediate consequence of Theorems 5.3 and 5.4.

Corollary 5.5 (The P. Lévy-Cramér continuity theorem). Let $\{\mu_k\}$ be a sequence of probability measures. The following conditions are equivalent:

(a) $\mu_k \to \mu$ weakly-*, where μ is a probability measure;

(b) $I_{\mu_k}(\phi) \to I_\mu(\phi)$ for all $\phi \in C_b$, where μ is a probability measure;

(c) $\hat{\mu}_k(x) \to F(x)$ for all $x \in \mathbb{R}^n$, where F is a function that is continuous at the origin.

REFERENCES

1. S. Bochner and K. Chandrasekharan, Fourier Transforms, Princeton University Press, Princeton, 1949.

2. Y. Katznelson, An Introduction to Harmonic Analysis, Wiley, New York, 1968.

3. H. Dym and H. P. Mc Kean, Fourier Series and Integrals, Academic Press, New York, 1972.

4. E. Hewitt and K. A. Ross, Abstract Harmonic Analysis, Springer-Verlag, New York, 1963-1970.

5. W. Rudin, Fourier Analysis on Groups, Wiley (Interscience), New York, 1963.

6. H. L. Royden, Real Analysis, MacMillan, New York, 1964.

Chapter 3

INVERSION THEORY AND HARMONIC FUNCTIONS

1. SUMMATION OF FOURIER INTEGRALS

In the previous chapter we introduced the Fourier transforms of integrable functions and finite Borel measures defined on $\mathbb{R}^n$. In both cases, the Fourier transforms lie in L^∞. The _inversion problem_ of determining f (or μ) given $\hat{f}$ (or $\hat{\mu}$) is particularly difficult in this context since in general $\hat{f}$ (and $\hat{\mu}$) $\notin L^1$ and so the _Fourier integral_

$$\int_{\mathbb{R}^n} \hat{f}(x) e^{2\pi i x \cdot t} dx \tag{1.1}$$

or the _Fourier-Stieltjes integral_

$$\int_{\mathbb{R}^n} \hat{\mu}(x) e^{2\pi i x \cdot t} dx \tag{1.1a}$$

need not exist.

Since the Fourier integral (1.1) is just the formula to get f from $\hat{f}$, as the case of good functions shows (see Exercise 1.4 of Chapter 2), the problem consists in finding a way to give meaning to Fourier integrals and to seek conditions to insure the convergence of the Fourier integral to the original function. This is what this section is about.

Definition 1.1. Given the integral $\int F(x)dx$, its Abel mean of order $\varepsilon > 0$, is the integral

$$A_\varepsilon(F) = A_\varepsilon = \int F(x)e^{-\varepsilon|x|}dx \tag{1.2}$$

For $F \in L^1$ it is evident that $A_\varepsilon(F)$ exists for every $\varepsilon > 0$ and that $\lim_{\varepsilon\to 0} A_\varepsilon(F) = \int F(x)dx$. But the interesting fact is that the Abel means may exist even if $F \notin L^1$. It exists, for instance, when F be bounded. Whenever $\lim_{\varepsilon\to 0} A_\varepsilon(F)$ exists, F is said to be Abel summable to $\ell = \lim_{\varepsilon\to 0} A_\varepsilon(F) = \lim_{\varepsilon\to 0} \int F(x)e^{-\varepsilon|x|}dx$.

Another method of summability of integrals is the Gauss method.

Definition 1.2. Given the integral $\int F(x)dx$, its Gauss mean of order $\varepsilon > 0$, is the integral

$$G_\varepsilon(F) = G_\varepsilon = \int F(x)e^{-\varepsilon|x|^2}dx \tag{1.3}$$

F is said to be Gauss summable to ℓ if

$$\lim_{\varepsilon\to 0} G_\varepsilon(F) = \lim_{\varepsilon\to 0} \int F(x)e^{-\varepsilon|x|^2}dx = \ell \quad \text{exists}$$

Abel and Gauss means are particular cases of Φ-means of $\int F(x)dx$, given by,

$$M_{\Phi,\varepsilon}(F) = \int F(x)\Phi(\varepsilon x)dx, \quad \varepsilon > 0 \tag{1.4}$$

for Φ such that $F(x)\Phi(\varepsilon x) \in L^1$ for every $\varepsilon > 0$, with $\lim_{\varepsilon\to 0} \Phi(\varepsilon x) = 1$.

F is said to be Φ-summable to ℓ if $\lim_{\varepsilon\to 0} M_{\Phi,\varepsilon}(F) = \ell$ exists. Our purpose is, given the Fourier transform $\hat{f}(x)$ and the Fourier integral $\int \hat{f}(x)e^{2\pi ix.t}dx$, which need not be convergent, to compute its Abel and Gauss means, in order to see if we can assure summability for (1.1). In these computations we are going to use the multiplication formula (Theorem 1.3 of Chapter 2) in order to change an integration involving $\hat{f}$ (which may be nonintegrable) into another involving the integrable f, so we shall have to compute the Fourier transforms of $\exp(-2\pi\varepsilon|t|)$ and $\exp(-4\pi\varepsilon^2|t|^2)$.

Both computations are very easy in the one dimensional case (we will carry them on for $\varepsilon = 1$ and then change variables):

$$\text{(a)} \quad \int_{\mathbb{R}^1} e^{-2\pi|t|}e^{-2\pi ix.t}dt = 2\int_0^\infty e^{-2\pi|t|}\cos 2\pi xt\, dt = \frac{1}{\pi(1+x^2)}$$

Let us now consider

$$I = \int_{\mathbb{R}^1} e^{-\pi|t|^2}e^{-2\pi ix.t}dt = \int_{-\infty}^{\infty} e^{-\pi(t^2+2\pi ixt)}dt$$

Multiplying and dividing the last integral by $\exp(\pi x^2)$ and calling $t + ix = z$, we get $I = e^{-\pi x^2}\int_C e^{-\pi z^2}dz$, where C is the line $\{t + ix : -\infty < t < +\infty\}$. By Cauchy's theorem on integration of analytic functions, the integral over C is equal to $\int_{-\infty}^{\infty} e^{-\pi x^2}dx = 1$. Thus,

$$\text{(b)} \quad \int_{\mathbb{R}^1} e^{-\pi|t|^2}e^{-2\pi ixt}dt = e^{-\pi|x|^2}$$

For $e^{-\pi|t|^2}$ it is very simple to pass from the case $n = 1$ to the case $n > 1$ since $f(t) = f_1(t_1) \dots f_n(t_n)$ implies $\hat{f}(x) = \hat{f}_1(x_1) \dots \hat{f}_n(x_n)$. For $t \in \mathbb{R}^n$, $n > 1$ it is $\exp(-\pi|t|^2) = \exp(-\pi(t_1^2+\dots+t_n^2)) = \exp(-\pi t_1^2)\dots(\exp(-\pi t_n^2)$, and we have that $(e^{-\pi|t|^2})\hat{}(x) = \prod_{k=1}^{n} (e^{-\pi x_k^2}) = e^{-\pi|x|^2}$, so we obtain <u>a function that is its own Fourier transform.</u>

Changing variables $(y = \sqrt{\varepsilon}t)$ we get

<u>Lemma 1.1.</u>

$$(e^{-\pi\varepsilon|t|^2})\hat{}(x) = \int e^{-\pi|t|^2} e^{-2\pi i x.t} dt = \varepsilon^{-n/2} e^{-\pi|x|^2/\varepsilon} \tag{1.5}$$

for every $\varepsilon > 0$.

The n-dimensional transform corresponding to (a) requires a more involved computation, since we cannot use $(\prod_{k=1}^{n} f_k)\hat{} = \prod_{k=1}^{n} \hat{f}_k$.

<u>Lemma 1.2.</u> For every $\varepsilon > 0$,

$$(e^{-2\pi\varepsilon|t|})\hat{}(x) = \int_{\mathbb{R}^n} e^{-2\pi\varepsilon|t|} e^{-2\pi i x.t} dt = C_n \frac{\varepsilon}{(\varepsilon^2+|x|^2)^{(n+1)/2}} \tag{1.6}$$

where

$$C_n = \Gamma(\frac{n+1}{2})\pi^{-(n+1)/2} \tag{1.6a}$$

<u>Proof.</u> By the change of variables $y = \varepsilon t$ it is enough to prove that

$$(e^{-2\pi|t|})\hat{}(x) = C_n(1+|x|^2)^{-(n+1)/2} \tag{1.7}$$

To get (1.7) we use formula (1.5), which is connected to our computation by the following formula, valid for all $s > 0$, that we assume now and prove later:

$$e^{-s} = \frac{1}{\sqrt{\pi}} \int_0^\infty \frac{e^{-u}}{\sqrt{u}} e^{-s^2/4u} du \qquad (1.8)$$

Taking $s = 2\pi|t|$,

$$\int_{\mathbb{R}^n} e^{-2\pi|t|} e^{-2\pi i x.t} dt = \int_{\mathbb{R}^n} \left(\frac{1}{\sqrt{\pi}} \int_0^\infty \frac{e^{-u}}{\sqrt{u}} e^{-\pi^2|t|^2/u} du\right) e^{-2\pi i x.t} dt$$

$$= \frac{1}{\sqrt{\pi}} \int_0^\infty \frac{e^{-u}}{\sqrt{u}} \left(\int_{\mathbb{R}^n} e^{-\pi^2|t|^2/u} dt\right) e^{-2\pi i x.t} du$$

(by Lemma 1.1 for $\varepsilon = \pi/\sqrt{u}$)

$$= \pi^{-(n+1)/2} \int_0^\infty e^{-u} u^{(n-1)/2} e^{-u|x|^2} du$$

$(s = (1 + |x|^2)u)$

$$= \pi^{-(n+1)/2} (1 + |x|^2)^{-(n+1)/2} \int_0^\infty e^{-s} s^{(n-1)/2} ds$$

$$= \pi^{-(n+1)/2} \Gamma\left(\frac{n+1}{2}\right)(1 + |x|^2)^{-(n+1)/2}$$

It is enough to establish the lemma to prove (1.8). This equality follows from two others:

$$\frac{1}{1+x^2} = \int_0^\infty e^{-(1+x^2)u} du \qquad (1.9)$$

and

$$e^{-s} = \frac{2}{\pi}\int_0^\infty \frac{\cos sx}{1+x^2}\,dx, \qquad s > 0 \tag{1.10}$$

where (1.9) is immediate and (1.10) is computed by applying the residues method to the function $\exp(isz).(1+z^2)^{-1}$. Thus,

$$e^{-s} = \frac{2}{\pi}\int_0^\infty \frac{\cos sx}{1+x^2}\,dx = \frac{2}{\pi}\int_0^\infty \cos sx\left(\int_0^\infty e^{-(1+x^2)u}\,du\right)dx$$

$$= \frac{2}{\pi}\int_0^\infty e^{-u}\left(\int_0^\infty \cos sx\, e^{-x^2u}\,dx\right)du$$

$$= \frac{1}{\pi}\int_0^\infty e^{-u}\left(\int_{-\infty}^\infty e^{isx}e^{-x^2u}\,dx\right)du$$

$(x = 2\pi t)$

$$= \frac{1}{\pi}\int_0^\infty e^{-u}\left(2\pi\int_{-\infty}^\infty e^{2\pi ist}e^{-4\pi^2t^2u}\,dt\right)du$$

(by Lemma 1.1 for $n = 1$, $\varepsilon = 4\pi u$)

$$= \frac{1}{\sqrt{\pi}}\int_0^\infty \frac{e^{-u}}{\sqrt{u}}e^{-s^2/4u}\,du$$

and the lemma is proved. ∇

In the following we denote by P and W the Fourier transforms of $e^{-2\pi\varepsilon|t|}$ and $e^{-4\pi^2\varepsilon|t|^2}$, $\varepsilon > 0$, respectively.

$$P(x,\varepsilon) = P_\varepsilon(x) = C_n\frac{\varepsilon}{(\varepsilon^2+|x|^2)^{(n+1)/2}}, \quad C_n = \Gamma\left(\frac{n+1}{2}\right)\pi^{-(n+1)/2} \tag{1.11}$$

and

$$W(x, \varepsilon) = W_\varepsilon(x) = (4\pi\varepsilon)^{-n/2} e^{-|x|^2/4\varepsilon} \tag{1.12}$$

$P(x, \varepsilon)$ is called the Poisson kernel and $W(x, \varepsilon)$ is called the Weierstrass kernel. Both functions appear in a number of problems in analysis.

We now consider a summation method given by the $M_{\Phi, \varepsilon}$ means defined by (1.4) of which Abel and Gauss methods are particular examples. Consider an integrable Φ and let $\phi = \hat{\Phi}$.

By the dilation relation (iii) of Section 1 of Chapter 2,

$$(\delta_\varepsilon \Phi)\hat{}(x) = \varepsilon^{-n}\phi(x/\varepsilon) = \phi_\varepsilon(x), \quad \text{for} \quad \varepsilon > 0 \tag{1.13}$$

Lemma 1.1 shows that when $\Phi(t) = e^{-4\pi^2|t|^2}$, then $\phi_\varepsilon(x) = W(x, \varepsilon^2)$, and Lemma 1.2 that when $\Phi(t) = e^{-2\pi|t|}$, then $\phi_\varepsilon(x) = P(x, \varepsilon)$.

Theorem 1.3. If $f, \Phi \in L^1$ and $\phi = \hat{\Phi}$ then

$$\int \hat{f}(x)e^{2\pi i x. t}\Phi(\varepsilon x)dx = \int f(x)\phi_\varepsilon(x - t)dx \tag{1.14}$$

for all $\varepsilon > 0$. In particular,

$$\int \hat{f}(x)e^{2\pi i x. t}e^{-2\pi\varepsilon|x|}dx = \int f(x)P(x - t, \varepsilon)dx = f * P_\varepsilon(t) \tag{1.15}$$

and

$$\int \hat{f}(x)e^{2\pi i x. t}e^{-4\pi^2\varepsilon|x|^2}dx = \int f(x)W(x - t, \varepsilon)dx = f * W_\varepsilon(t) \tag{1.16}$$

The second integral in (1.15) is called the Poisson integral of f and the corresponding one in (1.16), the Weierstrass integral of f. So, the Abel and Gauss means of the Fourier integral $\int \hat{f}(x)\exp(2\pi ix.t)dx$ are the Poisson and Weierstrass integrals of f, respectively.

Proof. By the multiplication formula applied to $f(x)$ and $\exp(2\pi ix.t)\Phi(\varepsilon x)$, we get, taking into account that (ii) of Section 1 of Chapter 2 and (1.13) give $(e^{2\pi ix.t}\delta_\varepsilon \Phi)^\wedge = \tau_t(\delta_\varepsilon \Phi)^\wedge = \tau_t\phi_\varepsilon$, the equality (1.14) for all $\varepsilon > 0$. Then (1.15) and (1.16) follow from 1.14, observing that $P_\varepsilon(x)$ and $W_\varepsilon(x)$ are radial functions of x. ∇

This theorem enables us to apply the results about convolution units obtained in Chapter 1 to the problem of summability of the Fourier integral. In order to do so, let us first show that the Abel and Gauss means give rise to convolution units.

Lemma 1.4. For all $\varepsilon > 0$, it is

(a) $\int_{\mathbb{R}^n} W(x,\varepsilon)dx = 1,$ (b) $\int_{\mathbb{R}^n} P(x,\varepsilon)dx = 1$

Proof. By changing variables it is immediate that

$$\int W(x,\varepsilon)dx = \int W(x,1)dx \quad \text{and} \quad \int P(x,\varepsilon)dx = \int P(x,1)dx$$

for all $\varepsilon > 0$

Thus, it suffices to prove (a) and (b) for $\varepsilon = 1$.

(a) follows, since

$$\int_{\mathbb{R}^n} W(x,1)dx = (4\pi)^{-n/2}\int e^{-|x|^2/4}dx$$

(continued)

$$= (4\pi)^{-n/2} \prod_{k=1}^{n} \int_{-\infty}^{\infty} e^{-x_k^2/4} \, dx_k$$

$$= (4\pi)^{-n/2} (2\sqrt{\pi})^n = 1$$

(b) means that

$$I = \int_{\mathbb{R}^n} \frac{dx}{(1+|x|^2)^{(n+1)/2}} = \frac{1}{C_n} = \frac{\pi^{(n+1)/2}}{\Gamma(\frac{n+1}{2})} \tag{1.17}$$

In order to prove this we take polar coordinates, where for $x \neq 0$, $r = |x|$, $x' = x/r \in \Sigma$. Recall (see Chapter 0, Section 2) that the surface area of $\Sigma = \Sigma_n$ is, for $n > 1$,

$$\omega_n = \int_{\Sigma} dx' = 2\pi^{n/2} (\Gamma(\tfrac{n}{2}))^{-1}$$

so (1.17) will be proved if the integral I is equal $\omega_{n+1}/2$, half of the surface area of the unit sphere of $\mathbb{R}^{n+1}$. In fact,

$$I = \int_{\mathbb{R}^n} \frac{dx}{(1+|x|^2)^{(n+1)/2}} = \int_0^{\infty} \int_{\Sigma} \frac{1}{(1+r^2)^{(n+1)/2}} r^{n-1} dr\, dx'$$

$$= \omega_n \int_0^{\infty} \frac{r^{n-1} dr}{(1+r^2)^{(n+1)/2}}$$

$(r = \tan \theta)$

$$= \omega_n \int_0^{\pi/2} \sin^{n-1}\theta \, d\theta$$

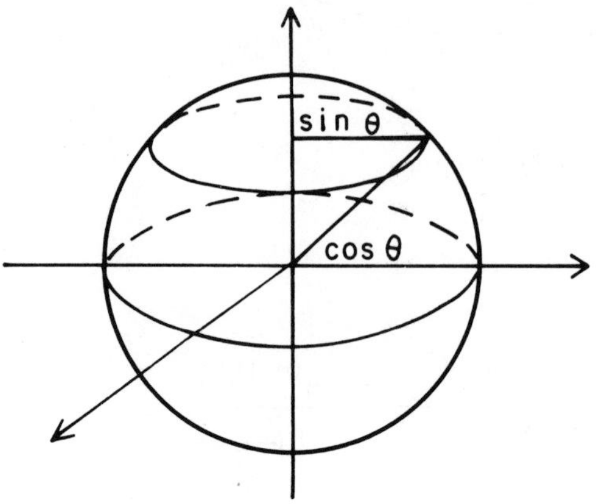

Figure 1

But $\omega_n \sin^{n-1}\theta$ is the surface area of the sphere of radius $\sin\theta$ obtained by intersecting Σ_{n+1}, the unit sphere of $\mathbb{R}^{n+1}$, by the hyperplane $x_n = \cos\theta$.

Thus, the surface area of the upper half of Σ_{n+1} can be computed by summing these (n-1) dimensional area for θ ranging from 0 to $\pi/2$; that is $I = \omega_n \int_0^{\pi/2} \sin^{n-1}\theta\, d\theta = \omega_{n+1}/2$. ∇

Exercise 1.1. Show that for every $\eta > 0$, $\int_{|x|>\eta} P(x, \varepsilon)dx \to 0$ as $\varepsilon \to 0$.

Since $P_\varepsilon(x)$ and $W_\varepsilon(x)$ are both positive functions of x, the following corollary is immediate.

Corollary 1.5. Given the Poisson kernel $P(x, \varepsilon) = P_\varepsilon(x)$ and the Weierstrass kernel $W(x, \varepsilon) = W_\varepsilon(x)$, for every $f \in L^p$, $1 \leq p \leq \infty$, it holds that the L^p norms of its Poisson and Weierstrass integrals are bounded by its L^p norm, i.e.,

$$\|P_\varepsilon * f\|_p \leq \|f\|_p \quad \text{and} \quad \|W_\varepsilon * f\|_p \leq \|f\|_p,$$

all $\varepsilon > 0$.

From Theorems 1.3 above and 1.4 of Chapter 1 we obtain now the following solution to the Fourier inversion problem:

Theorem 1.6. Let $\Phi \in L^1$ be such that $\phi = \hat{\Phi} \in L^1$ and $\int \phi = 1$. Then the Φ-means of the Fourier integral $\int \hat{f}(x)\exp(2\pi i x.t)dx$ converge to f in L^1. In particular, the Abel and Gauss means of the Fourier integral, i.e., the Poisson and Weierstrass integrals of f, converge to f in L^1.

Corollary 1.7 (Uniqueness). Given $f_1, f_2 \in L^1$ such that $\hat{f}_1(x) = \hat{f}_2(x)$, $x \in \mathbb{R}^n$, then $f_1(t) = f_2(t)$ a.e. $t \in \mathbb{R}^n$.

Proof. By Theorem 1.6, $\hat{f}(x) = 0$ for all x implies $f(t) = 0$ a.e. and we apply this result to $f = f_1 - f_2$. ∇

The Fourier inversion problem also admits a solution in the pointwise sense, complementing the solution in L^1 given by Theorem 1.6. We shall prove that the Φ-means of the Fourier integral of a given integrable function f converge to that function f at every point of the Lebesgue set $\mathscr{L}_f$, that is almost everywhere in $\mathbb{R}^n$, if Φ satisfies certain conditions that are satisfied by the functions corresponding to the Abel and Gauss means.

Theorem 1.8. Let $\Phi \in L^1$ be such that $\phi = \hat{\Phi} \in L^1$, $\int \phi = 1$ and $\psi \in L^1$ for $\psi(x) = \sup_{|y| \geq |x|} |\phi(y)|$. Then the Φ-means of the Fourier integral $\int \hat{f}(x)\exp(2\pi i x.t)dx$ converge to $f(t)$ whenever $t \in \mathscr{L}_f$, the Lebesgue set of f.

In particular, the Poisson and Weierstrass integrals of the integrable function f, $P_\varepsilon * f(t)$ and $W_\varepsilon * f(t)$ converge to $f(t)$ as $\varepsilon \to 0$ for almost every $t \in \mathbb{R}^n$.

Proof. By Theorem 2.1. of Chapter 1, we need only check that the hypothesis are fulfilled for $\phi(x) = P(x, 1)$ and $\phi(x) = W(x, 1)$. This is obvious since both P and W are radial functions of x, decreasing in $|x|$, and so in these cases $\phi(x)$ and $\psi(x)$ coincide. ∇

Corollary 1.9. If $f \in L^1$ and $\hat{f} \in L^1$ then

$$f(t) = \int \hat{f}(x) e^{2\pi i x.t} dx$$

for almost every $t \in \mathbb{R}^n$.

Proof. By Theorem 1.8, $\int \hat{f}(x) e^{2\pi i x.t} e^{-4\pi^2 \varepsilon |x|^2} dx$ converges a.e. to $f(t)$ when $\varepsilon \to 0$. If $\hat{f}$ is integrable, we apply Lebesgue's dominated convergence theorem to obtain the corollary. ∇

Remark 1.1. By Proposition 1.1(c) of Chapter 2 we know that $\hat{f}$ is continuous if f is integrable. If $\hat{f}$ is also integrable, the function defined by $f_0(t) = \int \hat{f}(x)\exp(2\pi i x.t)dx$ is also continuous ($f_0(t) = \hat{\hat{f}}(-t)$). By Corollary 1.9, $f_0(t) = f(t)$ a.e. so changing f in a set of measure zero we can obtain the equality $f(t) = \int \hat{f}(x)\exp(2\pi i x.t)dx$ for every t. In particular, if both f and $\hat{f}$ are integrable and f is continuous, the equality holds in every point, while it always holds at the points of continuity of f.

Remark 1.2. In the proof of Lemma 1.4, part (a) was much easier to prove than part (b). It is interesting to notice that Corollary 1.9 enables us to obtain (b) without computations. In fact, applying that result to the (integrable and continuous) function $f(t) = e^{-2\pi|t|}$, whose Fourier transform is, by Lemma 1.2, $P(x, 1)$, we get that

$$\int P(x,1) e^{2\pi i x.t} dx = e^{-2\pi|t|} \tag{1.18}$$

where equality holds for every t, as indicated in Remark 1.1. For $t = 0$, (1.18) becomes $\int P(x, 1)dx = 1$.

The following result is very useful in applications.

Corollary 1.10. Let $f \in L^1$ such that $\hat{f} \geq 0$. If furthermore f is continuous at $t = 0$, then $\hat{f} \in L^1$ and

$$f(t) = \int \hat{f}(x)e^{2\pi i x.t}(x)dx \qquad \text{a.e.}$$

In particular,

$$f(0) = \int \hat{f}(x)dx \tag{1.19}$$

Proof. Given $\hat{f}(x) \geq 0$, then $g_\varepsilon(x) = \hat{f}(x)\exp(-2\pi\varepsilon|x|) \geq 0$ for each $\varepsilon > 0$. As $\lim_{\varepsilon \to 0} \int \hat{f}(x)\exp(-2\pi\varepsilon|x|)\exp(ix.t)dx = f(t)$ for each t being a point of continuity of f, for $t = 0$ it is

$$\lim_{\varepsilon \to 0} \int \hat{f}(x)\exp(-2\pi\varepsilon|x|)dx = \lim_{\varepsilon \to 0} \int g_\varepsilon(x)dx = f(0)$$

Since $g_\varepsilon(x)$ is integrable for each $\varepsilon > 0$, applying Fatou's lemma to $\lim g_\varepsilon(x) = \hat{f}(x)$, we get

$$\int \hat{f}(x)dx \leq \lim_{\varepsilon \to 0} \int \hat{f}(x)\exp(-2\pi\varepsilon|x|)dx = f(0)$$

Therefore $0 \leq \hat{f} \in L^1$, and the thesis follows from Corollary 1.9 and again by the continuity of f at $t = 0$. ∇

Exercise 1.2. Prove that for every $\varepsilon > 0$,

(a) $\int W(x, \varepsilon)e^{2\pi i x.t}dx = \exp(-4\pi^2\varepsilon|x|^2)$

(b) $\int P(x, \varepsilon)e^{2\pi i x.t}dx = \exp(-2\pi\varepsilon|x|)$.

From this result, the uniqueness corollary and the fundamental Theorem 1.4 of Chapter 2, we obtain the semigroup properties of the Poisson and Weierstrass kernels:

Exercise 1.3. If ε_1 and ε_2 are positive real numbers, show that

(a) $W(x, \varepsilon_1 + \varepsilon_2) = W(x, \varepsilon_1) * W(x, \varepsilon_2)$,

(b) $P(x, \varepsilon_1+\varepsilon_2) = P(x, \varepsilon_1) * P(x, \varepsilon_2)$.

Accordingly, the Poisson and Weierstrass operators defined by convolution with such kernels verify the following composition relations, that follow from the associativity of convolution.

Corollary 1.11. Given $f \in L^1$ let $A_\varepsilon(f) = P_\varepsilon * f$ and $G_\varepsilon(f) = W_\varepsilon * f$. Then, if ε_1 and ε_2 are positive real numbers,

(a) $G_{\varepsilon_1} \circ G_{\varepsilon_2} = G_{\varepsilon_1+\varepsilon_2}$,

(b) $A_{\varepsilon_1} \circ A_{\varepsilon_2} = A_{\varepsilon_1+\varepsilon_2}$.

Summing up the results for the families $\{A_\varepsilon\}_{\varepsilon>0}$ of the Abel means and $\{G_\varepsilon\}_{\varepsilon>0}$ of the Gauss means of the Fourier integral of an integrable function, we have that they are two families of bounded operators in L^1 that form semigroup and converge, when $\varepsilon \to 0$, to the identity operator (in the sense of the norm and also in the pointwise sense a.e.). $\{A_\varepsilon\}$ is closely connected with the harmonic functions, as we shall see later on, since $u(x, y) = A_y(f)(x)$ is a harmonic function in the $n + 1$ variables (x, y), i.e., a function that verifies Laplace equation

$$\Delta u = \frac{\partial^2 u}{\partial x_1^2} + \dots + \frac{\partial^2 u}{\partial x_n^2} + \frac{\partial^2 u}{\partial y^2} = 0$$

In an analogous way, $\{G_\varepsilon\}$ appears as closely connected to the solution of the heat equation, since the $n + 1$ variables function $v(x, y) = G_y(f)(x)$ verifies it, i.e.,

$$\frac{\partial^2 v}{\partial x_1^2} + \dots + \frac{\partial^2 v}{\partial x_n^2} = \frac{\partial v}{\partial y}$$

Remark 1.3. Relationship between Abel and Gauss summation methods. If there existed 2m constants $C_1, \ldots, C_m, \lambda_1, \ldots, \lambda_m$ such that, for every $\varepsilon > 0$,

$$e^{-\varepsilon} = \sum_{k=1}^{m} C_k e^{-(\lambda_k \varepsilon)^2} \tag{1.20}$$

we would conclude that if one integral is Gauss summable then it must be Abel summable. Identity (1.20) does not hold, but we used (following [1], Chapter 1, Section 14) in the proof of Lemma 1.2 a somewhat "similar" identity. It gives $\exp(-\varepsilon)$ as a weighted average of the family $\{\exp(-\varepsilon^2/4u)\}_{0<u<\infty}$: for every $\varepsilon > 0$ we proved that

$$e^{-\varepsilon} = \frac{1}{\sqrt{\pi}} \int_0^\infty \frac{e^{-u}}{\sqrt{u}} e^{-\varepsilon^2/4u} du \tag{1.8}$$

So it is not entirely unexpected that if f has both Abel and Gauss means, $A_\varepsilon(f)$ and $G_\varepsilon(f)$, then $\lim_{\varepsilon \to 0} G_\varepsilon(f) = \ell$ implies that $\lim_{\varepsilon \to 0} A_\varepsilon(f)$ also exists and is equal to ℓ (see [1], Chapter 1).

Remark 1.4. Theorem 1.8 asserts that the Poisson integral of $f \in L^1$ converges to f(x) for each $x \in \mathscr{L}_f$. Actually a sharper result holds: $P_\varepsilon * f(x) \to f(x)$ when $\varepsilon \to 0$, for every x that belongs to the set of differentiability of f (see, [3], Theorem (5.9), Chapter 1).

Very similar arguments of those displayed above give rise to the Poisson-Stieltjes and Weierstrass-Stieltjes integrals and solve the Fourier inversion problem for finite Borel measures.

We shall specifically consider the Abel method of summation for the Fourier-Stieltjes integral $\int \hat{\mu}(x)e^{-2\pi ix.t}dx$, that need not converge for a given $\mu \in \mathcal{M}$. The arguments are analogous for the Gauss method.

For $\mu \in \mathcal{M}$, the Abel mean of order $\varepsilon > 0$ of the Fourier-Stieltjes integral of μ is

$$A_\varepsilon(\mu) = \int \hat{\mu}(x)e^{2\pi ix.t}e^{-2\pi\varepsilon|x|}dx \tag{1.21}$$

Lemma 1.12. For all $\varepsilon > 0$,

$$A_\varepsilon(\mu)(t) = (P_\varepsilon * \mu)(t) \tag{1.22}$$

where $P_\varepsilon(x)$ is the Poisson kernel.

Remark 1.5. $P_\varepsilon * \mu(t) = \int P_\varepsilon(t-x)d\mu(x)$ is called the Poisson-Stieltjes integral of μ.

Proof. By Proposition 2.2 of Chapter 2,

$$A_\varepsilon(\mu)(t) = \int \hat{\mu}(x)e^{2\pi ix.t}e^{-2\pi\varepsilon|x|}dx = \int \hat{\mu}_1(y)d\mu(y)$$

where $d\mu_1(x) = \exp(2\pi ix.t)\exp(-2\pi\varepsilon|x|)dx$. By definition,

$$\begin{aligned}\hat{\mu}_1(y) &= \int e^{-2\pi ix.y}e^{2\pi ix.t}e^{-2\pi\varepsilon|x|}dx \\ &= \int e^{-2\pi ix.(y-t)}e^{-2\pi\varepsilon|x|}dx \\ &= P_\varepsilon(y-t) = P_\varepsilon(t-y)\end{aligned}$$

Thus,

$$A_\varepsilon(\mu)(t) = \int P_\varepsilon(t-y)d\mu(y) = (P_\varepsilon * \mu)(t) \qquad \nabla$$

We shall first prove that the Abel means of the Fourier-Stieltjes integral of μ converge weakly-* to μ.

<u>Theorem 1.13</u>. For all $\mu \in \mathcal{M}$, $A_\varepsilon(\mu) \to \mu$ weakly-* as $\varepsilon \to 0$.

<u>Proof</u>. For every $\phi \in C_\infty$ we have, by Fubini's theorem, that

$$\begin{aligned}\int \phi(x)A_\varepsilon(\mu)(x)dx &= \int \phi(x)\left(\int P_\varepsilon(x - y)d\mu(y)\right)dx \\ &= \int\int \phi(x)P_\varepsilon(x - y)d\mu(y)dx \\ &= \int\left(\int P_\varepsilon(x - y)\phi(x)dx\right)d\mu(y) \\ &= \int P_\varepsilon * \phi(y)d\mu(y)\end{aligned}$$

By Theorem 1.4 of Chapter 1, $P_\varepsilon * \phi(y) \to \phi(y)$ uniformly when $\varepsilon \to 0$, so

$$\int \phi(x)A_\varepsilon(\mu)(x)dx = \int P_\varepsilon * \phi(y)d\mu(y) \to \int \phi(y)d\mu(y)$$

when $\varepsilon \to 0$, which proves the assertion. ∇

<u>Corollary 1.14 (Uniqueness of the Fourier transform of finite Borel measures)</u>. If $\mu_1, \mu_2 \in \mathcal{M}$ and $\hat{\mu}_1 = \hat{\mu}_2$, then $\mu_1 = \mu_2$.

<u>Proof</u>. $A_\varepsilon(\mu) \to \mu$ weakly-* when $\varepsilon \to 0$, and if $\hat{\mu}(x) \equiv 0$ then $A_\varepsilon(\mu)(x) \equiv 0$, so $\mu = 0$. Apply this to $\mu = \mu_1 - \mu_2$. ∇

There is also an analog of the pointwise convergence a.e. for Poisson-Stieltjes integrals.

<u>Theorem 1.15</u>. Let $\mu \in \mathcal{M}$ and f be its <u>Radon-Nikodym derivative</u>. Then,

$$A_\varepsilon(\mu)(x) \to f(x) \quad \text{a.e.}$$

as $\varepsilon \to 0$.

Proof. As before, $A_\varepsilon(\mu)(x) = (P_\varepsilon * \mu)(x)$ for every $\varepsilon > 0$. μ is a finite Borel measure so we can decompose it into a singular part ν and an absolutely continuous part $\mu_1 : \mu = \mu_1 + \nu$, where $d\mu_1 = fdx$ and $f \in L^1$ is the Radon-Nikodym derivative of μ. Thus,

$$A_\varepsilon(\mu)(x) = \int P_\varepsilon(x - t)d\mu_1(t) + \int P_\varepsilon(x - t)d\nu(t)$$

$$= \int P_\varepsilon(x - t)f(t)dt + \int P_\varepsilon(x - t)d\nu(t)$$

As $\varepsilon \to 0$, the first term of the sum tends to $f(x)$ a.e., since $f \in L^1$, and the second tends to zero a.e. This is shown by the same argument used for Poisson integrals of functions, taking into account the following fact valid for singular finite Borel measures:

$$\frac{1}{|S_x|}\int_{S_x} d\nu \to 0 \quad \text{when} \quad |S_x| \to 0$$

for almost every x, where $S_x = \{t \in \mathbb{R}^n : |x - t| \leq r\}$. (See [4], Chapter 4). Then,

$$A_\varepsilon(\mu)(x) \to f(x) + 0 = f(x) \qquad \text{a.e.}$$

whenever $\varepsilon \to 0$. ∇

2. FOURIER TRANSFORM IN L^2 AND THE PLANCHEREL THEOREM

The integral defining the Fourier transform of an integrable function (as in (1.1) of Chapter 2) is not defined as an ordinary Lebesgue integral for a general function belonging to $L^2(\mathbb{R}^n)$. Nevertheless, the Fourier transform has a natural definition in L^2 and its theory is particularly elegant in this space.

This is due to the fact that L^2 is a Hilbert space, and in what follows we shall rely on this additional structure. (See Chapter 0, Section 4.)

We want to study the Fourier transform of L^2 functions but we have defined it for L^1 functions. However, the Fourier transform is defined on $L^1 \cap L^2$ (because $L^1 \cap L^2 \subset L^1$) and

(i) $L^1 \cap L^2$ is a linear subspace (of L^1 and) of L^2,

(ii) $L^1 \cap L^2$ is a <u>dense</u> subspace (of L^1 and) of L^2.

The following is the basic result.

<u>Proposition 2.1</u>. If $f \in L^1 \cap L^2$, then $\hat{f} \in L^2$ and $\|\hat{f}\|_2 = \|f\|_2$.

<u>Proof</u>. Let $f^*(x) = \overline{f(-x)}$. If we consider $h = f * f^*$ then

(1) $h \in L^1$, by Theorem 1.1 of Chapter 1, since f and $f^* \in L^1$,

(2) h is continuous, by Exercise 1.5 of Chapter 1, since f and $f^* \in L^2$.

Furthermore, by Theorem 1.4 of Chapter 2, $\hat{h} = \hat{f}.\hat{f}^*$. But $(f^*)\hat{} = \overline{\hat{f}}$:

$$(f^*)\hat{}(x) = \int f^*(t)e^{-2\pi ix.t}dt = \int \overline{f(-t)}e^{-2\pi ix.t}dt = \int \overline{f(t)}e^{2\pi ix.t}dt = \overline{\hat{f}(x)}$$

since $e^{2\pi ix.t} = \overline{e^{-2\pi ix.t}}$. Thus, $\hat{h} = \hat{f}.\overline{\hat{f}} = |\hat{f}|^2 \geq 0$, and we may apply Corollary 1.10 to h, to obtain that $\hat{h} \in L^1$ and $h(0) = \int \hat{h}(x)dx$. From this,

$$\int |\hat{f}(x)|^2 dx = \int \hat{h}(x)dx = h(0) = \int f(t)f^*(0 - t)dt$$

$$= \int f(t)\overline{f(t)}dt = \int |f(t)|^2 dt$$

∇

Proposition 2.1 can be stated as $\mathscr{F} : L^1 \cap L^2 \to L^2$, $\mathscr{F}$ is a continuous linear operator on $L^1 \cap L^2$, a dense subset of

L^2. Furthermore, $\mathcal{F}$ is an isometry of $L^1 \cap L^2$, i.e., $\|\mathcal{F} f\|_2 = \|f\|_2$ for all $f \in L^1 \cap L^2$.

By Proposition 4.1 of Chapter 0, there exists a unique bounded extension to all L^2, that will also be called $\mathcal{F} : L^2 \to L^2$, such that $\|\mathcal{F} f\|_2 \leq \|f\|_2$ for all $f \in L^2$. (We will continue using, for $f \in L^2$ the notation $\mathcal{F} f = \hat{f}$.) This last estimate may be sharpened:

<u>Theorem 2.2.</u> $\mathcal{F}$ is an isometry of L^2, i.e., $\|\mathcal{F} f\|_2 = \|f\|_2$ for all $f \in L^2$.

<u>Proof</u>. For every $f \in L^2$ there is a sequence $\{f_k\} \subset L^1 \cap L^2$ such that $\|f_k - f\|_2 \to 0$ as $k \to \infty$. Hence $\|\hat{f}_k\|_2 \to \|\hat{f}\|_2$. (Since $|\,\|\hat{f}_k\|_2 - \|\hat{f}\|_2| \leq \|\hat{f}_k - \hat{f}\|_2 \leq \|f_k - f\|_2 \to 0$.) But $\|\hat{f}_k\|_2 = \|f_k\|_2 \to \|f\|_2$. By the uniqueness of the limits in L^2, $\mathcal{F}$ is an isometry. ▽

For a general square integrable function f, the definition of the Fourier transform gives $\hat{f}$ as the L^2 limit of a sequence $\{\hat{f}_k\}$, where $\{f_k\}$ is any sequence belonging to $L^1 \cap L^2$ with f as its L^2 limit. Given f, we may fix the sequence corresponding to it, for instance take $\{f_k\}$ such that $f_k(t) = f(t)$ when $|t| \leq k$ and zero when $|t| > k$ (we could have equally chosen any other sequence given by f restricted to other sets of finite measure dilating to the whole $\mathbb{R}^n$). Then $\hat{f}$ is the L^2 limit of the sequence $\{\hat{f}_k\}$ given by

$$\hat{f}_k(x) = \int_{\mathbb{R}^n} f_k(t) e^{-2\pi i x \cdot t} dt = \int_{|t| \leq k} f(t) e^{-2\pi i x \cdot t} dt$$

The multiplication formula extends immediately to L^2.

<u>Proposition 2.3.</u> For all $f, g \in L^2$,

$$\int \hat{f}(x) g(x) dx = \int f(x) \hat{g}(x) dx \tag{2.1}$$

Proof.

(1) If $f, g \in L^1 \cap L^2$ the result is Theorem 1.3 of Chapter 2.

(2) Let us fix $g \in L^1 \cap L^2$. For any $f \in L^2$, there exists $\{f_k\} \subset L^1 \cap L^2$ such that $\|f_k - f\|_2 \to 0$ when $k \to \infty$.

Then $\int f_k(x)\hat{g}(x)dx \to \int f(x)\hat{g}(x)dx$, since $\hat{g} \in L^2$ and convergence in the norm implies weak convergence.

But f_k and $g \in L^1 \cap L^2$, so $\int f_k(x)\hat{g}(x)dx = \int \hat{f}_k(x)g(x)dx$ and as $\int \hat{f}_k(x)g(x)dx \to \int \hat{f}(x)g(x)dx$, (2.1) follows in this case too.

(3) We repeat the process by approximating in the norm any $g \in L^2$ by $\{g_k\} \subset L^1 \cap L^2$. ∇

One of the main problems, as in the L^1 theory, is the inversion of the Fourier transform $\mathscr{F}$. But unitary operators in Hilbert spaces are always invertible (see Chapter 0, Section 4), so our immediate goal is to prove that $\mathscr{F}$ is a unitary operator in L^2. We have already proved that it is an isometry, we see now that $\mathscr{F}$ maps L^2 onto L^2.

Theorem 2.4. $\mathscr{F}$ is a unitary operator on L^2.

Proof. Let S be the image of L^2 by $\mathscr{F}$. Then

(i) S is subspace of L^2, since $\mathscr{F}$ is a linear operator,

and

(ii) S is a closed subspace of L^2, since $\mathscr{F}$ is an isometry: given $\{\hat{f}_k\} \subset S$ such that $\|\hat{f}_k - g\|_2 \to 0$ for $g \in L^2$, then $g \in S$, i.e., there is an $f \in L^2$ such that $g = \hat{f}$, since $\{f_k\}$ is a Cauchy sequence in L^2 ($\|f_k - f_m\|_2 = \|\hat{f}_k - \hat{f}_m\|_2$) and L^2 is complete.

Let us see that $S = L^2$. If the closed subspace S were not all of L^2, there would exist (by the theorem of the orthogonal in a Hilbert space) $g \in L^2$, $g \neq 0$, such that $(f, g) = \int f\bar{g} = 0$ for all $f \in S$, that is to say $(\hat{h}, g) = 0$ for all $h \in L^2$. But by the multiplication formula (2.1),

$$0 = (\hat{h}, g) = \int \hat{h}\bar{g} = \int h\hat{\bar{g}}, \quad \text{for all} \quad h \in L^2$$

In particular, taking $h = \hat{g} \in L^2$, $\|\hat{g}\|_2 = 0 = \|g\|_2$ and $g = 0$ a.e., contrary to the assumption $g \neq 0$. Thus $S = L^2$ and the thesis is proved. ∇

This result is the basic theorem of the L^2 theory of the Fourier transform:

Theorem 2.5 (Plancherel theorem).

If $f \in L^2$ and $\{f_k\} \subset L^1 \cap L^2$ converges to f in the L^2 norm, then $\{\hat{f}_k\}$ converges in the L^2 norm to a function $\hat{f} \in L^2$.

The linear operator $\mathscr{F} : f \to \hat{f}$, called the Fourier transform, is a unitary operator in L^2 and its inverse, $\mathscr{F}^{-1}$, can be obtained as

$$(\mathscr{F}^{-1}f)(t) = (\mathscr{F}f)(-t) \quad \text{for all} \quad f \in L^2 \tag{2.2}$$

In particular, for $f \in L^2$, if $\{h_k\}$ is given by

$$h_k(t) = \int_{|x| \leq k} \hat{f}(x)e^{2\pi i x \cdot t}dx \tag{2.3}$$

then $\|h_k - f\|_2 \to 0$ when $k \to \infty$.

Proof. The first part of the theorem corresponds to the definition we have given of $\mathscr{F}$ and Theorem 2.4. We have to check (2.2). This follows from the facts that $\mathscr{F}^{-1} = \mathscr{F}^*$ (the adjoint operator) in a Hilbert space and that $\mathscr{F}^*\hat{f}$ can be expressed as the L^2 limit of the sequence given in (2.3). In fact we shall show that this is true for $\hat{f} \in L^1 \cap L^2$; since $L^1 \cap L^2$ is dense in L^2, (2.2) for the general case will follow by continuity, taking into account that both $\mathscr{F}$ and $\mathscr{F}^*$ are isometries.

Let us consider then, for $\hat{f} \in L^1 \cap L^2$,

$$f^{\#}(t) = \int \hat{f}(x) e^{2\pi i x.t} dx = \lim_{k\to\infty} h_k(t) \quad (\text{in } L^2) = \mathscr{F}\hat{f}(-t)$$

Then, for every $g \in L^1 \cap L^2$, we have by Fubini's theorem,

$$(g, f^{\#}) = \int g(t) \overline{\left(\int \hat{f}(x) e^{2\pi i x.t} dx\right)} dt$$

$$= \int \left(\int g(t) e^{-2\pi i x.t} dt\right) \overline{\hat{f}(x)} dx$$

$$= (\mathscr{F} g, \hat{f}) = (g, \mathscr{F}^* \hat{f})$$

$\mathscr{F}$ is unitary, so it preserves the scalar product: $(g, f^{\#}) = (\mathscr{F} g, \mathscr{F} f) = (g, f)$ which implies $f^{\#} = f$ a.e.

That is, $\mathscr{F}\hat{f}(-t) = f(t)$ a.e., or $\hat{f}(t) = \mathscr{F}^{-1} f(t)$ a.e. as in (2.2). ∇

<u>Corollary 2.6 (Uniqueness)</u>. Given $f \in L^2$, if $\hat{f} \equiv 0$ then $f = 0$ a.e.

<u>Exercise 2.0.</u> Let $\mathscr{S} = \{f \in C^{\infty}(\mathbb{R}): \sup_x |x^{\alpha} \partial^{\beta} f(x)| < \infty$ for every $\alpha, \beta \in (\mathbb{Z}^+)^n\}$ be the Schwartz space. Prove that the Fourier operator $\mathscr{F}$ transforms $\mathscr{S}$ onto $\mathscr{S}$. (Hint: Use Propositions 1.4 and 1.5 of Chapter 1 and Theorem 2.5 above.)

By the Plancherel theorem, the inversion problem of the Fourier transform has a simple solution in L^2. Nevertheless, we may ask if there is also a solution of the inversion problem in L^2 involving summation techniques. More precisely, as $e^{-2\pi|x|}$ is a square integrable function, the Abel means of the Fourier antitransform are well defined for a function $f \in L^2$ and the question is: do they converge to f in L^2 and in the pointwise sense. The answer is affirmative as immediate consequence of Theorems 1.4 and 2.1 of Chapter 1, since formula (1.15) can be proved in L^2 using the multiplication formula (2.1).

We have defined the Fourier transform for functions in L^2 and in L^1 and we can now extend the definition to functions in L^p, $1 \leq p \leq 2$.

Let us consider first the class $L^1 + L^2 = \{f : f = f_1 + f_2, f_1 \in L^1, f_2 \in L^2\}$. In this class we define $\hat{f} = \hat{f}_1 + \hat{f}_2$, understanding by $\hat{f}_1$ the transform of f_1 given by integration in Definition 1.1 and by $\hat{f}_2$ the extended transform of f_2 given by Theorem 2.5. This new definition works, since if two functions coincide on $L^1 + L^2$ they belong to $L^1 \cap L^2$. In fact, if $f_1 + f_2 = g_1 + g_2$ with $f_1, g_1 \in L^1$ and $f_2, g_2 \in L^2$, then $f_1 - g_1 = g_2 - f_2$, such that the first term is an L^1 function and the second term, an L^2 function, so both belong to $L^1 \cap L^2$.

The two definitions of $\mathcal{F}$ coincide on $L^1 \cap L^2$, $\hat{f}_1 - \hat{g}_1 = \hat{g}_2 - \hat{f}_2$; hence $\hat{f}_1 + \hat{g}_1 = \hat{f}_2 + \hat{g}_2$.

Now observe that if a topological vector space $\mathcal{V}$ contains L^1 and L^2, as $L^1 + L^2$ does, it contains all the L^p, $1 \leq p \leq 2$. In fact, if $f \in L^p$, it can be decomposed as $f = f_1 + f_2$ where $f_2 = f$ when $|f| \leq 1$ and zero when $|f| > 1$ and $f_1 = f - f_2$. Then, for $1 \leq p \leq 2$,

$$\int |f_1| \leq \int |f_1|^p \leq \int |f|^p < \infty \quad \text{and so,} \quad f_1 \in L^1$$

$$\int |f_2|^2 \leq \int |f_2|^p \leq \int |f|^p < \infty \quad \text{and so,} \quad f_2 \in L^2$$

We define $\hat{f}$ for all $f \in L^p$, $1 \leq p \leq 2$, as above.

It is easy to see that the Fourier inversion problem can be solved also in this case in terms of the Abel or Gauss means of the Fourier integral.

The following extension holds for the fundamental theorem on convolutions.

Proposition 2.7. Given $f \in L^1$, $g \in L^p$, $1 \leq p \leq 2$, if $h = f * g$ then

$$\hat{h}(x) = \hat{f}(x).\hat{g}(x)$$

Exercise 2.1. Prove Proposition 2.7.

Exercise 2.2. Prove $(f.g)\hat{} = \hat{f} * \hat{g}$ for f and $g \in L^2$.

Remark 2.1. There is no extension of the Fourier transform for functions in L^p, $p > 2$, of the type of the one described above. All functions in L^p, $1 \leq p \leq \infty$, have Fourier transforms in the sense of distributions and these transforms are in general not functions but tempered distributions. If $1 \leq p \leq 2$, these distributions coincide with the functions defined as Fourier transforms in this chapter.

Remark 2.2. It was already remarked that the convolution operator K that has $k \in L^1$ as kernel is a bounded linear operator on L^p, $1 \leq p \leq \infty$, with norm $\| K \|^{(p)} \leq \|k\|_1$. For $p = 1$ we proved that $\| K \|^{(1)} = \|k\|_1$ (Exercise 1.11 of Chapter 1). For $p = 2$, $k \in L^1$ implies $\hat{k} \in L^\infty$ and we have that $\| K \|^{(2)} \leq \|\hat{k}\|_\infty$. More precisely, by Theorems 2.2 and 2.7, $\|K\|^{(2)} =$
$= \sup\{ \| K f\|_2 : \|f\|_2 = 1\} = \sup\{ \| \mathscr{F}(K f) \|_2 : \|f\|_2 = 1\} =$
$= \sup\{ \|\hat{k}.\hat{f}\|_2 : \|f\|_2 = 1\} = \|\hat{k}\|_\infty$. The question is whether $\| K \|^{(2)}$ can be expressed in terms of k. Provided that $k \geq 0$, by Corollary 1.10, $\|k\|_1 = |\hat{k}(0)| \leq \|\hat{k}\|_\infty \leq \|k\|_1$, thus $\| K \|^{(2)} =$
$= \|k\|_1$.

The M. Riesz interpolation theorem (see Chapter 4), together with the fact that K is a convolution operator, shows that the norm of operator $K : L^p \to L^p$, $1 \le p \le \infty$, can be expressed as $\| K \|^{(p)} = \|k\|_1$, provided $k \ge 0$.

3. HARMONIC FUNCTIONS

In this section D will denote a domain in $\mathbb{R}^n$ (i.e., an open connected subset of $\mathbb{R}^n$), $S(x_0, r) = \{x \in \mathbb{R}^n : |x - x_0| \le r\}$ the closed sphere of center x_0 and radius r, and $\mathbb{R}^{n+1}_+$ the open "upper half-space" of $\mathbb{R}^{n+1}$, $\mathbb{R}^{n+1}_+ = \{(x_1, \ldots, x_{n+1}) \in \mathbb{R}^{n+1} : x_{n+1} > 0\}$. $\mathbb{R}^{n+1}_+$ will be identified with $\mathbb{R}^n \times \mathbb{R}_+$ and the points $(x_1, \ldots, x_{n+1}) \in \mathbb{R}^{n+1}_+$ will be denoted by (x, y), $x = (x_1, \ldots, x_n) \in \mathbb{R}^n$, $y = x_{n+1} \in \mathbb{R}_+$. If $u(x, y)$ is defined in $\mathbb{R}^{n+1}_+$, letting $u_y(x) = u(x, y)$ we obtain a family $\{u_y(x)\}_{y \in \mathbb{R}_+}$ of functions defined in $\mathbb{R}^n$, so that it is the same to give a function u in $\mathbb{R}^{n+1}_+$ or a family $\{u_y\}_{y>0}$ of functions in $\mathbb{R}^n$.

In Section 1 we have associated to each $f \in L^p(\mathbb{R}^n)$ the Poisson integral $f_\varepsilon(x) = f * P_\varepsilon(x)$, $\varepsilon > 0$, which can be thought of either as a family $\{f_\varepsilon\}_{\varepsilon>0}$ of functions defined in $\mathbb{R}^n$ or as a function $u(x, \varepsilon) = f_\varepsilon(x)$ defined in $\mathbb{R}^{n+1}_+$. This function $u(x, \varepsilon)$, associated with f, determines f since we have already seen that $u(x, \varepsilon) = f_\varepsilon(x)$ converges to $f(x)$ as $\varepsilon \to 0$. In Section 1 we have studied the associated function u from the viewpoint of the family $\{f_\varepsilon\}$ of functions in $\mathbb{R}^n$. In Section 4 we shall consider u as a function in $\mathbb{R}^{n+1}_+$. It is easy to see that $u(x, \varepsilon) = f_\varepsilon(x)$ has the fundamental properties of being smooth and harmonic in $\mathbb{R}^{n+1}_+$. For precision, we recall some facts on harmonic functions.

Definition 3.1. A function u defined in a domain D is harmonic in D if $u \in C^2(D)$ and, for all $x \in D$,

$$\Delta u \equiv \sum_{k=1}^{n} \frac{\partial^2 u}{\partial x_k^2} = 0 \tag{3.1}$$

Exercise 3.1. If $n = 1$ then $u(x)$ is harmonic if and only if $u(x) = = ax + b$.

Exercise 3.2. Linear combinations of harmonic functions are harmonic. Also if $u(x)$ is harmonic in D, its translate $\tau_h u(x)$ is harmonic in $D + h = \{x + h : x \in D\}$. Similarly, harmonicity is preserved under dilations and rotations.

Exercise 3.3. For a fixed $t \in \mathbb{R}^n$, $u(x, y) = \exp(-2\pi(|t|y + ix.t))$, $(x, y) \in \mathbb{R}^n \times \mathbb{R}_+$, is harmonic in $\mathbb{R}_+^{n+1}$.

Exercise 3.4. The family of Abel means of the Fourier integral of an integrable function is harmonic in $\mathbb{R}_+^{n+1}$. (Hint: Use Exercise 3.3, applying the Lebesgue dominated convergence theorem as to differentiate under the integral sign.)

Exercise 3.5. The Poisson kernel

$$P(x, y) = P_y(x) = C_n y(|x|^2 + y^2)^{-(n+1)/2} \tag{3.2}$$

$(x, y) \in \mathbb{R}^n \times \mathbb{R}_+$ is harmonic in $\mathbb{R}_+^{n+1}$.

From this follows (as in Exercise 3.4) that the Poisson integral $u(x, y) = f * P_y(x)$ of an $f \in L^p(\mathbb{R}^n)$, $1 \leq p \leq \infty$, is harmonic in $\mathbb{R}_+^{n+1}$.

Remark 3.1. Observe that every partial derivative of an harmonic function is also harmonic (if the function is differentiable enough; but as we shall see later, all harmonic functions belong to C^∞). In particular, $(\partial/\partial y)P(x, y)$ is harmonic in $\mathbb{R}_+^{n+1}$, for P as in (3.2).

Through the connection with Poisson integrals, functions $f \in L^p(\mathbb{R}^n)$ of modern real analysis, that in general are neither differentiable nor even continuous, appear as boundary values of smooth harmonic functions u defined in $\mathbb{R}^{n+1}_+$. There is a close relationship between the Fourier theory for functions in $L^p(\mathbb{R}^n)$ and the theory of harmonic functions defined in $\mathbb{R}^{n+1}_+$.

It must be noted that although every $f \in L^p(\mathbb{R}^n)$ has an associated harmonic function u defined in $\mathbb{R}^{n+1}_+$, the converse is not true, i.e., not every harmonic function in $\mathbb{R}^{n+1}_+$ is the Poisson integral of an $f \in L^p(\mathbb{R}^n)$. To characterize those who are such is the aim of Section 4, and in this section we give some properties of harmonic functions of several variables necessary to this characterization.

By Exercise 3.1, it is enough to consider $D \subset \mathbb{R}^n$, $n \geq 2$. In the simplest and most basic case when $D = U = \{z \in \mathbb{C} : |z| < 1\}$ is the unit disc of $\mathbb{R}^2$, whose boundary is $\mathbb{T} = \{z \in \mathbb{C} : |z| = 1\}$, the study of harmonic functions u in U is closely related to the theory of analytic functions. In fact, as it is well-known, if $F = u + iv$ is analytic in U then F, u and v, are all harmonic in U, and conversely, every real harmonic function u in U is the real part of an analytic function F. In other words, if u is real and harmonic in U, there exists another real harmonic function v in U, such that $u + iv = F$ is analytic in U; such v is called a harmonic conjugate of u and is unique up to an additive constant. Since each $z \in U$ is of the form $z = r\exp(i\theta)$, $0 \leq r < 1$, $\exp(i\theta) \in \mathbb{T}$, it is the same to give a function $u(r\exp(i\theta))$ in U, or a family $\{u_r(\exp(i\theta)) = u(r\exp(i\theta))\}$, $0 \leq r < 1$, of functions in $\mathbb{T}$. As we may identify $\mathbb{T}$ with the interval $[0, 2\pi)$ and each function g defined in $\mathbb{T}$ with the periodic function $f(\theta) = g(\exp(i\theta))$ in $[0, 2\pi)$, we associate-in a way similar to that of Chapter 1-to each integrable function in $\mathbb{T}$, a harmonic function $u(z) = u(r\exp(i\theta))$ in U as follows. To each $f \in L^1[0, 2\pi)$ corresponds a Fourier series $f \sim \Sigma_{-\infty}^{\infty} c_n \exp(in\theta)$, $c_n = \hat{f}(n) = (2\pi)^{-1} \int f(t)\exp(-int)dt$, which in general is not convergent, but since $c_n \to 0$, the series $f_r(\theta) = \Sigma_{-\infty}^{\infty} c_n r^{|n|} \exp(in\theta)$ does converge absolutely for every $0 \leq r < 1$, and we get a function $u(r\exp(i\theta)) = f_r(\theta)$ defined in U. This associated function u is harmonic in U since for every $n \geq 0$ the function $r^n \exp(in\theta) + r^n \exp(-in\theta) = 2\mathrm{Re}\, z^n$ is harmonic. For each fixed $r < 1$ we have thus a function $f_r(\theta) = u(r\exp(i\theta))$ in $\mathbb{T}$, whose Fourier coefficients are $\{c_n r^{|n|}\}$. Therefore, if for each fixed r, $0 < r \leq 1$, we define

$$P_r(\theta) = \sum_{n=-\infty}^{\infty} r^{|n|} e^{in\theta} = \sum_{n=0}^{\infty} r^n e^{in\theta} + \sum_{n=1}^{\infty} r^n e^{-in\theta} \tag{3.3}$$

then the Fourier coefficients of P_r are $|r|^n$, and

$$f_r = f * P_r \tag{3.4}$$

$P_r(\theta)$ is the <u>Poisson kernel in $\mathbb{T}$</u>, and by (3.3),

$$P_r(\theta) = \sum_0^{\infty} z^n + \sum_1^{\infty} \bar{z}^n, \quad z = r\exp(i\theta), \qquad \text{so that}$$

$$P_r(\theta) = \frac{1-r^2}{1-2r\cos\theta+r^2} \tag{3.3a}$$

Thus to each $f \in L^1[0, 2\pi)$ we associate the harmonic function $u(z) = u(r\exp(i\theta)) = f_r(\theta)$, given by the Poisson integral

$$f_r(\theta) = u(re^{i\theta}) = f * P_r(\theta) = \frac{1}{2\pi}\int_0^{2\pi} f(t)P_r(\theta - t)dt \tag{3.5}$$

With an argument similar to that of Chapter 1 we obtain that $\{P_r\}$ is a convolution unit in $\mathbb{T}$, thus

$$f(\theta) = \lim_{r \to 1} u(re^{i\theta}) = \lim_{z \to \exp(i\theta)} u(z) \tag{3.6}$$

so that f is determined by u. Since there is a function v such that $F = u + iv$ is analytic, we see that there is a close connection between the theory of Fourier series of functions $f \in L^p(\mathbb{T}) \subset L^1(\mathbb{T})$ and the theory of harmonic and analytic function in U.

Similarly the theory of Fourier integrals of functions $f \in L^p(\mathbb{R})$ is related to that of harmonic and analytic functions in $\mathbb{R}^2_+$. (Observe that the map $z = (w - 1)/(w + 1)$ takes the half-plane $\mathbb{R}^2_+$ onto the unit disc U, the line $\mathbb{R}^1$ onto the circle $\mathbb{T}$, and transforms the Poisson kernel in $\mathbb{R}^1$ into that in $\mathbb{T}$.)

* The theory of harmonic functions in general domains originated in Mathematical Physics.

We will informally sketch the relation of harmonic functions to physics, to give a glimpse of the physical problems involved. However this motivation is not necessary to the study of the material of this and the following sections, and the reader who wants to proceed directly with the consideration of Poisson integrals can continue in Proposition 3.1.

Many (stationary) problems concerning mechanics or electromagnetic vibrations and heat diffusion lead to equations of the form

$$\sum_{j=1}^{n} \frac{\partial}{\partial x_j} \left(p \frac{\partial u}{\partial x_j}\right) + qu = F(x) \tag{3.7}$$

for $u = u(x_1, \ldots, x_n)$. For $p = \text{cst.}$, $q = 0$, (3.7) reduces to the Poisson equation

$$\Delta u = F \tag{3.8}$$

The corresponding homogeneous equation

$$\Delta u = 0 \tag{3.9}$$

is the Laplace equation. If u_0, u_1 are two solutions of (3.8) then $u_0 - u_1$ satisfies (3.9), so the study of certain physical problems reduces to the study of properties of harmonic functions (the solutions of (3.9)) and to find one particular solution u_0 of (3.8). Let us indicate briefly how this particular solution u_0 can be obtained in the case when f is a sufficiently "good" function (e.g., $f \in L^1(\mathbb{R}^n) \cap C^2(\mathbb{R}^n)$).

A locally integrable function E will be called a <u>fundamental solution of Δ</u> if

$$(\Delta\phi) * E = \phi, \quad \forall \phi \in C_0^\infty(\mathbb{R}^n) \tag{3.10}$$

<u>Remark 3.2</u>. If $E \in L^1 \cap C^2$, then $(\Delta\phi) * E = \Delta(\phi * E) = \phi * (\Delta E)$, and ΔE would be a convolution unit for C_0^∞. Thus a fundamental solution is a function E such that ΔE is, "in a generalized sense" a convolution unit in C_0^∞.

If E satisfies (3.10), the function

$$u_0 = U^f = f * E \tag{3.11}$$

is a particular solution of (3.8), i.e.,

$$\Delta(f * E) = f \tag{3.11a}$$

In fact, (3.11a) is equivalent to

$$\langle \Delta(f * E), \phi\rangle = \langle f, \phi\rangle, \quad \forall \phi \in C_0^\infty \tag{3.11b}$$

where $<g, \phi> = \int g(x)\phi(x)dx$. Since ϕ vanishes for large values of the variable, integration by parts gives $<\Delta g, \phi> = <g, \Delta\phi>$, so $<\Delta(f * E), \phi> = <f * E, \Delta\phi> = <f, \Delta\phi * E>$ and (3.10) implies (3.11a).

Similarly we prove that every locally integrable function g that belongs to C^2 can be written as

$$g = (\Delta g) * E \tag{3.12}$$

It is easy to see that condition (3.10) is equivalent to the following:

$$<\Delta\phi, E> = \phi(0), \quad \forall\phi \in C_0^\infty \tag{3.10a}$$

<u>Exercise 3.6.</u> Check the equivalence of (3.10) and (3.10a) as well as (3.12). Show that U^f is the only solution u_0 of (3.8) such that $u_0 * E$ exists.

Let now $E_n(x)$ be the function defined in $\mathbb{R}^n$, $n \geq 2$, as follows

$$E_n(x) = \begin{cases} -((n-2)\omega_n)^{-1}|x|^{2-n} & \text{if } n \geq 3 \quad (3.13) \\ (2\pi)^{-1}\log|x| & \text{if } n = 2 \quad (3.13a) \end{cases}$$

<u>Exercise 3.7.</u> Prove that $E = E_n(x)$ satisfies (3.10).

(Hint: For $\varepsilon > 0$ and $n > 2$, let $E^\varepsilon(x) = -((n-2)\omega_n)^{-1}(|x|^2 + \varepsilon^2)^{(2-n)/2}$, then $E \in L^1 \cap C^\infty(\mathbb{R}^n - \{0\})$, $E^\varepsilon \in C^\infty(\mathbb{R}^n)$ and $<E - E^\varepsilon, \Delta\phi> \to 0$ as $\varepsilon \to 0$, for $\phi \in C_0^\infty$. If E_j^ε are the derivatives of E^ε then $\psi^\varepsilon(x) \equiv \Delta E^\varepsilon(x) = \Sigma_1^n E_{jj}^\varepsilon(x) = \omega_n^{-1} n\varepsilon^2(|x|^2 + \varepsilon^2)^{-(n+2)/2}$ and $\psi^\varepsilon(x) = \varepsilon^{-n}\psi^1(x/\varepsilon)$, so that $<\psi^\varepsilon, \phi> = \int \psi^\varepsilon(-x)\phi(x)dx \to (\int \psi^1)\phi(0)$.)

Exercise 3.8. Prove that the Fourier transform of $E_n(t)$ is $-|x|^{-2}$ if $t \in \mathbb{R}^n$, $n \geq 3$, and deduce that $E_n(x)$ satisfies (3.10).

From Exercise 3.7 or 3.8 it follows that E_n is the fundamental solution of Δ in $\mathbb{R}^n$. In particular,

$$E_3(x) = -(4\pi|x|)^{-1} \tag{3.13b}$$

The required particular solution of (3.8) is given by

$$u_0 = U^f = f * E_n \tag{3.11b}$$

This function U^f is called the Newtonian potential of f.

More generally, if μ is a finite Borel measure, $U^\mu = \mu * E_n$ is called the Newtonian potential of μ.

Exercise 3.9. Prove that the fundamental solution $E_n(x)$ is harmonic in any domain D that does not contain the origin.

Remark 3.3. The formula (3.11b) has the following physical interpretation. If an ideal electrically charged body consists of a unitary negative charge at the origin, then it applies a force $t/|t|^3$ on each charged particle (of unitary charge) located at $t \in \mathbb{R}^3$, and $E_3 = -(4\pi|x|)^{-1}$ is just the potential function of this electrical field. In the case of a general body with charge given by a density f (or by a measure μ), the corresponding vector field is $\int |t - x|^{-3}(t - x)f(x)dx$ (or $\int |x - t|^{-3}(t - x)d\mu$), which has a potential function $U^f = \int f(x)|x - t|^{-1}dx$ (or $U^\mu = \mu * E_3$).

Remark 3.4. If S is a smooth (or piecewise smooth) surface, Φ a continuous function on S and μ the measure given by $I_\mu(\phi) = \langle\mu, \phi\rangle = \int_S \phi(x)\Phi(x)dS$, for all $\phi \in C_0^\infty$, then U^μ is called the

S-single layer potential with moment Φ. If μ is given by $<\mu, \phi> = \int_S \frac{\partial \phi}{\partial n}(x)\Phi(x)ds.$ for all $\phi \in C_0^\infty$, then U^μ is called the S-double layer potential with moment Φ.

The Poisson kernel $P_y(x)$ in $\mathbb{R}^n$ is a function defined in $\mathbb{R}_+^{n+1}$ and it is essentially equal to the derivative $\partial_{n+1}E_{n+1}$. As we shall see, the solution of the socalled Dirichlet problem for $\mathbb{R}_+^{n+1}$ is expressed through the kernel $P_y \sim \partial_{n+1}E_{n+1}$. On the other hand, the single and double layer potentials appear in the solution of the so called Neumann problem.

It can be proved that if $u \in C^2(\overline{D})$, $u(x) = 0$ for $x \notin \overline{D}$, then u is the sum of three potentials,

$$u = U^{\Phi_1} + U^{\Phi_2} + U^{\Phi_3}$$

where $\Phi_1 = \Delta u$, Φ_2 is the single layer potential on $S = \partial D$ with moment $\frac{\partial u}{\partial n}$, and Φ_3 is the double layer potential on $S = \partial D$ with moment $-((n-2)_{\omega_n})^{-1}u$. Therefore,

$$(n-2)\omega_n u(x) = -\int_D \frac{\Delta u(y)}{|x-y|^{n-2}}\,dy + \int_{\partial D} |x-y|^{2-n}\frac{\partial u}{\partial n}(y)dS$$
$$-\int_{\partial D} u(y)\frac{\partial}{\partial n_y}|x-y|^{2-n}\,dS \qquad (3.14)$$

The theorem that proves (3.14) is known as Green's theorem. This theorem allows to derive simply Theorem 3.3 below, as well as other properties of harmonic functions. However, we are not going to prove nor use formula (3.14) here. Instead, we shall repeatedly use the following simpler version of Green's theorem, as it is usually taught in Advanced Calculus courses:

Proposition 3.1. If $A \subset D$ is a subdomain with "sufficiently smooth" boundary (in our applications, ∂A will consist in one or more spherical surface) and $u, v \in C^2(A)$, then

$$\int_A (u\Delta v - v\Delta u)dx = \int_{\partial A} (u\frac{\partial v}{\partial n} - v\frac{\partial u}{\partial n})d\sigma \tag{3.15}$$

where $\partial u/\partial n$ indicates differentiation in the direction of the outward directed normal to ∂A and $d\sigma$ is the surface area element on ∂A.

Applying (3.15) to u harmonic and $v \equiv 1$, we get

Corollary 3.2. If u is harmonic in D and $A \subset D$ then

$$\int_{\partial A} \frac{\partial u}{\partial n} d\sigma = 0 \tag{3.16}$$

Definition 3.2. The mean value of $u(x)$ taken over the surface of the sphere of radius $r > 0$ centered in x is

$$\mathcal{M}u(x, r) = \frac{1}{\omega_n} \int_\Sigma u(x + rt')dt' \tag{3.17}$$

where dt' is the surface area element on Σ, unit sphere in $\mathbb{R}^n$.

Theorem 3.3 (Mean value theorem for harmonic functions). Let u be a harmonic function in the domain D. If $S(x, r_0) \subset D$ then for all $0 < r \leq r_0$,

$$\mathcal{M}u(x, r) = u(x) \tag{3.18}$$

Proof. By Exercise 3.2, we can assume $x = 0$. Let $0 < \varepsilon < r \leq r_0$ and consider the spheres $S_\varepsilon = S(0, \varepsilon)$ and $S_r = S(0, r)$, and $\Sigma_\varepsilon, \Sigma_r$ their surfaces. Applying Proposition 3.1 to the given u and $v = E_n$

(see (3.13); we write the proof for the case $n > 2$, the case $n = 2$ being analogous) in the domain $\mathcal{E}$ determined between S_ε and S_r, taking into account that $\Delta u = \Delta v = 0$, we obtain

$$0 = \left(\int_{\Sigma_r} - \int_{\Sigma_\varepsilon}\right)(2 - n)|x|^{1-n}u(x)d\sigma - \left(\int_{\Sigma_r} - \int_{\Sigma_\varepsilon}\right)|x|^{2-n}\frac{\partial u}{\partial n}\,d\sigma$$

But $|x| = r$ on Σ_r and $|x| = \varepsilon$ on Σ_ε, and u being harmonic in $D \supset S(0, r_0) \supset S_r \supset S_\varepsilon$, the two last integrals vanish by Corollary 3.2, and we get

$$\varepsilon^{1-n}\int_{\Sigma_\varepsilon} u d\sigma = r^{1-n}\int_{\Sigma_r} u d\sigma$$

Thus,

$$u(0, r) = \frac{1}{\omega_n}\int_{\Sigma} u(x + rt')dt'$$

$$= \frac{1}{\omega_n r^{n-1}}\int_{\Sigma_r} u d\sigma$$

$$= \frac{1}{\omega_n \varepsilon^{n-1}}\int_{\Sigma_\varepsilon} u d\sigma \xrightarrow[\varepsilon \to 0]{} u(0) \qquad \nabla$$

The mean value property characterizes harmonic functions completely, as the following result shows.

<u>Proposition 3.4.</u> Let $u \in C^2(D)$ be such that $\mathcal{M}u(x, r) = u(x)$ if $S(x, r) \subset D$. Then u is harmonic in D.

<u>Proof</u>. Let $x \in D$ be fixed. Since $\mathcal{M}u(x, r) = \mathcal{M}(r)$ is constant by hypothesis, it is $\mathcal{M}''(0) = 0$, so it suffices to prove that $n\,\mathcal{M}''(0) = \Delta u$. In fact,

$$\frac{d}{dr}\mathcal{M}(r) = \frac{1}{\omega_n}\int_\Sigma \sum_{j=1}^{n} u_j(x + rt')t'_j\, dt'$$

and

$$\frac{d^2}{dr^2}\mathcal{M}(r) = \frac{1}{\omega_n}\int_\Sigma \sum_{j,k=1}^{n} u_{jk}(x + rt')t'_j t'_k\, dt' \qquad (3.19)$$

where the differentiation under the integral sign is justified by the fact that $u \in C^2$. Thus (3.19) yields

$$\mathcal{M}''(0) = \frac{1}{\omega_n}\sum_{j,k=1}^{n}\left(\int_\Sigma t'_j t'_k\, dt'\right)u_{jk}(x)$$

But $\int_\Sigma t'_j t'_k\, dt' = 0$ if $j \neq k$, while $\int_\Sigma (t'_j)^2 dt' = \omega_n/n$ is independent of j, $j = 1, \ldots, n$, as can be easily checked, so

$$\mathcal{M}''(0) = \frac{1}{n}\sum_{j=1} u_{jj}(x) = \frac{1}{n}\Delta u \qquad \nabla$$

Proposition 3.4 enables us to prove the following theorem that, together with the mean value theorem asserts that <u>every function harmonic on a domain</u> D <u>belongs to</u> $C^\infty(D)$.

<u>Theorem 3.5</u>. Let u be a locally integrable function in D such that $\mathcal{M}u(x, r) = u(x)$ if $S(x, r) \subset D$. Then u is harmonic in D and $u \in C^\infty(D)$.

<u>Proof</u>. As the problem is a local one, it is sufficient to restrict it to a sphere which closure is contained in D. Let us assume directly that D is such a sphere and that $u \in L^1(\overline{D})$. We may extend u to the whole of $\mathbb{R}^n$ by making it zero outside $\overline{D}$.

Let $\phi \in C^\infty(\mathbb{R}^n)$ be a radial function such that $\int \phi = 1$ and $\operatorname{supp} \phi \subset S(0,1)$ and consider the convolution unit $\{\phi_\varepsilon\}_{\varepsilon>0}$, $\phi_\varepsilon(x) = \varepsilon^{-n}\phi(x/\varepsilon)$. Calling $u_\varepsilon(x) = u * \phi_\varepsilon(x)$ we have $u_\varepsilon \in C^\infty(\mathbb{R}^n)$. Furthermore,

$$u_\varepsilon(x) = \int_{R^n} u(x-y)\phi_\varepsilon(y)dy = \int_0^\varepsilon \phi_\varepsilon(r)\left(\int_\Sigma u(x-ry')dy'\right)r^{n-1}dr$$

$$= \int_0^\varepsilon \phi_\varepsilon(r)(\omega_n \mathscr{M}u(x,r))r^{n-1}dr$$

If ε is less than the distance from x to ∂D then $\mathscr{M}u(x,r) = u(x)$ for $0 < r \leq \varepsilon$, so

$$u_\varepsilon(x) = u(x)\omega_n \int_0^\varepsilon \phi_\varepsilon(r)r^{n-1}dr = u(x) \tag{3.20}$$

In particular, if the sphere D is centered at x_0 and has radius 2ρ, $u(x) = u_\rho(x)$ for all x such that $|x - x_0| < \rho$. Thus for each point $x_0 \in D$ there is a neighborhood where u coincides with a C^∞ function and the thesis follows by Theorem 3.4. ∇

Corollary 3.6. Let $\{u_k\}$ be a sequence of harmonic functions in D, converging uniformly to a function u in each bounded subdomain A, $\bar{A} \subset D$. Then u is harmonic in D.

Proof. The uniform limit in A of continuous functions is continuous and, furthermore, locally integrable. Applying Theorem 3.5 to $u(x)$ it is only necessary to show that u satisfies the mean value condition (3.18). Let $x_0 \in D$ and r such that $S(x_0, r) \subset D$. For each k, $u_k(x_0) = \mathscr{M}u_k(x_0, r)$. As $u_k(x) \to u(x)$ uniformly on the surface of $S(x_0, r)$, $\mathscr{M}u_k(x_0, r) \to \mathscr{M}u(x_0, r)$. Moreover, $u_k(x_0) \to u(x_0)$, so $u(x_0) = \lim u_k(x_0) = \lim \mathscr{M}u_k(x_0, r) = \mathscr{M}u(x_0, r)$. ∇

Remark 3.5. We state some stronger versions of the last results and leave them as exercises to the reader:

Proposition 3.7. If $u \in C(D)$ and $<\Delta u, \phi> = \int u\Delta\phi = 0$ for all $\phi \in C_0^\infty(D)$, then $u \in C^\infty(D)$ and u is harmonic.

Proposition 3.8. Let $\{u_k\}$ be a sequence of harmonic function in D which converges weakly to a function $u \in C(D)$, i.e.,

$$<u_k, \phi> \to <u, \phi> \quad \text{for all} \quad \phi \in C_0^\infty(D)$$

Then u is harmonic in D.

In particular, Proposition 3.8 immediately implies Corollary 3.6.

We now show that harmonic functions on a bounded domain are characterized by their boundary values.

Corollary 3.9 (Maximum principle for harmonic functions). Given a (real valued) harmonic function u in D such that $M = \sup_{x \in D} u(x) < \infty$, either $u(x) < M$ for all $x \in D$ or $u(x)$ is a constant function.

Proof. Suppose $u(x) = M$ for some $x \in D$. Then there exists $r_0 > 0$ such that $M = u(x) = \mathcal{M}u(x, r)$ for $0 < r \leq r_0$. Since u is continuous and less than or equal M, it must be $u \equiv M$ on $\Sigma_r(x) = \{y \in \mathbb{R}^n : |x - y| = r\}$ for all $0 < r \leq r_0$. That is, $u \equiv M$ in a neighborhood of x, so the set $\{y \in \mathbb{R}^n : u(y) = M\}$ is open. The continuity of u implies, on the other hand, that this set is closed. Since D is connected it must be $\{y \in \mathbb{R}^n : u(y) \equiv M\} = D$, that is $u(x) \equiv M$ in D. ∇

Applying this result to $-u$ we obtain the Minimum principle for harmonic functions: given a harmonic function u in D such that $m = \inf u(x) > -\infty$, either $u(x) > m$ for all $x \in D$ or $u(x)$ is a constant in D.

An equivalent form of these statements is the following:

Corollary 3.10. If u is harmonic in a bounded domain D and continuous in $\bar{D} = D \cup \partial D$, u attains its maximum (minimum) only on the boundary ∂D, provided u is not a constant.

Applying this to $u = u_1 - u_2$ we obtain

Corollary 3.11 (Uniqueness of the values of an harmonic function in the boundary). Given two functions u_1, u_2 continuous on the closure $\bar{D}$ of a bounded domain D and harmonic in D, if $u_1(x) = u_2(x)$ for all $x \in \partial D$, then $u_1(x) = u_2(x)$ for all $x \in \bar{D}$.

Theorem 3.12 (Louiville's theorem for harmonic functions). If u is a harmonic function in $\mathbb{R}^n$ that is bounded throughout, then u is a constant function.

Proof. By the mean value property (3.18) we have that, for $x \in \mathbb{R}^n$ and $t > 0$,

$$u(x) = u(x) \frac{n}{t^n} \int_0^t r^{n-1} dr = \frac{n}{t^n} \int_0^t u(x, r) r^{n-1} dr$$

$$= \frac{n}{\omega_n} \frac{1}{t^n} \int_0^t \left(\int_\Sigma u(x + ry') dy' \right) r^{n-1} dr$$

$$= \frac{1}{\Omega_n t^n} \int_{|y| \leq t} u(x + y) dy$$

Thus,

$$u(x_1) - u(x_2) = \frac{1}{\Omega_n t^n} \left(\int_{S(x_1, t)} u(x) dx - \int_{S(x_2, t)} u(x) dx \right)$$

and

$$|u(x_1) - u(x_2)| \leq \frac{\|u\|_\infty}{\Omega_n t^n} \int_{S(x_1, t) \Delta S(x_2, t)} dx$$

But the measure of the symmetric difference $S(x_1, t) \Delta S(x_2, t)$ divided by t^n clearly tends to zero when $t \to \infty$, and the thesis follows. ∇

4. POISSON INTEGRALS

A classical problem on harmonic functions on a domain is the Dirichlet problem:

Let D be a domain with compact closure $\bar{D}$, and f a continuous function defined on $\partial D = \bar{D} - D$. Does there exist a (unique) function u, continuous on $\bar{D}$, harmonic when restricted to D, and that agrees with f on ∂D?

This problem is of great importance in the theory of harmonic functions and in its applications. Corollary 3.11 shows that uniqueness holds. The solution for the special case when $n = 2$ and D is the unit sphere (actually, the unit disc) was outlined at the beginning of Section 3, by means of the Poisson kernel in $\mathbb{T}$. Through conformal mappings the solution for the problem on the unit disc gives the solution on other bounded domains in $\mathbb{R}^2$ as well. In the case $n > 2$, the solution of the Dirichlet problem on a sphere also involves the Poisson kernel in the unit sphere, $P(x, w) = (1 - |x|^2)|x - w|^{-n}$, that can be written as before as $P(x, w) = P_r(\theta) = (1 - r^2)(1 - 2r \cos \theta + r^2)^{-n/2}$ if $r = |x| < 1 = |w|$ and θ is the angle between x and w.

Proposition 4.1. Let D be the interior of the sphere $S(x_0, R)$ and f a continuous function on $\partial D = \{x \in \mathbb{R}^n : |x - x_0| = R\}$.

The function u defined by

$$u(x) = \frac{R^{n-2}}{\omega_n} = \int_\Sigma f(x_0 + Rs) \frac{R^2 - |x-x_0|^2}{|x-x_0-Rs|^n} d\sigma \tag{4.1}$$

if $x \in D$ and by $u(x) = f(x)$ if $x \in \partial D$, is harmonic in D and continuous in $\bar{D} = D \cup \partial D$.

Proof. Since harmonicity is preserved under translations and dilations (see Exercise 3.2) it will be enough to consider $S(0,1)$, the unit sphere of $\mathbb{R}^n$. Since $P(x,u)$ is harmonic for $x \in D$, the harmonicity of $u(x) = \omega_n^{-1} f * P(x)$ follows from the standard argument of differentiation under the integral sign (as was already done in Section 3).

To prove the continuity of u in the closed sphere $\bar{D}$ it will be enough to prove it on its boundary Σ.

Let be $w \in \Sigma$ and $x = rw$, $0 < r < 1$. By the mean value property $\omega_n^{-1} \int_\Sigma P(x,w)dw = P(0,w) = 1$, and $u(w) = f(w)$, so

$$\begin{aligned} |u(x) - u(w)| &\le \omega_n^{-1} \int_\Sigma |f(s) - f(w)| P(x,s)ds \\ &= \omega_n^{-1} \int_{|s-w|<\delta} + \omega_n^{-1} \int_{|s-w|>\delta} \end{aligned} \tag{4.2}$$

Choosing $\delta > 0$ conveniently, the first integral in (4.2) can be made arbitrarily small since f is continuous on Σ, and for such δ, the last integral is majorized by a constant times $\int_{|s-w|>\delta} P(rw,s)ds$ which tends to zero uniformly in w when $r \to 1$. ∇

Let us consider now a variant of the Dirichlet problem that arises when ∂D coincides with $\mathbb{R}^n$, so that its solution provides a tool for the study of functions defined in $\mathbb{R}^n$.

We shall work in the upper half-space $\mathbb{R}^{n+1}_+$ instead of D, so $\bar{D}$ is no longer compact. Now $\overline{\mathbb{R}^{n+1}_+} = \{(x, y) : x \in \mathbb{R}^n, y \in \mathbb{R}, y \geq 0\}$, and we identify $\partial\mathbb{R}^{n+1}_+ = \{(x, y) : x \in \mathbb{R}^n, y = 0\}$ with $\mathbb{R}^n$. The problem becomes:

Given f(x) continuous on $\mathbb{R}^n$, does there exist a (unique) function u(x, y) harmonic in $\mathbb{R}^{n+1}_+$ and continuous in $\overline{\mathbb{R}^{n+1}_+}$ such that $u(x, 0) = f(x)$ for $x \in \mathbb{R}^n$?

Considering the case $u(x, y) = y$, we already see that other conditions must be imposed in order to obtain uniqueness, since $v(x, y) \equiv 0$ is such that both u and v are harmonic in $\mathbb{R}^{n+1}_+$ and coincide on $\mathbb{R}^n$. This fact does not contradict Corollary 3.11 since $D = \mathbb{R}^{n+1}_+$ is not bounded.

We do have the following result that will entail uniqueness in the $\mathbb{R}^{n+1}_+$ case.

Proposition 4.2. Let u(x, y) be continuous in $\overline{\mathbb{R}^{n+1}_+}$, harmonic in $\mathbb{R}^{n+1}_+$ and $u \equiv 0$ on $\mathbb{R}^n$. If u is bounded in $\mathbb{R}^{n+1}_+$ then $u \equiv 0$ throughout $\mathbb{R}^{n+1}_+$.

This proposition is an immediate consequence of the Liouville theorem (Theorem 3.12) and the Reflection Principle.

Proposition 4.3 (The Reflection Principle). $D \subset \mathbb{R}^{n+1}$ is a domain symmetric with respect to $\mathbb{R}^n$ (i.e., $(x, y) \in D$ implies $(x, -y) \in D$) and u(x, y) is a continuous function defined on D, such that $u(x, y) = -u(x, -y)$. If u is harmonic in $D^+ = \{(x, y) \in D : y > 0\}$ then u is harmonic in all D.

Proof. Since u is harmonic in D^+ and $\Delta u(x, -y) = -\Delta u(x, y)$, u is also harmonic in $D^- = \{(x, y) \in D : y < 0\}$. It remains to prove harmonicity in a neighborhood of each point of $\{(x, y) \in D : y = 0\}$. If $x_0 = (x_0, 0)$ is such a point, let $S(x_0, R)$ be contained in D and let

$$v(x, y) = \frac{R^{n-1}}{\omega_{n+1}} \int_{\Sigma_{n+1}} u(x_0 + Rs, Rt) \frac{R^2 - |(x-x_0, y)|^2}{|(x-x_0-Rs, y-Rt)|^{n+1}} \, d\sigma \tag{4.3}$$

where $\sigma = (s, t) \in \Sigma_{n+1}$ for $(x, y) \in S(x_0, R)$. By Proposition 4.1, v is a harmonic function coinciding with u on $\partial S(x_0, R)$. The integral in (4.3) vanishes with y, since $u(x_0 + Rs, Rt) = -u(x_0 + Rs, -Rt)$, so it is $v(x, 0) \equiv 0$, and by continuity, $u(x, 0) \equiv 0$ whenever $(x, 0) \in D$. So v coincides with u on $T^+ = \{(x, y) \in \overline{\mathbb{R}^{n+1}_+} : |x - x_0|^2 + |y|^2 = R^2\}$, a set including $S(x_0, R) \cap \mathbb{R}^n$. Since both u and v are harmonic in the interior of T^+, they must coincide in it (by the uniqueness Corollary 3.11). The same will be true in $\mathbb{R}^{n+1}_-$, that is $u \equiv v$ in the interior of and on $T^- = \{(x, y) \in \overline{\mathbb{R}^{n+1}_-} : |x - x_0|^2 + |y|^2 = R^2\}$. Thus $u \equiv v$ on $S(x_0, R)$ and, since v is harmonic, the thesis is proved. ∇

We can now give the solution for the Dirichlet problem in $\mathbb{R}^{n+1}_+$.

Theorem 4.4. Given $f \in C_\infty(\mathbb{R}^n)$ there exists a unique bounded function $u(x, y)$ defined in $\overline{\mathbb{R}^{n+1}_+}$, such that

(a) $u(x, y)$ is harmonic in $\mathbb{R}^{n+1}_+$,

(b) $u(x, y)$ is continuous in $\overline{\mathbb{R}^{n+1}_+}$ and $u \in C_\infty(\overline{\mathbb{R}^{n+1}_+})$,

(c) $u(x, 0) = f(x)$.

Furthermore, u is the Poisson integral of f,

$$u(x, y) = \int f(t) P(x - t, y) dt \tag{4.4}$$

Remark 4.1. Considering C_∞ instead of L^∞, this result may be viewed as a substitute for $p = \infty$ of the next theorem, and we shall prove both together.

Theorem 4.4A. Given $f \in L^p(\mathbb{R}^n)$, $1 \leq p < \infty$, there exists a function $u(x, y)$ defined in $\mathbb{R}_+^{n+1}$, such that

(a) $u(x, y)$ is harmonic in $\mathbb{R}_+^{n+1}$,

(b') $(\int_{\mathbb{R}^n} |u(x, y)|^p dx)^{1/p} \leq \|f\|_p$ for all $y > 0$,

(c') $\|u(., y) - f\|_p \to 0$ when $y \to 0$,

and also

$u(x, y) \to f(x)$ when $y \to 0$ for a.e. $x \in \mathbb{R}^n$.

Furthermore, u is the Poisson integral of f,

$$u(x, y) = \int f(t)P(x - t, y)dt$$

Proof. We shall prove that $u(x, y) = f * P(x, y)$ satisfies conditions (a), (b), (c) for $f \in C_\infty$ and (a), (b'), (c') for $f \in L^p$, $1 \leq p < \infty$.

In both cases, (a) follows from the harmonicity of the Poisson kernel $P(x, y)$, as in Exercise 3.5.

Condition (b'), which is valid for $1 \leq p \leq \infty$, corresponds to Theorem 1.1 of Chapter 1 and, similarly, condition (c') corresponds to Theorems 1.4 and 2.1 of Chapter 1.

If $f \in C_\infty$, the same Theorem 1.4 asserts that $u(x, y) \to f(x)$, uniformly as $y \to 0$. Extending $u(x, y)$ to $\overline{\mathbb{R}_+^{n+1}}$ by defining $u(x, 0) = f(x)$ we get a continuous function in $\overline{\mathbb{R}_+^{n+1}}$ that fulfills condition (c). To complete condition (b) for $f \in C_\infty$ it remains to prove that $u(x, y)$ tends to zero whenever $|(x, y)| \to \infty$.

First notice that given $\varepsilon > 0$, there is $y_0 > 0$ such that $|u(x, y)| < \varepsilon$ for all x and all $y \geq y_0$. In fact, let $A > 0$ be such that $|f(t)| < \varepsilon/2$ if $|t| > A$, then

$$|u(x, y)| \leq \int |f(t)|P(x - t, y)dt = \int_{|t| \leq A} + \int_{|t| > A}$$

(continued)

$$\leq C_n y^{-n} \int_{|t| \leq A} |f(t)| dt + \frac{\varepsilon}{2} \int P(x - t, y) dt < \frac{\varepsilon}{2} + \frac{\varepsilon}{2} = \varepsilon$$

if

$$y \geq y_0 = \left(\frac{2C_n}{\varepsilon} \int_{|t| \leq A} |f| dt\right)^{1/n}$$

For $y < y_0$ there exists $K > 0$ such that $|u(x, y)| < \varepsilon$ for $|x| > K$. In fact,

$$|u(x, y)| \leq \int |f(t)| P(x - t, y) dt \leq \int_{|x-t| \leq B} + \int_{|x-t| > B}$$

$$\leq \max_{|x-t| \leq B} |f(t)| \int P(t) dt + C_n y \|f\|_\infty \int_{|s| > B} s^{-n-1} ds$$

and we may take B large enough as to have the second term of the sum less than $\varepsilon/2$, and for that B, we take K such that for $|x| > K$ it is $|f(t)| < \varepsilon/2$ whenever $|x - t| \leq B$.

Thus $u(x, y) \in C_\infty(\overline{\mathbb{R}^{n+1}_+})$ whenever $f \in C_\infty(\mathbb{R}^n)$.

In this case the uniqueness of u follows from the fact that $u = f * P_y$ is bounded, so Proposition 4.2 applies. ∇

Like other results involving integrable functions, Theorem 4.4A admits an extension to finite Borel measures.

Theorem 4.4B. Given $\mu \in \mathcal{M}(\mathbb{R}^n)$, if $u(x, y) = \int P(x - t, y) d\mu(t)$ is its Poisson-Stieltjes integral, then

(a) $u(x, y)$ is harmonic in $\mathbb{R}^{n+1}_+$,

(b'') $\|u(., y)\|_1 \leq \|\mu\|$,

(c'') $u(x, y)$ tends weakly-* to μ when $y \to 0$.

Proof. (a) can be proved similarly to (a) in Theorems 4.2 and 4.2A, (b'') corresponds to Proposition 3.2(c) of Chapter 1, and (c'') was already proved as Theorem 1.13. ∇

The condition of L^p boundedness in x, for all $y > 0$, (condition (b') of Theorem 4.4A) characterizes the Poisson integrals. To obtain the converse to Theorem 4.4 we prove now two lemmas.

Lemma 4.5. Let $u(x, y)$ be a harmonic function in $\mathbb{R}_+^{n+1}$ such that there exist a constant C, $0 < C < \infty$, and a number p, $1 \leq p \leq \infty$, with $\|u(.,y)\|_p < C$ for all $y > 0$. Then there exists a constant $A = A(n, p)$ such that $\|u(.,y)\|_\infty = \sup_{x\in\mathbb{R}^n} |u(x, y)| \leq ACy^{-n/p}$.

In particular, u is a bounded function in each proper subhalf-space $\mathbb{R}_+^{n+1}(y_0) = \{(x, y) \in \mathbb{R}_+^{n+1} : y > y_0 > 0\} \subset \mathbb{R}_+^{n+1}$.

Proof. Let us apply the mean value theorem in $\mathbb{R}^{n+1}$.

$$u(x, y) =$$

$$= u(x, y) \cdot \frac{n+1}{(y/2)^{n+1}} \int_0^{y/2} r^n dr = \frac{(n+1)2^{n+1}}{y^{n+1}} \int_0^{y/2} \mathscr{M}u((x, y), r)r^n dr$$

$$= \frac{(n+1)2^{n+1}}{y^{n+1}} \frac{1}{\omega_{n+1}} \int_0^{y/2} (\int_{\Sigma_{n+1}} u(x_1 + rt_1', \ldots, x_n + rt_n', y + rt_{n+1}')dt')r''dr$$

(here $t = (t_1, \ldots, t_n, t_{n+1})$)

$$= \frac{2^{n+1}}{y^{n+1}} \frac{1}{\Omega_{n+1}} \int_{|t|\leq y/2} u(x_1 + t_1, \ldots, x_n + t_n, y + t_{n+1})dt$$

Hence

$$|u(x, y)| \leq \left(\frac{2}{y}\right)^{n+1} \frac{1}{\Omega_{n+1}} \int_{|(x, y)-(\xi, \eta)| \leq y/2} |u(\xi, \eta)| d\xi \, d\eta$$

where $(\xi, \eta) = (\xi_1, \ldots, \xi_\eta, \eta) = (x_1 + t_1, \ldots, x_n + t_n, y + t_{n+1})$.

By Hölder's inequality,

$$|u(x, y)| \leq \left(\frac{2}{y}\right)^{n+1} \frac{1}{\Omega_{n+1}} \left(\int_{|(x, y)-(\xi, \eta)| \leq y/2} |u/\xi, \eta)|^p d\xi \, d\eta\right)^{1/p} \cdot$$

$$\left(\Omega_{n+1}\left(\frac{y}{2}\right)^{n+1}\right)^{1/p'}$$

$$\leq \left(\frac{2}{y}\right)^{(n+1)/p} \Omega_{n+1}^{-1/p} \left(\int_{y/2}^{3y/2} \left(\int_{\xi \in \mathbb{R}^n} |u(\xi, \eta)|^p d\xi\right) d\eta\right)^{1/p}$$

$$\leq \left(\frac{2}{y}\right)^{(n+1)/p} \Omega_{n+1}^{-1/p} \left(\int_{y/2}^{3y/2} C^p d\eta\right)^{1/p}$$

$$= \left(\frac{2^{n+1}}{\Omega_{n+1}}\right)^{1/p} C y^{-n/p}$$

that is the thesis with $A = (2^{n+1}/\Omega_{n+1})^{1/p}$. ∇

Lemma 4.6. For all $u(x, y)$ harmonic in $\mathbb{R}_+^{n+1}$ and bounded in each proper subhalf-space,

$$u(x, y_1+y_2) = \int u(t, y_1) P(x-t, y_2) dt$$

Proof. For $y_0 > 0$ fixed, let $v(x, y) = u(x, y + y_0)$ for all $y \geq 0$, and let $w(x, y) = \int u(t, y_0) P(x - t, y) dt$ for all $(x, y) \in \mathbb{R}_+^{n+1}$.

We must prove that v coincides with w.

The function v is harmonic in $\mathbb{R}_+^{n+1}$ and continuous and bounded in $\mathbb{R}_+^{n+1}$. As in Theorem 4.4, we can extend $w(x, y)$ to all $\overline{\mathbb{R}^{n+1}_+}$ by letting $w(x, 0) = u(x, y_0)$ and so we obtain a bounded continuous function that is harmonic in $\mathbb{R}^{n+1}_+$. So $v - w = U$ is a harmonic function in $\mathbb{R}^{n+1}_+$, bounded and continuous in $\overline{\mathbb{R}^{n+1}_+}$, that vanishes on $\mathbb{R}^n$: $U(x, 0) = 0$. Thus by Proposition 4. 2, $v \equiv w$ in all $\mathbb{R}^{n+1}_+$. ∇

<u>Theorem 4. 7</u>. Let $u(x, y)$ be a harmonic function in $\mathbb{R}^{n+1}_+$ such that there exist $C > 0$ and p, $1 \le p \le \infty$, with $\|u(., y)\|_p \le C$ for all $y > 0$.

(i) If $1 < p \le \infty$, then $u(x, y)$ is the Poisson integral of an $f \in L^p(\mathbb{R}^n)$.

(ii) If $p = 1$, then $u(x, y)$ is the Poisson-Stieltjes integral of a $\mu \in \mathcal{M}(\mathbb{R}^n)$. Furthermore, if $\{u(., y)\}_{y>0}$ is Cauchy in L^1 as $y \to 0$, then $u(x, y)$ is the Poisson integral of an $f \in L^1(\mathbb{R}^n)$.

<u>Proof</u>. For $1 < p \le \infty$, as $\|u(., y)\|_p \le C$ for all $y > 0$, there exist a sequence $\{y_k\}$ tending to zero and an $f \in L^p$, such that $\{u(., y_k)\}$ converges weakly to f, i. e., for each $g \in L^{p'}$, $1/p + 1/p' = 1$, $\int u(x, y_k)g(x)dx \to \int f(x)g(x)dx$.

Analogously, if $p = 1$, there exist $\{y_k\}$ and a $\mu \in \mathcal{M}$, such that $\{u(., y_k)\}$ converges weakly-* to μ, i. e., for each $g \in C_\infty$, $\int u(x, y_k)g(x)dx \to \int g(x)d\mu(x)$.

(Both facts follow from the weak-* compactness of the unit sphere of the dual of any Banach space, the Alaoglu-Bourbaki theorem, see Chapter 0, Section 4, and by $(L^p)' = L^{p'}$, $1 \le p < \infty$, $(C_\infty)' = \mathcal{M}$.)

For each $y > 0$, the Poisson kernel $P(., y) \in L^{p'}$ for any $1 \le p' \le \infty$ and also $P(., y) \in C_\infty$, so

$$\int P(x - t, y)u(t, y_k)dt \to \int P(x - t, y)f(t)dt = v(x, y)$$

for $1 < p \leq \infty$, and

$$\int P(x - t, y)u(t, y_k)dt \to \int P(x - t, y)d\mu(t) = v(x, y)$$

for $p = 1$. It remains to show that, in each case, $v(x, y) = u(x, y)$. But this follows from Lemma 4.6, since

$$\int P(x - t, y)u(t, y_k)dt = u(x, y + y_k)$$

and $u(x, y + y_k) \to u(x, y)$ when $k \to \infty$.

To show that the last part of (ii) holds, suppose $\{u(., y)\}_{y>0}$ is Cauchy in L^1. By the completeness of L^1, there exists an $f \in L^1$ for which $\|u(., y) - f\|_1 \to 0$ as $y \to 0$. Thus, $\int u(x, y)g(x)dx \to \int f(x)g(x)dx$ for all $g \in L^\infty$, and we can reproduce the initial argument and get again $u = P * f$. ∇

REFERENCES

1. S. Bochner and K. Chandrasekharan, Fourier transforms, Princeton University Press, Princeton, 1949.
2. G. Weiss, Análisis Armónico en varias variables, Teoría de los espacios H^p. Cursos y seminarios de Matemáticas, Universidad de Buenos Aires, fasc. 9, Buenos Aires, 1960.
3. E. M. Stein and G. Weiss, Introduction to Fourier Analysis on Euclidean Spaces, Princeton University Press, Princeton, 1971.
4. S. Saks, Theory of the Integral, Hafner Publ. Co., New York, 1938.

Chapter 4

INTERPOLATION OF OPERATORS IN L^p SPACES

1. THE M. RIESZ-THORIN CONVEXITY THEOREM

Let $(\mathscr{X}, \mu)$ be a measure space and L^p the corresponding complex (or real) Lebesgue spaces of all complex (real) valued measurable functions f with $\|f\|_p < \infty$.

For every functions f and every $\lambda > 0$ we set

$$f_\lambda(x) = f(x) \quad \text{if} \quad |f(x)| \leq \lambda \quad \text{and zero otherwise} \tag{1.1}$$

$$f^\lambda(x) = f(x) \quad \text{if} \quad |f(x)| > \lambda \quad \text{and zero otherwise} \tag{1.1a}$$

Then

$$f = f_\lambda + f^\lambda \quad \text{and} \quad |f|^p = |f_\lambda|^p + |f^\lambda|^p \tag{1.2}$$

and

$$|f_\lambda| \leq \lambda, \quad |f_\lambda| \leq |f|, \quad |f^\lambda| \leq |f|$$

Hence,

$$|f_\lambda|^{p_1} \leq \lambda^{p_1 - p} |f_\lambda|^p \leq \lambda^{p_1 - p} |f|^p \quad \text{if} \quad p_1 \geq p \tag{1.2a}$$

and also

$$|f^\lambda|^{p_0} \le \lambda^{p_0 - p} |f^\lambda|^p \le \lambda^{p_0 - p} |f|^p \quad \text{if} \quad p_0 \le p \tag{1.2b}$$

$L^{p_0} + L^{p_1}$ is the vector space of the functions of the form $f = g + h$, $g \in L^{p_0}$, $h \in L^{p_1}$, then

<u>Lemma 1.1</u>. Let be $1 \le p_0 \le p \le p_1 \le \infty$ and $\lambda > 0$.

(a) If $f \in L^p$ and f_λ, f^λ are as in (1.1), (1.1a), then $f_\lambda \in L^{p_1}$, $f^\lambda \in L^{p_0}$.

(b) $L^p \subset L^{p_0} + L^{p_1}$.

(c) If $f \in L^{p_0} \cap L^{p_1}$ and $p_1 \neq \infty$, then $\|f\|_p^p \le \|f\|_{p_0}^{p_0} + \|f\|_{p_1}^{p_1}$.

<u>Proof</u>. (a) and (b) follow immediately from (1.2)-(1.2b), and so does (c), taking $\lambda = 1$, since $|f|^p = |f_1|^p + |f^1|^p \le |f_1|^{p_0} + |f^1|^{p_1} \le |f|^{p_0} + |f|^{p_1}$. ▽

We fix now two measure spaces $(\mathscr{X}, \mu)$ and $(\mathscr{Y}, \nu)$ and consider linear operators T, from μ-measurable functions on $\mathscr{X}$ to ν-measurable functions on $\mathscr{Y}$.

If $1 \le p_0 \le p_1 < \infty$ and T is a linear operator on $L^{p_0} + L^{p_1}$, then T is defined on the (nonlinear) set $L^{p_0} \cup L^{p_1}$ and is linear on L^{p_0} and L^{p_1}. Conversely, if T is defined on $L^{p_0} \cup L^{p_1}$ and is linear on L^{p_0} and L^{p_1} then there exists a (unique) linear operator T_1 on $L^{p_0} + L^{p_1}$ such that $T_1 = T$ on $L^{p_0} \cup L^{p_1}$ (given by $T_1(g + h) = Tg + Th$ for $g \in L^{p_0}$, $h \in L^{p_1}$). Clearly, it is the same to give a linear operator T on $L^{p_0} + L^{p_1}$ or to give an operator on $L^{p_0} \cup L^{p_1}$ which is linear on L^{p_0} and L^{p_1}; in this case T extends to all L^p for $p_0 \le p \le p_1$.

An operator T is called quasi-linear if $T(f + g)$ is uniquely defined whenever Tf and Tg are defined, and if $|T(f + g)| \leq K(|Tf| + |Tg|)$, where K is a constant independent of f and g. If $K = 1$, T is called sublinear.

Definition 1.1. A linear (or sublinear or quasilinear) operator T defined in the vector space $L \subset L^p(\mathscr{X}, \mu)$ is of type (p, q) on L, with constant $M_{pq} < \infty$, if $Tf \in L^q(\mathscr{Y}, \nu)$ and

$$\|Tf\|_q \leq M_{pq}\|f\|_p \quad \text{for all } f \in L \tag{1.3}$$

If $L = L^p$ then we simply say that T is of type (p, q) with constant M_{pq}, which is equivalent to saying that T is a bounded operator from $L^p(\mathscr{X}, \mu)$ to $L^q(\mathscr{Y}, \nu)$ with norm $\|T\|_{pq} \leq M_{pq}$.

Definition 1.2. The type diagram of T is the set $\tau(T)$ of all points $(\alpha, \beta) \in \mathbb{R}^2$ such that $L^{1/\alpha}$ is contained in the domain of definition of T and T is of type $(1/\alpha, 1/\beta)$.

We deal with operators of type (p, q) with $1 \leq p, q \leq \infty$ and are interested in type diagrams contained in the unit square or type square $\{(x, y) \in \mathbb{R}^2 : 0 \leq x \leq 1, 0 \leq y \leq 1\}$. For example, if T is of type (p, q), the set $\tau(T)$ contains the point $(1/p, 1/p)$ on the main diagonal of Figure 1, and if T is of type (p, p'), $1/p + 1/p' = 1$, $\tau(T)$ contains the point $(1/p, 1/p')$ on the secondary diagonal of Figure 2.

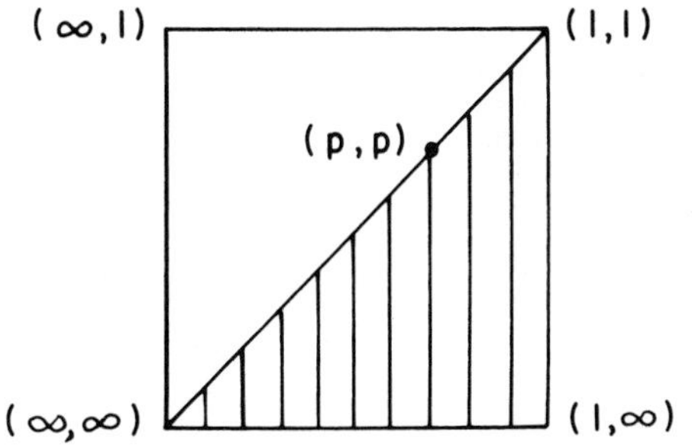

Figure 1

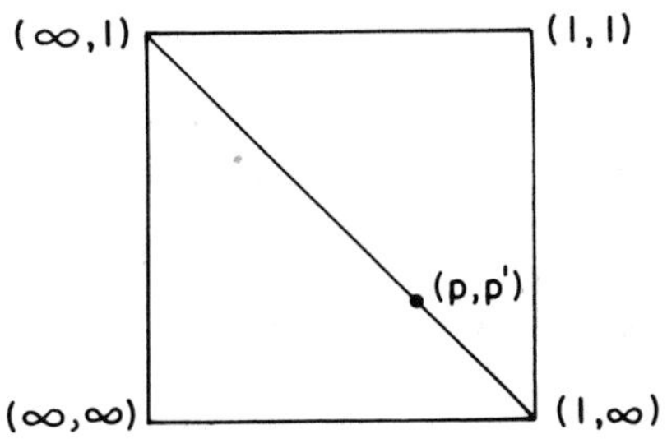

Figure 2

In the figures we mark the points by the inverses of the corresponding coordinates. Note that if $p \leq q$, the corresponding point $(1/p, 1/q)$ belongs to the lower triangle that appears shaded in Figure 1.

Exercise 3.1. Let $Tf(x) = a(x).f(x)$, $a \in L^r$. Then T is of type (p, q) if $1/q \geq 1/r + 1/p$. If $a \in L^r$ for $1/r \geq 1/r_0$ and $a \notin L^r$ for $1/r < 1/r_0$ then T is of type (p, q) if and only if $1/q \geq 1/r_0 + 1/p$.

Exercise 3.2. Prove that if T is of type (p, q) with $p < 1$, $q > p$ then $T \equiv 0$. (Hint: Prove that $T(\chi_A) = 0$ for every set A of finite measure. If $A = A_1 + \ldots + A_n$, $\mu(A_1) = \ldots = \mu(A_n)$ then $\|T\chi_{A_j}\|_q \leq Mm^{-1/p}\mu(A)^{1/p}$, $i = 1, \ldots, m$.)

Exercise 3.3. An operator T is said to operate from L^p into L^q if T is defined for every $f \in L^p$ and assigns to it a function $Tf \in L^p$. An operator T is said to be positive if $f(x) \geq 0$ a.e. implies $Tf(x) \geq 0$ a.e. Prove that if T is linear, operates from L^p into L^q and is positive, then T is continuous from L^p into L^q. (Hint: If the opposite were true, there would exist a sequence $0 \leq f_k \in L^p$, $\|f_k\|_p \leq 1$ with $\|Tf_k\|_q \geq 2^k k$. Thus $g = \Sigma_{k=1}^{\infty} 2^{-k} f_k \in L^p$ and $\|Tg\|_q \geq k$ for all k.)

Let S be a dense linear subspace of $L^p = L^p(\mathscr{X}, \mu)$ for all $1 \leq p < \infty$, and let T be a linear operator on S. By the extension theorem (Chapter 0, Proposition 4.1), if $1 \leq p < \infty$, $1 \leq q < \infty$, T is of type (p, q) on S if and only if there exists a unique operator T_1 defined on L^p such that T_1 is of type (p, q) and $T_1 = T$ on S.

Generally we shall take S as one of the following two classes of "easily manageable" functions: (i) the class of all simple functions in $\mathscr{X}$, (ii) the class of all continuous (or smooth) functions of compact support, in case $\mathscr{X} = \mathbb{R}^n$.

Lemma 1.2. Let S be as (i) or (ii), T be a linear operator in S and $1 \le p_0, p_1 < \infty$, $1 \le q_0, q_1 \le \infty$. Then T is of type (p_0, q_0) and of type (p_1, q_1) on S, with constants M_0 and M_1, if and only if T extends to an operator in $L^{p_0} + L^{p_1}$ whose restriction to L^{p_0} is of type (p_0, q_0) and whose restriction to L^{p_1} is of type (p_1, q_1), with the same constants M_0 and M_1.

Proof. If T is of type (p_i, q_i) on S, $i = 0, 1$, then T extends to an operator T_0, defined on L^{p_0}, of type (p_0, q_0) and to an operator T_1, defined on L^{p_1}, of type (p_1, q_1). By the remark preceding Definition 1.1, we have only to show that $f \in L^{p_0} \cap L^{p_1}$ implies $T_0 f = T_1 f$. If S is as in (ii) and $f \in L^{p_0} \cap L^{p_1}$ then, by Theorem 1.4 of Chapter 1 there is a convolution unit $\{f_m\} \subset S$ with $f_m \to f$ in L^{p_0} and in L^{p_1}, hence $T_i f = \lim T_i f_m = \lim Tf_m$ in L^{p_i}, $i = 0, 1$, and passing to a subsequence we may get $T_i f = \lim Tf_m$ a.e. for $i = 0, 1$; thus $T_0 f = T_1 f$.

If S is as in (i), we may assume $f \ge 0$ and there is a sequence $\{f_m\} \subset S$ such that $0 \le f_1 \le \dots \le f_m \le \dots \uparrow f$ a.e. Hence $|f - f_m|^{p_i} \to 0$ a.e. and $|f - f_m|^{p_i} \le 2|f|^{p_i}$, so $f_m \to f$ in L^{p_i}, $i = 0, 1$ and, as above, $T_0 f = T_1 f$.

Given $1 \le p_0, p_1, q_0, q_1 \le \infty$, we define, for each $t \in [0, 1]$, the intermediate p_t and q_t by

$$\frac{1}{p_t} = \frac{1-t}{p_0} + \frac{t}{p_1}, \qquad \frac{1}{q_t} = \frac{1-t}{q_0} + \frac{t}{q_1} \tag{1.4}$$

It is immediate that the set of all such p_t (resp. q_t) coincides with the interval which has p_0, p_1 (resp. q_0, q_1) as endpoints, that

$$p = p_t \quad \text{if and only if} \quad t = \frac{p_0 - p}{p_0 - p_1} \cdot \frac{p_1}{p} \tag{1.4a}$$

and that replacing p_0, p_1 by their conjugate numbers p_0', p_1', we get as intermediate p_t'.

In this section we are concerned with the following classical theorem.

Theorem 1.3 (The M. Riesz-Thorin interpolation or convexity theorem).

(a) Let S be the set of all complex valued simple functions in $\mathscr{X}$ and T a linear operator in S. If T is of type (p_0, q_0) and of type (p_1, q_1) on S, with constants M_0, M_1, where $1 \leq p_0, p_1, q_0, q_1 \leq \infty$, then, for all $t \in [0, 1]$, T is of type (p_t, q_t) on S, with constant M_t satisfying

$$M_t \leq M_0^{1-t} M_1^t \tag{1.5}$$

Moreover T extends to a linear operator in the vector space spanned by $\bigcup_{p_t<\infty} L^{p_t}$, whose restriction to L^{p_t} is a bounded operator from L^{p_t} to L^{q_t} of norm $\|T\|_{p_t q_t} \leq M_0^{1-t} M_1^t$ whenever $p_t < \infty$.

(b) If T is a linear operator in the complex space $L^{p_0} + L^{p_1}$ whose restriction to L^{p_i} is of type (p_i, q_i), $i = 0, 1$, then the restriction of T to L^{p_t} is of type (p_t, q_t) for all $t \in [0, 1]$ and (1.5) holds.

From this follows immediately

Corollary 1.4. If S is the class of all continuous (smooth) complex functions of compact support in $\mathscr{X} = \mathbb{R}^n$ and if T is of type (p_i, q_i) on S, $i = 0, 1$, where $1 \leq p_0, p_1 \leq \infty$, $1 \leq q_0, q_1 \leq \infty$, then T extends to a linear operator in $L^{p_0} + L^{p_1}$ whose restriction to L^{p_t} is a bounded operator from L^{p_t} to L^{q_t} with norm $\|T\|_{p_t q_t} \leq M_t$ satisfying (1.5) for all $t \in [0, 1]$.

Part (b) of Theorem 1.3 can be restated as follows, justifying the name of "convexity theorem": under the hypothesis of Theorem 1.3,

(i) $\tau(T)$ is a convex set in the type square

and

(ii) if $(\alpha_0, \beta_0), (\alpha_1, \beta_1) \in \tau(T)$ and $1/p_t = (1-t)\alpha_0 + t\beta_0$, $1/q_t = (1-t)\alpha_1 + t\beta_1$, then $\log \|T\|_{p_t q_t}$ is a convex function of t.

The proof of Theorem 1.3 as well as a more general theorem will be given in Section 2.

In this section we give some applications of the theorem, following a few remarks.

Remark 1.1. Counter-examples are known which show that (1.5) does not hold in the case of real L^p spaces, but the theorem remains true in that case replacing (1.5) by $M_t \le 2M_0^{1-t}M_1^t$. However, even for real L^p spaces, (1.5) holds if $p_i \le q_i$, $i = 0, 1$, (i.e., if the type diagram of the operator lies in the lower triangle), as well as if T is a positive operator (Cfr. [1]).

Theorem 1.3 was first proved by M. Riesz [1] in 1926 for real spaces, in the case $p_i \le q_i$, by real variable methods. It was extended as stated here for complex spaces in 1938 by Thorin [2] using complex methods, and further by Thorin [3] and Tamarkin and Zygmund [4]. Calderón and Zygmund [5] extended the theorem for the case $\alpha_i, \beta_i > 1$ and sublinear operators, while Weiss and Stein [6] extended it for spaces with different measures. For the interpolation theory of compact and positive operators and applications to integral equations, as well as for some sufficient conditions for a set in $\mathbb{R}^2$ to be the type diagram of some T, see [7].

Remark 1.2. In most applications of the theorem, the interesting case is $p_i \le q_i$, $i = 0, 1$, as the following example, due to Hörmander, shows. If T is a convolution operator of type (p, q)

$p > q$, then $T = 0$. In fact, as T is linear and commutes with translations, $\tau_h(Tf) = T(\tau_h f)$,

$$\|\tau_h(Tf) + Tf\|_q = \|T(\tau_h f + f)\|_q \leq M_{pq}\|\tau_h f + f\|_p \tag{1.6}$$

Since $\|\tau_h \psi + \psi\|_r \to 2^{1/r}\|\psi\|_r$ when $|h| \to \infty$ for $\psi \in L^r$, (see Chapter 0, Section 3), letting $|h| \to \infty$ in (1.6) we obtain $\|Tf\|_q \leq 2^{1/p-1/q}M_{pq}\|f\|_p$, which contradicts the type (p, q) condition since $M_{pq}2^{1/p-1/q} < M_{pq}$ for $p > q$. Then $M_{pq} = \|T\|_{pq} = 0$ and the operator is zero.

Theorem 1.3 and Corollary 1.4 find many applications in analysis. Certain operators T are studied and it is important to know the values (p, q) for which T is a bounded operator from L^p to L^q. Generally T is defined by an analytic expression which makes sense only for f in an "easily manageable" class S (e.g., the Fourier transform is defined by a formula which makes sense if $f \in L^1$ but not if $f \in L^p$, $p \neq 1$). If two pairs (p_0, q_0), (p_1, q_1) can be found such that T is of type (p_i, q_i), $i = 0, 1$, on S then Theorem 1.3 or Corollary 1.4 allows to define T on L^p for a whole interval of values $p = p_t$, $t \in [0, 1]$, as a bounded operator acting from L^p to L^q, $q = q_t$.

On L^2, the Plancherel theorem holds and there are powerful methods to check if a given T acts on L^2 (is of type (2, 2)). Usually, the problem is also simpler for L^1 or L^∞, since the L^1 norm is additive for positive functions ($\|f + g\|_1 = \|f\|_1 + \|g\|_1$ for $f, g \geq 0$) and L^∞ is an algebra under the pointwise product, etc. So, if by particular means one is able to prove that T is of type (2, 2) and of type (∞, ∞), then the convexity theorem tells us that T is of type (p, p) for $2 \leq p \leq \infty$; and, similarly, if T can be proved to be of type (2, 2) and of type (1, 1), it will be of type (p, p) for $1 \leq p \leq 2$. Furthermore, the convexity theorem asserts that the set of p's such that T is of type (p, p) is an interval.

In the preceding chapters we considered the Fourier transform operator $\mathscr{F}$ as acting on $L^1(\mathbb{R}^n, dx)$ and $L^2(\mathbb{R}^n, dx)$, and extended it to $L^p(\mathbb{R}^n, dx)$, $1 \leq p \leq 2$.

More precisely, we proved that $\mathscr{F} : L^1 \to L^\infty$ continuously, and that $\|\hat{f}\|_\infty \leq \|f\|_1$, i.e., that $\mathscr{F}$ is of type $(1, \infty)$ with constant less than or equal to 1. On the other hand, the Plancherel theorem asserts that $\mathscr{F} : L^2 \to L^2$ continuously and as an isometry, $\|\hat{f}\|_2 = \|f\|_2$, i.e., that $\mathscr{F}$ is of type $(2, 2)$ with constant less than or equal to 1. Which, then, will be the q such that $\mathscr{F}$ is of type (p, q), $1 \leq p \leq 2$, if any?

To this question answers the following theorem, which proof is immediate as a corollary of the M. Riesz-Thorin theorem.

Theorem 1.5 (Hausdorff-Young inequality in $\mathbb{R}^n$). If $f \in L^p(\mathbb{R}^n)$ for $1 \leq p \leq 2$, then $\mathscr{F}f = \hat{f} \in L^{p'}(\mathbb{R}^n)$, $1/p + 1/p' = 1$, and $\|\hat{f}\|_{p'} \leq \|f\|_p$.

Proof. As $\mathscr{F}$ is of type $(1, \infty)$ and $(2, 2)$ with norms M_0 and $M_1 \leq 1$, by Theorem 1.3, $\mathscr{F}$ is of type (p_t, q_t) with

$$\frac{1}{p_t} = \frac{1-t}{1} + \frac{t}{2} = 1 - \frac{t}{2} \quad \text{and} \quad \frac{1}{q_t} = \frac{1-t}{\infty} + \frac{t}{2} = \frac{t}{2}$$

and constant $M_t \leq M_0^{1-t} M_1^t = 1$. As $\frac{1}{p_t} = 1 - \frac{1}{q_t}$, $q_t = p_t'$ (see Figure 3). ▽

In the periodic case, $\mathscr{F}$ acts on $(\mathscr{X}, \mu)$, $\mathscr{X} = \mathbb{T}$, $d\mu = \frac{dx}{2\pi}$, as $\mathscr{F} : f \to \{c_n\}$ where $c_n = \int_0^{2\pi} f(x)e^{-inx} d\mu$ is the n-th Fourier coefficient of (periodic) f, so $\mathscr{F}f$ is a function in the measure space $(\mathscr{Y}, \nu)$, $\mathscr{Y} = \mathbb{Z}$, ν discrete. Bessel's inequality

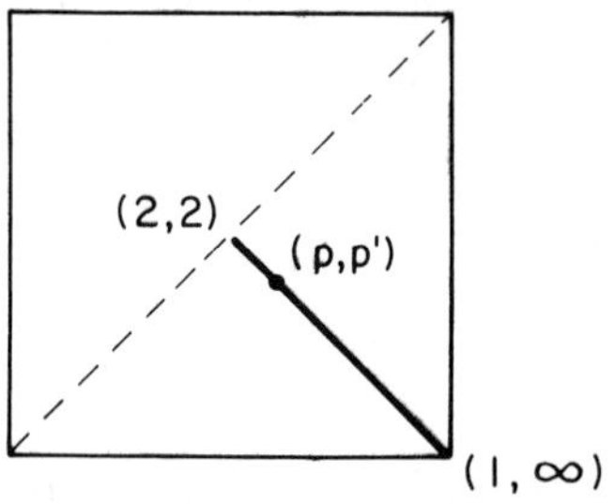

Figure 3

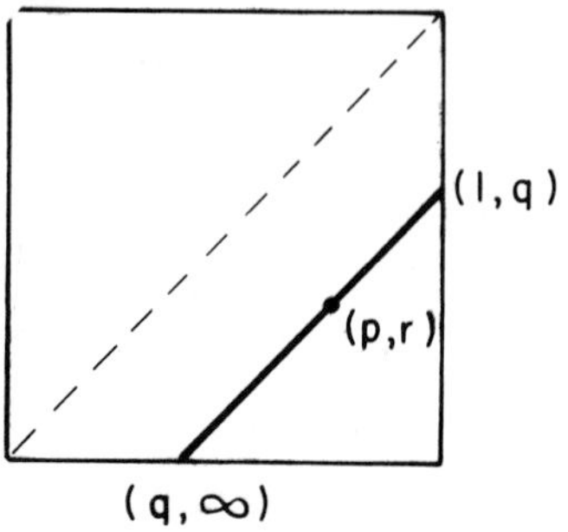

Figure 4

$\sum_n |c_n|^2 \le \int_0^{2\pi} |f(x)|^2 d\mu$, gives the type (2, 2) of $\mathcal{F}$ in this case, as $\sup_n |c_n| \le \int_0^{2\pi} |f(x)| d\mu$ gives the type $(1, \infty)$; so, as before, $\mathcal{F}$ is of type (p, p') and we get the (original) version of the result that is a generalization of Bessel's inequality.

<u>Theorem 1.5A (Hausdorff-Young's inequality for Fourier series).</u> If $f(x) \sim \sum_{n=-\infty}^{\infty} c_n e^{inx}$ then, for $1/p + 1/p' = 1$,

$$\left(\sum_{n=-\infty}^{\infty} |c_n|^{p'}\right)^{1/p'} \le \left(\frac{1}{2\pi} \int_0^{2\pi} |f(x)|^p \, dx\right)^{1/p}$$

for $1 \le p \le 2$.

<u>Remark 1.3</u>. Both Theorems 1.5 and 1.5A are not valid for $p > 2$. In fact, every $f \in L^p$, $1 \le p \le \infty$ has a Fourier transform defined as a tempered distribution that coincides with an L^p function if $1 \le p \le 2$, but an $f \in L^p$, $p > 2$, can be constructed such that the corresponding Fourier transform distribution cannot be expressed as a function.

As a first application of interpolation results to convolutions, let us obtain an immediate proof of Theorem 1.2 of Chapter 1. For any fixed $k \in L^1$, the corresponding convolution operator

$K : f \to k * f$ is of type $(1, 1)$ with constant equal to $\|k\|_1$. Furthermore, it is evident that K is of type (∞, ∞) with constant less than or equal to $\|k\|_1$, thus, if $p = p_t$, K is of type (p, p) $1 \le p \le \infty$, with norm less than or equal to $\|k\|_1^{1-t} \|k_1\|^t = \|k\|_1$.

We can obtain similarly a new result. Let now $k \in L^q$ be a fixed kernel and $1 \le q \le \infty$. By the preceding remark, the convolution operator $K : f \to k * f$ is of type $(1, q)$ with constant less than or equal to $\|k\|_q$.

Furthermore, we have already noticed that, by Hölder's inequality, $\|k * f\|_\infty \le \|k\|_q \|f\|_{q'}$, for $1/q + 1/q' = 1$, so K is of type (q', ∞) with constant less than or equal to $\|k\|_q$. Since K is defined in L^1 and in $L^{q'}$ and is a linear operator, it can be properly extended to L^p for $1 \le p \le q'$. Looking for an r such that K will be of type (p, r) (see Figure 4) we get a simple proof of

<u>Theorem 1.6 (Young's generalized inequality)</u>. If $f \in L^p$ and $g \in L^q$, $1 \le p, q \le \infty$, where $1/p + 1/q \ge 1$, then $f * g \in L^r$, for $1/r = 1/p + 1/q - 1$ and

$$\|f * g\|_r \le \|f\|_p \cdot \|g\|_q$$

<u>Proof</u>. As for a fixed $g \in L^q$, $K : f \to f * g$ is of type $(1, q)$ and of type (q', ∞) with constants less than or equal to $\|g\|_q$, then by the M. Riesz-Thorin theorem, K is of type (p_t, r_t), $t \in [0, 1]$, where

$$\frac{1}{p_t} = \frac{1-t}{1} + \frac{t}{q'} = 1 - \frac{t}{q'}, \quad \frac{1}{r_t} = \frac{1-t}{q} + \frac{t}{\infty} = \frac{1-t}{q} = \frac{1}{q} + \left(1 - \frac{t}{q'}\right) - 1 = \frac{1}{q} + \frac{1}{p_t} - 1$$

with constant less than or equal to $\|g\|_q$. If $p = p_t$ and $r = r_t$ we have the thesis. ∇

2. PROOF OF M. RIESZ-THORIN THEOREM BY THE COMPLEX METHOD

Let be $L^p = L^p(\mathscr{X}, \mu)$ and $1 \leq p_0 < p < p_1 \leq \infty$. Theorem 1.3 relates norms of an operator T in different L^p spaces so, to prove it, we will express the p-norm in terms of the p_i-norms, $i = 0, 1$. A simple way to do it is as follows: if $0 \leq f \in L^p$ then $\int f^p = \int (f^{p/p_0})^{p_0} = \int (f^{p/p_1})^{p_1} < \infty$, so that

$$f_0 = f^{p/p_0} \in L^{p_0}, \quad f_1 = f^{p/p_1} \in L^{p_1} \quad \text{and} \quad \|f\|_p^p = \|f_0\|_{p_0}^{p_0} = \|f_1\|_{p_1}^{p_1}$$

which is the seeked expression. More generally if we let $f_{0+iy} = f^{(1+i)p/p_0}$, $f_{1+iy} = f^{(1+i)p/p_1}$, then since $|f_{0+iy}| = f^{\mathrm{Re}(1+i)p/p_0} = f_0$, we have also $\int f^p = \int |f_{0+iy}|^{p_0}$. Thus $\|f\|_p^p = \|f_{iy}\|_{p_0}^{p_0} = \|f_{1+iy}\|_{p_1}^{p_1}$, for all $y \in \mathbb{R}$.

We drop now the assumption $f \geq 0$, and extend this relation for complex valued simple function $f \in S$, modifying slightly the definition of f_{0+iy}.

Let $D = \{z = x + iy \in \mathbb{C};\ 0 \leq x \leq 1\}$ be the strip in the complex plane bounded laterally by the lines $\Delta_0 = \{z = iy;\ y \in \mathbb{R}\}$ and $\Delta_1 = \{z = 1 + iy;\ y \in \mathbb{R}\}$. Given $1 \leq p_0 < p_1 \leq \infty$, for each $z \in D$ let p_z be defined by $1/p_z = 1-z/p_0 + z/p_1$, so that $p_z = p_0p_1/zp_0+(1-z)p_1$ and $p_z = p_0/1-z$ if $p_1 = \infty$. Fix $t \in (0,1)$ and let $p_t = p$. Given $f = \Sigma a_k\chi_{A_k}(x) \in S$, denoting the argument of $f(x)$ by $\phi(x)$, $f(x) = |f(x)| \exp i\phi(x)$, we define, for $z \in D$, $f_z(x) = |f(x)|^{p/p_z} \cdot \exp i\phi(x)$, so that

$$f_z(x) = \Sigma |a_k|^{p/p_z} \arg a_k \chi_{A_k}(x) = \Sigma a_k(z) b_k(x) \tag{2.1}$$

where $a_k(z)$ is a (numerical) analytic function bounded in D, $b_k \in S$ and

$$|f_z(x)| = |f(x)|^{\operatorname{Re} p/p_z} \tag{2.2}$$

$\operatorname{Re} p/p_z = p/p_0$ if $z = iy \in \Delta_0$ and $\operatorname{Re} p/p_z = p/p_1$ if $z = 1 + iy \in \Delta_1$, so by (2.2) we have that

$$\int |f_{iy}|^{p_0} d\mu = \int |f|^p d\mu \quad \text{for every} \quad iy \in \Delta_0 \tag{2.3}$$

$$\int |f_{1+iy}|^{p_1} d\mu = \int |f|^p d\mu \quad \text{for every} \quad 1 + iy \in \Delta_1 \tag{2.3a}$$

and we obtain the desired expression,

$$\|f\|_p^p = \|f_{iy}\|_{p_0}^{p_0} = \|f_{1+iy}\|_{p_1}^{p_1} \quad \text{for all} \quad y \in \mathbb{R} \tag{2.3b}$$

Let us observe these formulae more closely. Recall first that an L^p-valued function $F : D \to L^p$, which assigns to each $z \in D$ a function $F(z) = F_z \in L^p$, is called analytic in D if for every simple function ψ, the numerical function $\langle F(z), \psi \rangle = \int F_z(x)\psi(x)dx$ is analytic in the usual sense.

Defining now $\Phi(z) = f_z(x)$ for $z \in D$, we obtain from (2.1) and (2.3b) that

(a) $\Phi : z \to \Phi(z) \in S \subset L^{p_0} \cap L^{p_1}$,

so that Φ can be thought of as an L^{p_i}-valued function $\Phi : D \to L^{p_i}$, $i = 0, 1$, and these functions are bounded and analytic in D,

(b) $\Phi(t) = f$, i.e., $f_t(x) = f(x)$,

(c) defining

$$|||\Phi||| = \sup_{y\in\mathbb{R}} \max\{\|\Phi(iy)\|_{p_0}, \|\Phi(1+iy)\|_{p_1}\} \tag{2.4}$$

we have that

$$|||\Phi||| = \max\{\|f\|_p^{p/p_0}, \|f\|_p^{p/p_1}\} \tag{2.4a}$$

so that

$$|||\Phi||| < 1 \quad \text{whenever} \quad \|f\|_p < 1 \tag{2.4b}$$

and

$$|||\Phi||| = 1 \quad \text{if} \quad \|f\|_p = 1 \tag{2.4c}$$

Definition 2.1. Given $f \in L^{p_0} \cap L^{p_1}$, $0 \leq t \leq 1$, let

$$\mathscr{F}_f(p_0, p_1) = \mathscr{F}_f = \{\Phi : D \to L^{p_0} \cap L^{p_1};\ \Phi \text{ satisfies (a) and (b)}\}$$

For $\Phi \in \mathscr{F}_f$, $|||\Phi|||$ is as given in (2.4).

We have already seen that $\mathscr{F}_f$ is not empty if f is simple, and for general $f \in L^{p_0} \cap L^{p_1}$, this will be shown in Proposition 2.2 below. We prove now the converse of (2.4b).

Proposition 2.1. If $\Phi \in \mathscr{F}_f(p_0, p_1)$ and $|||\Phi||| < 1$ then $\|f\|_{p_t} < 1$. Furthermore, if $\|\Phi(iy)\|_{p_0} \leq M_0$ and $\|\Phi(1+iy)\|_{p_1} \leq M_1$ for all $y \in \mathbb{R}$ then $\|f\|_{p_t} \leq M_0^{1-t} M_1^t$.

Proof. Let $p_t = p$. It suffices to show that for every simple function g with $\|g\|_{p'} \leq 1$ $(1/p + 1/p' = 1)$, $|\int f(x)g(x)d\mu| \leq 1$ (or $\leq M_0^{1-t}M_1^t$). Replacing p_0, p_1, p and f by p_0', p_1', p' and g, we get the existence of $\Phi^* \in \mathscr{F}_g(p_0', p_1')$ such that $\Phi^*(t) = g$ and $\|\Phi^*(iy)\|_{p_0'} < 1$, $\|\Phi^*(1 + iy)\|_{p_1'} < 1$ for all $y \in \mathbb{R}$. The numerical function $F(z) = \int_{\mathscr{X}} \Phi(z)\Phi^*(z)d\mu(x)$ is analytic in D, since $\Phi(z)\Phi^*(z)$ is of the form $f_z(x)\Sigma_k c_k(z)d_k(x)$, c_k analytic, and so $F(z) = \Sigma_k c_k(z)\int_{\mathscr{X}} f_z(x)d_k(x)d\mu(x)$, and each integral is an analytic function of z. By Hölder's inequality,

$$|F(iy)| \leq \|\Phi(iy)\|_{p_0} \cdot \|\Phi^*(iy)\|_{p_0'} < 1 \text{ (or } < M_0) \quad \text{for all } y \in \Delta_0$$

and

$$|F(1+iy)| \leq \|\Phi(1+iy)\|_{p_1} \cdot \|\Phi^*(1+iy)\|_{p_1'} < 1 \text{ (or } < M_1) \text{ for all } 1+iy \in \Delta_1$$

Then the Three Lines theorem (see Chapter 0, Section 5) implies that $|F(t)| < 1$ (or $< M_0^{1-t}M_1^t$) for $0 \leq t \leq 1$, and by condition (b),

$$|F(t)| = |\int \Phi(t)\Phi^*(t)d\mu| = |\int f(x)g(x)d\mu| < |(\text{or } < M_0^{1-t}M_1^t) \qquad \nabla$$

Proof of the M. Riesz-Thorin theorem. Let T be a linear operator defined on S, satisfying $\|Tf\|_{q_0} \leq M_0\|f\|_{p_0}$, $\|Tf\|_{p_1} \leq M_1\|f\|_{p_1}$ for all $f \in S$, and let $p = p_t$. Given $f = \Sigma a_k \chi_{A_k} \in S$, with $\|f\|_p = 1$, we set $\Phi(z) = f_z$. We know that $\Phi \in \mathscr{F}_f(p_0, p_1)$ and $|||\Phi||| = \|f\|_p$ by (2.4c).

Since $g = Tf \in L^{q_0} \cap L^{q_1}$, letting $\Psi(z) = T(\Phi(z))$, we have by (2.1), $\Psi(z) = \Sigma a_k(z)T(b_k)$, so that Ψ is an analytic function with values in $L^{q_0} \cap L^{q_1}$. Moreover, $\Psi(t) = T(\Phi(t)) = T(f) = g$, and

$$\|\Psi(iy)\|_{q_0} \le M_0 \|\Phi(iy)\|_{p_0} \le M_0 |||\Phi||| = M_0 \|f\|_p$$

$$\|\Psi(1+iy)\|_{q_1} \le M_1 \|\Phi(1+iy)\|_{p_1} \le M_1 |||\Phi||| = M_1 \|f\|_p$$

This shows that $\Psi \in \mathscr{F}_g(q_0, q_1)$ and, by Proposition 2.1,

$$\|g\|_q = \|g\|_{q_t} \le (M_0 \|f\|_p)^{1-t}(M_1 \|f\|_p)^t = M_0^{1-t} M_1^t \|f\|_p$$

which is the thesis. ∇

* We conclude this section with a generalization of the M. Riesz-Thorin theorem due to E. M. Stein, and the converse of Proposition 2.1 which motivates the introduction of the abstract complex method of interpolation of operators that is outlined in Appendix B.

<u>Theorem 2.2 (M. Riesz-E. M. Stein's interpolation theorem)</u>. Let $(\mathscr{X}, \mu)$ and $(\mathscr{Y}, \nu)$ be measure spaces, S be the subspace of simple functions on $(\mathscr{X}, \mu)$, T_z be an operator that transforms each $f \in S$ into a $T_z f$ measurable in $(\mathscr{Y}, \nu)$, for each $z \in D$ (unit strip in the complex plane) and $1 \le p_0, q_0, p_1, q_1 \le \infty$, fixed numbers. If $T_z f$ is an analytic bounded function in D, such that $T_{iy} f \in L^{q_0}(\mathscr{Y}, \nu)$ and

$$\|T_{iy} f\|_{q_0} \le M_0 \|f\|_{p_0} \quad \text{for all } f \in S \text{ and all } iy \in \Delta_0 \tag{2.5}$$

and

$$T_{1+iy} f \in L^{q_1}(\mathscr{Y}, \mu) \text{ and } \|T_{1+iy} f\|_{q_1} \le M_1 \|f\|_{p_1}$$
$$\text{for all } f \in S \text{ and all } 1 + iy \in \Delta_1 \tag{2.5a}$$

then, for all $t \in [0,1]$, $T_t f \in L^{q_t}(\mathscr{Y}, \mu)$ and

$$\|T_t f\|_{q_t} \leq M_0^{1-t} M_1^t \|f\|_{p_t} \quad \text{for all} \quad f \in S \tag{2.6}$$

where $1/p_t = (1-t)/p_0 + t/p_1$ and $1/q_t = (1-t)/q_0 + t/q_1$.

Proof. It suffices to prove (2.6) for every $f \in S$ such that $\|f\|_{p_t} = 1$. As before, let $\Phi(z) = f_z(x) = |f(x)|^{p_t/p_z} \exp(i \arg f(x))$ and let $\Psi(z) = T_z(f_z(x)) = T_z(\Sigma_k\, a_k(z) b_k(x)) = \Sigma_k\, a_k(z) T_z(b_k(x))$. So $\Psi(z)$ is an analytic function and $\Psi(t) = T_t(f_t(x)) = T_t f$. Furthermore, by hypothesis (2.6) and (2.5a) and by (2.3) and (2.3a), it is

$$\|\Psi(iy)\|_{q_0} = \|T_{iy} f_{iy}\|_{q_0} \leq M_0 \|f_{iy}\|_{p_0} = M_0 \|f\|_{p_t}^{p_t/p_0} = M_0$$

for all $iy \in \Delta_0$ and, similarly,

$$\|\Psi(1+iy)\|_{q_1} \leq M_1$$

for all $1 + iy \in \Delta_1$.

So, $\Psi \in \mathscr{F}_{T_t f}(q_0, q_1)$ and, by Proposition 2.1,

$$\|T_t f\|_{q_t} \leq M_0^{1-t} M_1^t = M_0^{1-t} M_1^t \|f\|_{p_t} \qquad \nabla$$

Remark 2.1. This theorem includes the M. Riesz-Thorin theorem as the particular case when $T_z \equiv T$ for all $z \in D$.

Proposition 2.3. Given $f \in L^{p_0} \cap L^{p_1}$ and $t \in [0,1]$,

$$\|f\|_{p_t} = \inf\{|||\Phi||| : \Phi \in \mathscr{F}_f(p_0, p_1)\} \tag{2.7}$$

for $\mathscr{F}_f(p_0, p_1)$ as in Definition 2.1.

Proof. Formula (2.7) holds whenever both condition (2.4b) and Proposition 2.1 are satisfied. This is the case if f is simple. For a general $f \in L^{p_0} \cap L^{p_1}$, $\|f\|_{p_t} < 1$, Proposition 2.1 holds but we have to prove condition (2.4b). Choose $\{f_k\} \subset S$ such that $\|f - f_k\|_{p_t} \to 0$ as well as $\|f - f_k\|_{p_0} \to 0$ and $\|f - f_k\|_{p_1} \to 0$ when $k \to \infty$ (as $f \in L^{p_0} \cap L^{p_1}$ and at least $p_0 < \infty$, this can always be done, even if $p_1 = \infty$).

Condition (2.4b) holds for every f_k simple so, for each k, there exists $\Phi^{(k)} \in \mathscr{F}_{f_k}$ with $|||\Phi^{(k)}||| < 1$. Thus if $\Phi(z) = \Phi^{(k)}(z) + f - f_k$ we get $\Phi(t) = \Phi^{(k)}(t) + f - f_k = f_k + f - f_k = f$, i.e., $\Phi \in \mathscr{F}_f$. Furthermore,

$$\begin{aligned} \|\Phi(iy)\|_{p_0} &\le \|\Phi^{(k)}(iy)\|_{p_0} + \|f - f_k\|_{p_0} \\ &\le \|f_k\|_{p_0} + \|f - f_k\|_{p_0} \\ &< \|f_k\|_{p_0} + \varepsilon_k < 1 \end{aligned}$$

and, similarly, $\|\Phi(1 + iy)\|_{p_1} < 1$, so Condition (2.4b) holds for a general $f \in L^{p_0} \cap L^{p_1}$ and the thesis follows. ∇

Remark 2.2. Instead of the Lebesgue spaces L^{p_0} and L^{p_1}, consider any two Banach spaces E_0 and E_1 contained in a topological vector space V. For $f \in E_0 \cap E_1$ we can define the classes $\mathscr{F}_f$ as in Definition 2.1 and the norm $\|f\|_{p_t}$ as in Formula (2.7). Completing $E_0 \cap E_1$ in this norm $\|.\|_{p_t}$ we obtain a Banach space E_t, called the intermediate space between E_0 and E_1. Thus a scale of spaces E_t, analogous to the one of Lebesgue spaces L^{p_t} can be constructed and it can be proved that the M. Riesz-Thorin interpolation theorem holds for such abstract scale. This is done in the general interpolation theory through the complex method due to Calderón and S. Krein, which is presented in Appendix B.

3. DISTRIBUTION FUNCTION AND WEAK TYPE OPERATORS

Given a point $P = (1/p, 1/q)$, $1 \leq p, q \leq \infty$, in the type square, we shall say that an operator T is of type at P if T is of type (p, q). The examples presented in Section 1 show that operators T of type at P for every point P on a closed segment ℓ of the type square arise frequently. In such cases, we shall say that T is of type on ℓ. For instance, if T is the convolution operator given by an integrable kernel, then T is of type on ℓ, for ℓ = principal diagonal of the type square.

The M. Riesz-Thorin interpolation theorem asserts--and this is why it is useful--that an operator T is of type on ℓ if and only if it is of type at the endpoints P_0 and P_1 of ℓ. As we shall see in examples given in the next section, operators T of type at P for all points P belonging to the open segment ℓ^o but not of type at its endpoints arise also. We say that these operators are of type on ℓ^o, but from the M. Riesz-Thorin theorem we can only deduce the following characterization: T is of type on ℓ^o if and only if there exist sequences of points P_0^n, $P_1^n \in \ell$ such that $P_0^n \to P_0$, $P_1^n \to P_1$, and T is of type at P_i^n for every $n = 1, 2, \ldots$ and $i = 0, 1$. Since we would have to check the type condition at infinitely many points to apply the theorem, this method is of little value in this case.

Now, operators T of type on ℓ^o, but not of type at the endpoints P_0, P_1, may still satisfy certain weaker type conditions at P_0, P_1. To each space L^p we can associate a larger space

$L^p_* \supset L^p$, in a way that will be defined later, and we shall say that T is of "weak type" at $P = (1/p, 1/q)$ if it transforms continuously L^p into L^q_*. Since we assume $L^q \subset L^q_*$ continuously, if T is of type at P it will also be of weak type at P, but the reverse will not necessarily be true. Many operators T, of type on ℓ^o, are of weak type at the endpoints of ℓ.

An important example is the Hardy-Littlewood maximal operator (see Chapter 5). It is of weak type but not of type (1, 1). The same will be true for the singular integrals considered in Chapter 6.

In Section 5 we shall prove the Marcinkiewicz interpolation theorem which asserts that if T is of weak type at P_0 and P_1, endpoints of ℓ, then T is of type on ℓ^o whenever ℓ satisfies certain restrictions on its position in the type square. With that theorem, to establish the type of T on ℓ^o, it will suffice to prove its weak type at just two points, P_0 and P_1. Thus the Marcinkiewicz theorem is a generalization of the M. Riesz-Thorin theorem that is useful in instances where the latter is not. Nevertheless, it does not replace the M. Riesz-Thorin theorem entirely, since its hypotheses impose restrictions on ℓ and, furthermore, it gives weaker estimates for the norms (a weaker convexity inequality).

The above considerations can be restated as follows. If the operator T is of type on ℓ^o but not at $P_1 = (1/p_1, 1/q_1)$, then $T(L^{p_1})$ is not contained in L^{q_1}. To make T continuous on L^{p_1} we have to look for a larger space $L^{q_1}_*$ containing $T(L^{p_1})$.

This leads to the introduction of new spaces L^q_* and to the Marcinkiewicz extension of the M. Riesz-Thorin interpolation theorem to operators of type (L^p, L^q_*). In this Marcinkiewicz theorem the space L^q_* will actually be larger than L^q if $1 \le q < \infty$ but will be equal to L^q if $q = \infty$. In Chapter 5 we will present another important extension of the Lebesgue spaces for the case of L^∞, namely the space BMO.

Thus, to get a certain type condition at the endpoint $P_1 = (1/p_1, 1/q_1)$ we leave unchanged the domain L^{p_1} of T and extend its range. Alternatively, we could leave the range fixed and look for a smaller domain for T. That is, we could associate to each L^p a smaller space $L^p_\# \subset L^p$, so that many operators T of type on ℓ^o will be continuous from $L^{p_1}_\#$ into L^{q_1} or from $L^{p_0}_\#$ into L^{q_0}.

In Section 6 we will give an example of such a smaller space for the case of L^1. There $L^1_\#$ will be the Zygmund space $L \log^+ L$ (for definitions, see Chapter 0, Section 1). Other important examples are the Hardy spaces $H^p \subset L^p$. These will not be treated here. (For reference see [9].)

These procedures for modifying the Lebesgue spaces give rise to many important scales of spaces and to the need for applicable interpolation theorems. The development of general interpolation theories can be studied in [10]. Indications on these methods will be given in some remarks to this and next sections and in Appendix B.

Before defining the spaces L^p_* we introduce the distribution function.

<u>Definition 3.1</u>. Let $(\mathscr{X}, \mu)$ be a measure space with μ positive and f a μ-measurable function defined on $\mathscr{X}$.

For every $\alpha > 0$ the set

$$E_\alpha = E_\alpha(f) = \{x \in \mathscr{X} : |f(x)| > \alpha\} \tag{3.1}$$

is measurable and the function

$$f_*(\alpha) = \mu(E_\alpha) \tag{3.2}$$

defined from $\mathbb{R}^+$ to $\mathbb{R}^+$ is called the <u>distribution function</u> of f.

(Usually μ will be taken σ-finite).

Example 3.1. Let $\mathscr{X} = \mathbb{R}$, $d\mu = dx$, $f(x) = x$. Then $f_*(\alpha) = \infty$ for every $\alpha > 0$.

Example 3.2. Let $\mathscr{X} = \mathbb{R}$, $d\mu = dx$, $f(x) \equiv 1$. Then $f_*(\alpha) = \infty$ for $\alpha < 1$ and $f_*(\alpha) = 0$ for $\alpha \geq 1$.

Example 3.3. Let $A \subset \mathscr{X}$ be such that $\mu(A) = a < \infty$ and let $f(x) = c\chi_A(x)$, $c \geq 0$. Then $f_*(\alpha) = a$ for $\alpha < c$ and $f_*(\alpha) = 0$ for $\alpha \geq c$. (See Figure 5 for this case when $\mathscr{X} = \mathbb{R}$ and A is an interval.)

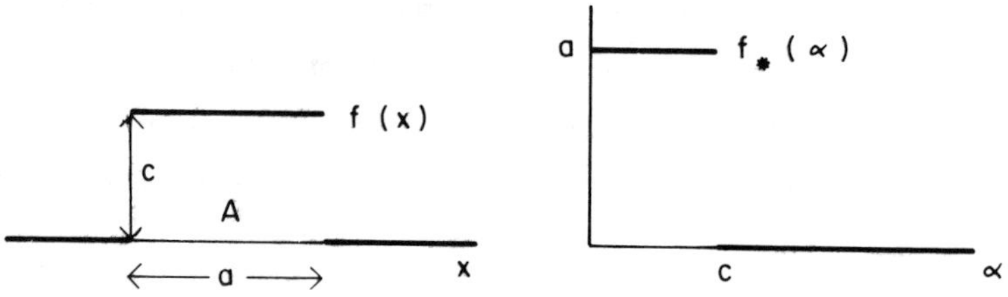

Figure 5

Example 3.4. Let $\{A_k\}$, $k = 1, \ldots, m$, be a collection of disjoint sets with $\mu(A_k) = a_k < \infty$, and $\{c_k\}$ such that $c_1 \geq c_2 \geq \ldots \geq c_m \geq$ ≥ 0. Let $f(x) = \Sigma_{k=1}^m c_k \chi_{A_k}(x)$. If $\alpha \geq c_1$, $E_\alpha = \phi$ and $f_*(\alpha) = 0$. If $c_2 \leq \alpha < c_1$, $E_\alpha = A_1$, $f_*(\alpha) = a_1$. If $c_3 \leq \alpha < c_2$, $E_\alpha = A_1 \cup A_2$, $f_*(\alpha) = a_1 + a_2$, and so on. Then,

$$f_*(\alpha) = a_1\chi_{[c_2, c_1)}(\alpha) + (a_1 + a_2)\chi_{[c_3, c_2)}(\alpha) + \ldots + \\ + (a_1 + \ldots + a_n)\chi_{[0, c_m)}(\alpha) \tag{3.3}$$

Proposition 3.1. For $f : \mathscr{X} \to \mathscr{Y}$ measurable function with respect to $\mu \geq 0$, the distribution function $f_* : \mathbb{R}^+ \to \mathbb{R}^+$ has the following properties:

(a) f_* is nonincreasing and continuous to the right (but not necessarily continuous);

(b) if $|f(x)| \leq |g(x)|$, then $f_*(\alpha) \leq g_*(\alpha)$, $\alpha > 0$;

(c) if $\{f\}$ are positive μ-measurable functions such that $f_1 \leq f_2 \leq \leq \ldots \uparrow f$, then $(f_n)_* \uparrow f_*$;

(d) if $|f(x)| \leq |g(x)| + |h(x)|$, then $f_*(\alpha) \leq g_*(\alpha/2) + h_*(\alpha/2)$ (3.4)

Proof. (a) As $E_\alpha \subset E_\beta$ for $\alpha \geq \beta$ and $\mu \geq 0$, f_* is nonincreasing. If $\alpha_n \downarrow \alpha$, $E_\alpha = \bigcup_n E_{\alpha_n}$ where $\{E_{\alpha_n}\}$ is an increasing sequence of sets. Then $\lim f_*(\alpha_n) = f_*(\alpha)$.

(b) $|f(x)| \leq |g(x)|$ implies $E_\alpha(f) \subset E_\alpha(g)$ for all $\alpha > 0$.

(c) Now $E(f) = \bigcup_n E(f_n)$ where $E(f_n)$ is an increasing sequence of sets. Then $\lim (f_n)_*(\alpha) = f_*(\alpha)$.

(d) If for a given x, $|f(x)| > \alpha$, then $|g(x)| > \alpha/2$ or else $|h(x)| > \alpha/2$. Then $E_\alpha(f) \subset E_{\alpha/2}(g) \cup E_{\alpha/2}(h)$ and $f_*(\alpha) \leq g_*(\alpha/2) + h_*(\alpha/2)$. ∇

Proposition 3.2. Let f and f_* be as in Proposition 3.1.

(a) (Chebyshev's inequality) For every $0 < p < \infty$

$$f_*(\alpha) \leq \alpha^{-p} \int_{E_\alpha} |f(x)|^p \, d\mu \tag{3.5}$$

for all $\alpha > 0$, where E_α is as in (3.1).

(b) If $f \in L^p(\mathscr{X})$, $1 \leq p < \infty$, for every $\alpha > 0$, $f_*(\alpha)$ is finite and

$$\sup_{\alpha > 0} \alpha^p f_*(\alpha) \leq \|f\|_p^p \tag{3.6}$$

(c) If $f \in L^p(\mathscr{X})$, $1 \leq p < \infty$, then $\alpha^p f_*(\alpha) \to 0$ as $\alpha \to 0$.

(d) If $\int_0^\infty \alpha^{p-1} f_*(\alpha) d\alpha < \infty$ then $\alpha^p f_*(\alpha) \to 0$ as $\alpha \to \infty$ and as $\alpha \to 0$.

Proof. (a) and (b): As $x \in E_\alpha$ iff $|f(x)| > \alpha$,

$$\int_{\mathscr{X}} |f(x)|^p d\mu \geq \int_{E_\alpha} |f(x)|^p d\mu \geq \alpha^p \int_{E_\alpha} d\mu = \alpha^p f_*(\alpha)$$

(c) From (b), $f_*(\alpha) < \infty$ for all α and $\mu(E_\alpha) = f_*(\alpha) \to 0$ when $\alpha \to \infty$; hence, by the continuity of the integral, $\int_{E_\alpha} |f|^p d\mu \to 0$ as $\alpha \to \infty$ and from (3.5), $\alpha^p f_*(\alpha) \to 0$ as $\alpha \to \infty$.

For fixed $\beta > 0$ and $\alpha < \beta$, we have

$$\lim_{\alpha \to 0} \alpha^p f_*(\alpha) = \lim_{\alpha \to 0} \alpha^p (f_*(\alpha) - f_*(\beta)) = \lim_{\alpha \to 0} \alpha^p (\mu(E_\alpha) - \mu(E_\beta))$$

$$\leq \int_{|f| \leq \beta} |f|^p d\mu$$

and since $\beta > 0$ is arbitrary, $\lim_{\alpha \to 0} \alpha^p f_*(\alpha) = 0$.

(d) $\int_{\alpha/2}^{\alpha} \eta^{p-1} f_*(\eta) d \geq f_*(\alpha)\alpha^p(1 - 2^{-p})$

and the integral tends to zero as $\alpha \to 0$ or $\alpha \to \infty$. ∇

Observe that if $f_*(\alpha)$ is finite for all positive α (as in the case when $f \in L^p$, $1 \leq p < \infty$) then, by Proposition 3.1(a), $-df_*(\alpha)$ is a positive measure in $\mathbb{R}$. The following proposition expresses $\|f\|_p$ in terms of f_* and is very useful.

Proposition 3.3. If f is a measurable function in $(\mathscr{X}, \mu)$ and $1 \leq p < \infty$, then

$$\int_{\mathscr{X}} |f(x)|^p d\mu = p\int_0^\infty \alpha^{p-1} f_*(\alpha)d\alpha \tag{3.7}$$

If in addition f is a finite function and $f_*(\alpha) < \infty$ for all $\alpha > 0$, then also

$$\int_{\mathscr{X}} |f(x)|^p d\mu = -\int_0^\infty \alpha^p df_*(\alpha) \tag{3.7a}$$

More generally, for any differentiable function $\Phi : \mathbb{R}^+ \to \mathbb{R}^+$,

$$\int_{\mathscr{X}} \Phi(|f(x)|)d\mu = \int_0^\infty \Phi'(\alpha)f_*(\alpha)d\alpha \tag{3.7b}$$

<u>Proof</u>. If $f_*(\alpha)$ is finite for all positive α and f is also finite, then the integral at the right in Formula (3.7a) makes sense, while the integral at the left can be expressed by the Lebesgue sums as follows. Taking an ε-subdivision, $0 < \varepsilon < 2\varepsilon < \ldots < m\varepsilon < \ldots$ and letting $\mathscr{X}_j = \{x \in \mathscr{X} : (j-1)\varepsilon \le f(x) < j\varepsilon\}$, we have $\mu(\mathscr{X}_j) = f_*((j-1)\varepsilon) - f_*(j\varepsilon)$ and

$$\int_{\mathscr{X}} |f(x)|^p d\mu = \lim_{\varepsilon \to 0} \sum_j (j\varepsilon)^p \mu(\mathscr{X}_j) =$$

$$= -\lim_{\varepsilon \to 0} \sum_j (j\varepsilon)^p (f_*(j\varepsilon) - f_*((j-1)\varepsilon)) = -\int_0^\infty \alpha^p df_*(\alpha) \tag{3.8}$$

This proves (3.7a). If both integrals in (3.7) are infinite, then (3.7) holds. If one of these integrals is finite then $f_*(\alpha) < \infty$ for all $\alpha > 0$, and f is finite a.e. so that (3.7a) holds and, by Proposition 3.2, $\alpha^p f_*(\alpha)$ tends to zero as $\alpha \to \infty$ or $\alpha \to 0$, so integrating by parts,

$$-\int_0^\infty \alpha^p df_*(\alpha) = p\int_0^\infty \alpha^{p-1} f_*(\alpha)d\alpha - \alpha^p f_*(\alpha)\Big|_0^\infty = p\int_0^\infty \alpha^{p-1} f_*(\alpha)d\alpha$$

From the last equality and (3.7a) we get (3.7). The proof of (3.7b) is quite similar and is left as an exercise.

Exercise 3.1. Prove (3.7) directly for simple functions $f = \sum_j c_j \chi_{A_j}$, $c_1 > c_2 > \dots c_m > \dots > 0$, by using Example 3.4. Then deduce (3.7) for general f.

Two functions f and g (which may be defined over different measure spaces) are called equimeasurable if their distribution functions coincide. Formula (3.7) tells us that two equimeasurable functions have the same L^p norm, $p < \infty$. This holds also for $p = \infty$, since $\|f\|_\infty = \inf\{\alpha : f_*(\alpha) = 0\}$. In general, if we know f_*, (3.7) allows us to learn also about the "size" of f as an L^p function If $\|f\|_p = M < \infty$ then, by (3.6), $\sup_\alpha \alpha f_*(\alpha)^{1/p} \leq M < \infty$, but the converse does not hold; this is the case for instance of $f(x) = x^{-1/p}$, $x \in (0,1)$, since $f_*(\alpha) < \alpha^{-p} (x^{-1/p} > \alpha$ implies $x < \alpha^{-p})$ and $af_*(\alpha)^{1/p} < 1$, but $f \notin L^p (|f(x)|^p = |x|^{-1}$ is not integrable in $(0,1))$. This suggests the following definitions.

Definition 3.2. For $1 \leq p < \infty$, f belongs to the Marcinkiewicz class L_*^p (or satisfies Marcinkiewicz condition p) if

$$[f]_p = \sup_{\alpha>0} \alpha f_*(\alpha)^{1/p} < \infty \tag{3.9}$$

For $p = \infty$, let $L_*^\infty = L^\infty$.

Note that, by Chebyshev's inequality (3.5), $[f]_p \leq \|f\|_p$, ($\|f\|_p$ can be infinite), thus $L^p \subset L_*^p$. Note also that $[\cdot]_p$ is not a norm for the class L_*^p: however, as $(f + g)_*(\alpha) \leq f_*(\alpha/2) + g_*(\alpha/2)$

by Proposition 3.1(d), and $(a + b)^{1/p} \leq a^{1/p} + b^{1/p}$ for $p \geq 1$,

$$[f + g]_p \leq 2([f]_p + [g]_p)$$

and L^p_* becomes a quasi-normed vector space under the quasinorm $[\cdot]_p$. But it can be proved that for $1 < p < \infty$, $[\cdot]_p$ is equivalent to a norm and that, with that norm, L^p_* becomes a Banach space. It can also be shown that L^1_* is a complete but nonnormable space. (See exercises at the end of this section.)

<u>Definition 3.3</u>. An operator T defined in $L^p = L^p(\mathscr{X}, \mu)$ with values in the class of measurables functions of $(\mathscr{Y}, \nu)$ is of <u>weak type</u> (p, q) with constant A_{pq}, for $1 \leq p \leq \infty$, $1 \leq q < \infty$, if

$$(Tf)_*(\alpha) \equiv \nu\{y \in \mathscr{Y} : |Tf(y)| > \alpha\} \leq (A_{pq}\|f\|_p/\alpha)^q \tag{3.10}$$

for all $\alpha > 0$.

In other words, T is of weak type (p, q) if $Tf \in L^q_*$ and

$$[Tf]_q \leq A_{pq}\|f\|_p \tag{3.10a}$$

If $q = \infty$, by definition, weak type (p, q) coincides with type (p, q) and this extends Definition 3.3 to $1 \leq p, q \leq \infty$. By (3.10a) an operator T of weak type (p, q) is a bounded operator from L^p to L^q_*. For this reason the Marcinkiewicz class L^q_* is often called the <u>weak</u> L^q space.

Another useful expression of the preceding matters is through a function equimeasurable with a given f, namely, the "nonincreasing rearrangement" of f.

<u>Definition 3.4</u>. Let f be a measurable function in $(\mathscr{X}, \mu)$ and f_* its distribution function. The <u>nonincreasing rearrangement</u> of f is the function $f^* : \mathbb{R}^+ \to \mathbb{R}^+$ defined by

$$f^*(t) = \inf\{\alpha : f_*(\alpha) \le t\} \tag{3.11}$$

Properties (a) and (e) of the following proposition justify the name "nonincreasing rearrangement," while (b) and (c) show that f^* is essentially the inverse function of f_* but for discontinuities and intervals of constancy.

Proposition 3.4. f^* satisfies the following properties:

(a) f^* is nonincreasing and continuous to the right;

(b) $f^*(f_*(\alpha)) \le \alpha < f^*(f_*(\alpha) - \varepsilon)$ for all $0 < \varepsilon < f_*(\alpha) < \infty$;

(c) if f^* is continuous at $t = f_*(\alpha)$ then $f^*(f_*(\alpha)) = \alpha$;

(d) $(fg)^*(t_1 + t_2) \le f^*(t_1)g^*(t_2)$;

(e) f and f^* are equimeasurable functions, i.e., $(f^*)_* = f_*$;

(f) if $f \in L^p$, $\|f\|_p = \|f^*\|_p$.

Proof. (a) Follows from the definition of f^* and Proposition 3.1(a).

(b) Evidently, $\alpha \in \{\eta : f_*(\eta) \le f_*(\alpha)\}$, so $f^*(f_*(\alpha)) = \inf\{\eta : f_*(\eta) \le f_*(\alpha)\} \le \alpha$. Analogously, $\alpha \notin \{\eta : f_*(\eta) \le f_*(\alpha) - \varepsilon\}$ and this is a closed half-line.

(c) Follows from (b).

(d) $f^*(t_1) = \inf\{\alpha : f_*(\alpha) \le t_1\}$, $g^*(t_2) = \inf\{\alpha : g_*(\alpha) \le t_2\}$ and $(fg)^*(t_1 + t_2) = \inf\{\alpha : (fg)_*(\alpha) \le t_1 + t_2\}$. As the distribution functions are nonincreasing, if $\alpha \ge f^*(t_1)$ then $f_*(\alpha) \le t_1$ and if $\beta \ge g^*(t_2)$ then $g_*(\beta) \le t_2$. Since $E_\alpha(fg) \subset E_\alpha(f) \cup E_\beta(g)$, $(fg)_*(\alpha\beta) \le f_*(\alpha) + g_*(\beta)$, so $(fg)_*(\alpha\beta) \le t_1 + t_2$ for all α, β such that $\alpha \ge f^*(t_1)$, $\beta \ge g^*(t_2)$. This implies $(fg)^*(t_1 + t_2) \le \alpha\beta$ for all such α, β and the result.

(e) From Definition 3.4, $f^*(t) > \alpha$ if and only if $t < f_*(\alpha)$ or $\{t : f^*(t) > \alpha\} = [0, f_*(\alpha))$. Thus, $(f^*)_*(\alpha) = |\{t : f^*(t) > \alpha\}| = f_*(\alpha)$ for all $\alpha > 0$.

(f) Follows from (e), with $\|f\|_p^p = \int_{\mathscr{X}} |f(x)|^p d\mu = \int_0^\infty f^*(t)^p dt =$ $= \|f^*\|_p^p$. ∇

Exercise 3.2. Show that $(f+g)^*(t) \leq f^*(t) + g^*(t)$ does not hold in general.

(Hint: Try $f(x) = \chi_{(0,1)}(x)$, $g(x) = \chi_{(1,2)}(x)$, $x \in \mathbb{R}$.)

By Proposition 3.4(b), condition $\alpha f_*(\alpha)^{1/p} \leq A$ for all $\alpha > 0$ is equivalent to $f^*(t)t^{1/p} \leq A$ for all $t > 0$, therefore Definitions 3.2 and 3.3 can be restated in terms of the nonincreasing rearrangements as:

$$f \in L_*^p,\ 1 \leq p < \infty, \quad \text{if and only if} \quad \sup_{t>0} t^{1/p} f^*(t) < \infty \tag{3.9a}$$

T is of weak type (p, q), $q < \infty$, if and only if, for all $f \in L^p$,

$$\sup_{t>0} t^{1/q}(Tf)^*(t) \leq A_{pq} \|f\|_p \tag{3.10b}$$

Since by Proposition 3.4 (f),

$$\int |f|^p d\mu = \int_0^\infty (f^*(t)t^{1/p})^p \frac{dt}{t}$$

this suggests the introduction, for $1 \leq p < \infty$, $1 \leq q < \infty$, of the expressions

$$I(f;p,q) = \int_0^\infty (f^*(t)t^{1/p})^q \frac{dt}{t})^{1/q} \tag{3.12}$$

and

$$I(f;p,\infty) = \sup_{t>0} f^*(t)t^{1/p} \tag{3.12a}$$

<u>Definition 3.5</u>. For $1 \leq p < \infty$ and $1 \leq q \leq \infty$ the (p, q) <u>Lorentz space</u>, denoted L_{pq}, is the class of measurable functions f such that $I(f;p, q) < \infty$. Letting $L_{\infty q} = L^\infty$ for all q, L_{pq} is defined for all $1 \leq p, q \leq \infty$.

The L_{pq} are vector spaces for all p, q and it can be proved that $I(\cdot\,;p, q)$ is equivalent to a norm if $p > 1$.

Obviously, $L_{pp} = L^p$, with $I(f;p, p) = \|f\|_p$, for all $1 \leq p < \infty$, and $L_{p\infty} = L^p_*$, with $I(f;p, \infty) = [f]_p$, for all $1 \leq p < \infty$; thus, both the (strong) Lebesgue spaces and the weak Lebesgue spaces can be considered as particular cases of Lorentz spaces.

<u>Exercise 3.3</u>. Prove that L_{pq} is a complete space for $1 < p < \infty$, $1 \leq q \leq \infty$ and for $p = q = 1$ or ∞. (Hint: Prove first that if $0 \leq f_n \uparrow f$ and $\sup_n I(f_n;p, q) < \infty$, then $\lim_n I_n(f_n;p, q) = I(f;p, q)$.)

<u>Exercise 3.4</u>. Prove that $I(f + g;p, q) \leq I(f;p, q) + I(g;p, q)$ if and only if $1 \leq q \leq p < \infty$ or $p = q = \infty$.

The fact that $L^p \subset L^p_*$ for $1 \leq p < \infty$ can be restated in terms of Lorentz spaces as $L_{pp} \subset L_{p\infty}$ for all p. This is a special case of the following result.

<u>Exercise 3.5</u>. If $1 \leq q_1 \leq q_2 \leq \infty$ then $L_{pq_1} \subset L_{pq_2}$ for all $1 < p < \infty$ (Hint: Prove it first for $q_2 = \infty$ using the fact that $t^{1/p} f^*(t) \leq$ $\leq c_p \int_0^t s^{1/p} f^*(s)\, ds/s$, $t > 0$.)

<u>Exercise 3.6</u>. Let $f^{**}(t) = t^{-1} \int_0^t f^*(s)ds$. Prove that $f^{**}(t)$ is a nonincreasing function of positive t and that

$$(f + g)^{**}(t) \leq f^{**}(t) + g^{**}(t)$$

Remark 3.1. Defining $J(f;p,q) = I(f^{**};p,q)$ for $1 < p, q < \infty$ it can be proved that I and J are equivalent, i.e., there exist two positive constants C_1 and C_2 such that $C_1 I(f;p,q) \le J(f;p,q) \le$ $\le C_2(f;p,q)$ for all $f \in L_{pq}$. It can also be proved that $J(.;p,q)$ is a norm in L_{pq} for $1 < p, q < \infty$. For this and other facts concerning Lorentz spaces see [11] and [12].

4. THE MARCINKIEWICZ INTERPOLATION THEOREM: DIAGONAL CASE

As announced in Section 3, we are going to state and prove a theorem that insures the type of an operator T on certain open segments whenever T is of weak type at the endpoints. We consider first the particular but important case when the endpoints of the segment belong to the main diagonal of the type square, and follow Marcinkiewicz' original idea of the proof, and will deal with the general case in Section 5.

Let $(\mathscr{X}, \mu)$ and $(\mathscr{Y}, \nu)$ be two measure spaces, S be the class of simple function in $(\mathscr{X}, \mu)$ and T be a sublinear operator that assigns to every $f \in S$ a function Tf defined in $\mathscr{Y}$ and ν-measurable. Let be $1 \le p_0 < p_1 \le \infty$ and, for each $t \in (0,1)$, let $p = p_t$ be given by (1.4), i.e., $1/p_t = (1-t)/p_0 + t/p_1$.

Theorem 4.1 (The Marcinkiewicz interpolation theorem). Assume that the sublinear operator T, defined in S, is of weak type (p_i, p_i) with constant M_i, for $i = 0, 1$, i.e.,

$$(Tf)_*(\alpha) \equiv \nu(\{|Tf| > \alpha\}) \le (M_0 \|f\|_{p_0} / \alpha)^{p_0} \tag{4.1}$$

and

$$(Tf)_*(\alpha) \equiv \nu(\{|Tf| > \alpha\}) \leq (M_1 \|f\|_{p_1}/\alpha)^{p_1} \tag{4.1a}$$

Then, for all $0 < t < 1$,

(a) T is of (strong) type (p, p), $p = p_t$, i.e.,

$$\|Tf\|_p \leq M_t \|f\|_p, \quad \text{for all} \quad f \in S,$$

where

(b) $M_t \leq KM_0^{1-t}M_1^t$ and

$$K = 2(p/(p - p_0) + p/(p_1 - p))^{1/p} \tag{4.2}$$

Remark 4.1. The constant K in (4.2) is bounded for $0 < \varepsilon < t < < 1 - \varepsilon$, ε fixed, but tends to infinity when $t \to 0$ or $t \to 1$. Part (a) of Theorem 4.1 implies the type condition of the M. Riesz-Thorin theorem for the case $p_i = q_i$, but for the convexity inequality (1.5) for the norm. Thus, as condition (b) of Theorem 4.1 is weaker than (1.5) since the factor K varies with t, the Marcinkiewicz theorem does not include the M. Riesz-Thorin theorem.

Remark 4.2. The statement and the sketch of the proof of Theorem 4.1 was sent by J. Marcinkiewicz (1910-1940) to A. Zygmund in a personal letter written in 1939, while he was a war prisoner, and appeared in [13]. Zygmund published the Marcinkiewicz theorem, as stated in Theorem 5.2 below, in 1956 [14], giving important applications.

Before giving the proof of Theorem 4.1 we explain its basic idea. Since we want to estimate $\|Tf\|_p$, $p = p_t$, it is natural to use Formula (3.7) which expresses $\|Tf\|_p$ in terms of $(Tf)_*(\alpha)$ and then use the estimates (4.1) and (4.1a). But doing so directly we would get either

$$\|Tf\|_p^p = p\int_0^\infty \alpha^{p-1}(Tf)_*(\alpha)d\alpha \le pM_0^{p_0}\left(\int_0^\infty \alpha^{p-p_0-1}d\alpha\right)\|f\|_{p_0}^{p_0} \qquad (4.3)$$

or

$$\|Tf\|_p^p = p\int_0^\infty \alpha^{p-1}(Tf)_*(\alpha)d\alpha \le pM_1^{p_1}\left(\int_0^\infty \alpha^{p-p_1-1}d\alpha\right)\|f\|_{p_1}^{p_1} \qquad (4.3a)$$

and none of these inequalities gives the desired result since they bound $\|Tf\|_p$ in terms of $\|f\|_{p_0}$ or $\|f\|_{p_1}$ and not of $\|f\|_p$ and, more essentially, since $\int_0^\infty \alpha^r d\alpha$ cannot be convergent for any value of r. These two difficulties will be avoided by replacing, for each α in (4.2), f by $f_\alpha + f^\alpha$ and by using the following lemma.

Lemma 4.2. Let $f \in L^p$, f_α, f^α be defined as in (1.1), (1.1a) and $1 \le p_0 < p_1 < \infty$. Then for all $\alpha > 0$, $f_\alpha \in L^{p_1} \cap L^p$ and $f^\alpha \in L^{p_0} \cap L^p$ and, moreover,

$$\|f\|_p^p = (p - p_0)\int_0^\infty \alpha^{p-p_0-1}\|f^\alpha\|_{p_0}^{p_0} d\alpha \qquad (4.4)$$

and

$$\|f\|_p^p = (p_1 - p)\int_0^\infty \alpha^{p-p_1-1}\|f_\alpha\|_{p_1}^{p_1} d\alpha \qquad (4.4a)$$

Proof. We already know, by Lemma 1.1, that $f_\alpha \in L^{p_1} \cap L^p$ and $f^\alpha \in L^{p_0} \cap L^p$. Furthermore,

$$\int_0^\infty \alpha^{p-p_0-1} \|f^\alpha\|_{p_0}^{p_0} d\alpha = \int_0^\infty \alpha^{p-p_0-1} \left(\int_{\mathscr{X}} |f^\alpha|^{p_0} d\mu\right) d\alpha$$

$$= \int_0^\infty \alpha^{p-p_0-1} \left(\int_{\{|f|>\alpha\}} |f|^{p_0} d\mu\right) d\alpha$$

$$= \int_{\mathscr{X}} |f|^{p_0} \left(\int_0^{|f|} \alpha^{p-p_0-1} d\alpha\right) d\mu$$

$$= (p - p_0)^{-1} \int_{\mathscr{X}} |f|^{p_0} |f|^{p-p_0} d\mu$$

$$= (p - p_0)^{-1} \|f\|_p^p$$

which is (4.3). Similarly (4.3a) is obtained. ∇

Proof of the Marcinkiewicz theorem. Given $\alpha > 0$, let us decompose $f \in L^p$ as $f = f^\alpha + f_\alpha$, where f^α, f_α are as in (1.1), (1.1a). Thus, $|Tf(x)| \leq |Tf^\alpha(x)| + |Tf_\alpha(x)|$ and, by Proposition 3.1(d),

$$(Tf)_*(\lambda) \leq (Tf^\alpha)_*(\lambda/2) + (Tf_\alpha)_*(\lambda/2)$$

for all $\lambda > 0$.

Case $p_1 < \infty$: As by Lemma 4.2, $f^\alpha \in L^{p_0}$ and $f_\alpha \in L^{p_1}$ we may apply to Tf^α and Tf_α the weak type hypotheses to obtain the estimate

$$\int_0^\infty \alpha^{p-1}(Tf)_*(\alpha)d\alpha \leq \int_0^\infty \alpha^{p-1}((Tf^\alpha)_*(\alpha/2) + (Tf_\alpha)_*(\alpha/2))d\alpha$$

$$\leq \int_0^\infty \alpha^{p-1}\left(\left(\frac{2M_0}{\alpha} \|f^\alpha\|_{p_0}\right)^{p_0} + \left(\frac{2M_1}{\alpha} \|f_\alpha\|_{p_1}\right)^{p_1}\right)d\alpha$$

$$= (2M_0)^{p_0} \int_0^\infty \alpha^{p-p_0-1} \|f^\alpha\|_{p_0}^{p_0} d\alpha$$

$$+ (2M_1)^{p_1} \int_0^\infty \alpha^{p-p_1-1} \|f_\alpha\|_{p_1}^{p_1} d\alpha$$

$$= \left(\frac{(2M_0)^{p_0}}{p-p_0} + \frac{(2M_1)^{p_1}}{p_1-p}\right) \|f\|_p^p \tag{4.5}$$

Formula (4.5), that follows from (4.4) and (4.4a), gives the type condition (p, p), $p_0 < p < p_1$, for operator T. To obtain the estimate (4.2) let us first observe that

$$\frac{(2M_0)^{p_0} p}{p-p_0} + \frac{(2M_1)^{p_1} p}{p_1-p} \le \left(\frac{p}{p-p_0} + \frac{p}{p_1-p}\right) \max((2M_0)^{p_0}, (2M_1)^{p_1})$$

$$= C \max((2M_0)^{p_0}, (2M_1)^{p_1})$$

and, second, that replacing T by aT for $a > 0$, in (4.5), we get

$$a^p \|Tf\|_p^p \le C \max((2aM_0)^{p_0}, (2aM_1)^{p_1}) \|f\|_p^p \tag{4.6}$$

Taking a such that

$$(2aM_0)^{p_0} = (2aM_1)^{p_1} \quad (\text{or } a^{p_0-p_1} = 2^{p_1-p_0} M_0^{-p_0} M_1^{p_1})$$

(4.6) becomes

$$\|Tf\|_p^p \le C2^{p_0} a^{p_0-p} M_0^{p_0} \|f\|_p^p$$
$$= C2^p (M_0^{-p_0} M_1^{p_1})^{(p_0-p)/(p_0-p_1)} M_0^{p_0} \|f\|_p^p \tag{4.7}$$

Since by (1.4), $1/p = (1-t)/p_0 + t/p_1$ for $0 < t < 1$, $(1-t)p + tp = p_0 p_1$, and by an elementary computation,

$$(M_0^{-p_0} M_1^{p_1})^{(p_0-p)/(p_0-p_1)} M_0^{p_0} = (M_0^{1-t} M_1^t)^p$$

so (4.7) becomes (4.2), where $K = 2C^{1/p}$ depends only on p_0, p_1 and t, is bounded for fixed $t \in (0, 1)$ and tends to infinity for $t \to 0$ or $t \to 1$.

Case $p_1 = \infty$: Replacing T by cT we may always assume that $M_1 = 1$. Then, as weak type (∞, ∞) coincides with type (∞, ∞), and $f_\alpha = f$ for $|f| \le \alpha$ and $f_\alpha = 0$ otherwise, $\|Tf_\alpha\|_\infty \le 1$. $\|f_\alpha\|_\infty \le \alpha$ and $(Tf_\alpha)_*(\lambda) = 0$ for every $\lambda \ge \alpha$. So by Proposition 3.1(d) and (4.4),

$$\|Tf\|_p^p = p\int_0^\infty \alpha^{p-1}(Tf)_*(\alpha)d\alpha = 2^p p\int_0^\infty \alpha^{p-1}(Tf)_*(2\alpha)d\alpha$$

$$\le 2^p p\int_0^\infty \alpha^{p-1}(Tf^\alpha)_*(\alpha)d\alpha$$

$$\le 2^p p M_0^{p_0} \int_0^\infty \alpha^{p-p_0-1} \|f^\alpha\|_{p_0}^{p_0} d\alpha$$

$$= \frac{2^p p M_0^{p_0}}{p-p_0} \|f\|_p^p$$

and

$$\|Tf\|_p \le 2\left(\frac{p}{p-p_0}\right) M_0^{p_0/p} \|f\|_p, \quad \text{for} \quad p_0 < p < \infty$$

which is the desired type inequality, since $M_1 = 1$ and $p_0/p = 1 - t$.

∇

An immediate application of the M. Riesz-Thorin theorem given in Section 1 was the Hausdorff-Young theorem for Fourier series and integrals. We shall now give an application of the Marcinkiewicz theorem to Fourier analysis that is obtained considering $\mathbb{R}^n$ as two different measure spaces given by two different measures in it.

Theorem 4.3 (The Hardy-Littlewood-Paley theorem in $\mathbb{R}^n$). Let $\hat{f}$ denote the Fourier transform of f.

(a) If $1 < p \le 2$, there exists a constant M_p such that, for every $f \in L^p(\mathbb{R}^n)$,

$$\left(\int_{\mathbb{R}^n} |\hat{f}(x)|^p |x|^{(p-2)n} dx\right)^{1/p} \le M_p \left(\int_{\mathbb{R}^n} |f(x)|^p dx\right)^{1/p} \tag{4.8}$$

(b) If $2 \le q < \infty$ there exists M_q such that

$$\left(\int_{\mathbb{R}^n} |f(x)|^q dx\right)^{1/q} \le M_q \left(\int_{\mathbb{R}^n} |f(x)|^q |x|^{(q-2)n} dx\right)^{1/q} \tag{4.9}$$

Remark 4.3. Part (a) provides a generalization of the classical Bessel inequality, which corresponds to the case $p = 2$, $M_p = 1$ in (4.8). The result does not hold for $p = 1$ or $q = \infty$, since $M_p \to \infty$ as $p \to 1$ and $M_q \to \infty$ as $q \to \infty$. The fact that (a) does not hold for $p = 1$ can be seen by considering

$$f(x) \sim \sum_{n=2}^{\infty} \frac{\cos 2\pi nx}{\log n}$$

Proof. (a) Consider the spaces $(\mathscr{X}, \mu) = (\mathbb{R}^n, dx)$, $(\mathscr{Y}, \nu) = (\mathbb{R}^n, |x|^{-2n}dx)$, where dx is the ordinary Lebesgue measure in $\mathbb{R}^n$, and let T be the operator defined by $Tf(x) = \hat{f}(x). |x|^n$, where f is defined in $(\mathscr{X}, \mu)$ and Tf in $(\mathscr{Y}, \nu)$. By the Plancherel theorem,

$$\int_{\mathscr{Y}} |Tf(x)|^2 d\nu = \int_{\mathbb{R}^n} |\hat{f}(x)|^2 dx = \int_{\mathbb{R}^n} |f(x)|^2 dx = \int_{\mathscr{X}} |f(x)|^2 d\mu$$

thus, T is of type (2, 2) with constant $M_0 = 1$.

We prove now that T is also of weak type (1, 1), i.e., that there is M_1 such that $(Tf)_*(\alpha) \le M_1 \|f\|_1/\alpha$ for all $\alpha > 0$, where $(Tf)_*(\alpha) = \nu(E_\alpha) = \nu\{x \in \mathbb{R}^n \cdot |Tf(x)| > \alpha\} = \nu\{x : |\hat{f}(x)|\,|x|^n > \alpha\}$. For $x \in \mathbb{R}^n$, $|\hat{f}(x)| \le \|f\|_1$; for each $x \in E_\alpha$, $\alpha < |\hat{f}(x)|. |x|^n \le \|f\|_1 |x|^n$, that is, $|x| > (\alpha/\|f\|_1)^{1/n} = b$. Thus $E_\alpha \subset B = \{x : |x| > b\}$ and

$$\nu(E_\alpha) = \int_{E_\alpha} d\nu \le \int_B d\nu = \int_{|x|>b} |x|^{-2n} dx$$

$$= \int_\Sigma \int_b^\infty r^{n-1} r^{-2n} dr\, dx' = \omega_n \|f\|_1/\alpha \tag{4.10}$$

and (4.10) in the weak type (1, 1) inequality with $M_1 = \omega_n$.

Hence T is of type (2, 2) with constant 1 and of weak type (1, 1) with constant ω_n. By the Marcinkiewicz Theorem 4.1, T is of type (p, p) for $1 < p \le 2$, i.e.,

$$\left(\int_{\mathscr{Y}} |Tf(x)|^p d\nu\right)^{1/p} \le M_p \left(\int_{\mathscr{X}} |f(x)|^p d\mu\right)^{1/p}$$

or

$$\left(\int_{\mathbb{R}^n} |\hat{f}(x)|^p |x|^{np} |x|^{-2n} dx\right)^{1/p} \leq M_p \left(\int_{\mathbb{R}^n} |f(x)|^p dx\right)^{1/p}$$

which is the thesis (4. 8).

(b) Let now be $2 \leq q < \infty$ and call $p = q'$. Thus $1 < p \leq 2$ and (a) is satisfied for this p. If $f \in L^p$, by the Hausdorff-Young Theorem 1. 5, $\hat{f} \in L^q$ and, by the F. Riesz representation theorem, there exists $g \in L^{q'}$, $\|g\|_{q'} = 1$, such that $\|\hat{f}\|_q = |\int \hat{f} g|$. By the multiplication formula for Fourier transforms (Theorem 1. 3 of Chapter 2) and by Hölder's inequality,

$$\|\hat{f}\|_q = |\int \hat{f} g| = |\int f\hat{g}|$$

$$= |\int f(x)|x|^{n(q-2)/q} \hat{g}(x)|x|^{-n(q-2)/q} dx|$$

$$\leq \left(\int |f(x)|^q |x|^{n(q-2)} dx\right)^{1/q} \left(\int |\hat{g}(x)|^{q'} |x|^{n(q'-2)} dx\right)^{1/q'}$$

Applying (a) to $q' = p$, we get

$$\|\hat{f}\|_q \leq M_{q'} \|g\|_{q'} \left(\int |f(x)|^q |x|^{n(q-2)} dx\right)^{1/q}$$

which is Thesis (4. 9), since $\|g\|_{q'} = 1$. ∇

*<u>Exercise 4. 1</u>. Let be $1 \leq p_0 < p_1 \leq \infty$, $p = p_t$ as in (1. 4), $f \in L^p$. Let D_f be the set of pairs of functions of parameter λ, (h^λ, h_λ), such that, for each $\lambda > 0$, it is $f(x) = h^\lambda(x) + h_\lambda(x)$, $h^\lambda \in L^{p_0}$, $h_\lambda \in L^{p_1}$. Define in D_f the norm

$$|||(h^\lambda, h_\lambda)||| = \max \{(p - p_0)\int_0^\infty \lambda^{p-p_0-1} \|h^\lambda\|_{p_0}^{p_0} d\lambda \qquad (4.11)$$

$$(p_1 - p)\int_0^\infty \lambda^{p-p_1-1} \|h_\lambda\|_{p_1}^{p_1} d\lambda\}$$

and let

$$(|||f|||_p)^p = \inf\{|||(h^\lambda, h_\lambda)||| : (h^\lambda, h_\lambda) \in D_f\} \qquad (4.12)$$

Prove that $|||f|||_p \leq \|f\|_p$. (Cfr. Lemma 4.2.)

*Exercise 4.2. Let D_f^* be the set of pairs of functions (h^λ, h_λ) such that for every $\lambda > 0$ it is $f(x) = h^\lambda(x) + h_\lambda(x)$, $h^\lambda \in L_*^{p_0}$, $h_\lambda \in L_*^{p_1}$ and define $|||(h^\lambda, h_\lambda)|||_*$ by replacing in (4.11) $\|h^\lambda\|_{p_0}^{p_0}$ by $[h^\lambda]_{p_0}^{p_0}$ (as in (3.9)), idem $\|h_\lambda\|_{p_1}^{p_1}$.
If $(|||f|||_{p^*})^p = \inf\{|||(h^\lambda, h_\lambda)|||_* : (h^\lambda, h_\lambda) \in D_f^*\}$, prove that $|||f|||_{p^*} \leq |||f|||_p \leq \|f\|_p \leq c|||f|||_{p^*} \leq c|||f|||_p$ for $c = c(p, p_0, p_1)$, so that the norms $\|f\|_p$, $|||f|||_p$ and $|||f|||_{p^*}$ are equivalent.

*Exercise 4.3. From Exercise 4.2, prove part (a) of Theorem 4.1.

*Remark 4.4. The formulation of Exercises 4.1 and 4.2 can be done also in an abstract context. Instead of the Lebesgue spaces L^{p_j}, i = 0, 1, let E_0, E_1 be two normed spaces, contained in the same vector space V, in a way that for each pair $g \in E_0$, $h \in E_1$, we have $g + h = f \in V$ and we can define $E_0 + E_1 = \{f \in V : f = g + h, g \in E_0, h \in E_1\}$. For a fixed $f \in E_0 + E_1$, let D_f be the set of pairs (g^λ, h_λ) of parameter λ, such that for every $\lambda > 0$ it is $g^\lambda + h_\lambda = f$, $g^\lambda \in E_0$, $h_\lambda \in E_1$ and with $\|g^\lambda\|_{E_0}$, $\|h_\lambda\|_{E_1}$ measurable functions in λ. For each $(g^\lambda, h_\lambda) \in D_f$ and $p_0 < p < p_1$ fixed numbers, we define

$$\begin{aligned} |||(g^\lambda, h_\lambda)||| = \max\{&(p - p_0)\int_0^\infty \lambda^{p-p_0-1} \|g^\lambda\|_{E_0}\, d\lambda, \\ &(p_1 - p)\int_0^\infty \lambda^{p-p_1-1} \|g_\lambda\|_{E_1}\, d\lambda\} \end{aligned} \qquad (4.11a)$$

and

$$(|||f|||_p)^p = \inf \{ |||(g^\lambda, h_\lambda)||| : (g^\lambda, h_\lambda) \in D_f \} \tag{4.12a}$$

Setting $E_{p,p_0,p_1} = E_p = \{f \in E_0 + E_1 : |||f|||_p < \infty\}$ we get a new space between E_0 and E_1. The spaces E_p, p ranging between p_0 and p_1, form a scale of spaces. If $E_i = L^{p_i}$ then $E_p = L^p$ and $|||\cdot|||_p$ is equivalent to the norm $\|\cdot\|_p$. The analog of the result in Exercise 4.3 is the

<u>Abstract Interpolation Theorem</u>. Let $(E_0, E_1) \subset V$ and $(E'_0, E'_1) \subset V'$ be two pairs of normed spaces, let $p_0 < p = p_t < p_1$ be fixed numbers and let E_p, E'_p be the corresponding intermediate spaces. If $T : E_0 + E_1 \to E'_0 + E'_1$ is a sublinear operator such that $\|Tg\|_{E_i} \le M_i \|g\|_{E_i}$ for all $g \in E_i$, $i = 0, 1$, then

$$\|Tf\|_{E'_p} \le K M_0^{1-t} M_1^t \|f\|_{E_p} \quad \text{for all } f \in E_p$$

This abstract theorem stems from the <u>real method</u> of interpolation due to Lions-Peetre. For a study of this topic see [10] and [15].

*5. THE MARCINKIEWICZ THEOREM: GENERAL CASE

We have seen in Lemma 1.1 that if $p_0 < p < p_1$ and $f \in L^p$ then $f^\lambda \in L^{p_0}$, $f_\lambda \in L^{p_1}$ and, in Lemma 4.2, that by knowing the norms $\|f^\lambda\|_{p_0}$, $\|f_\lambda\|_{p_1}$ for all λ, we may reconstruct the norm $\|f\|_p$. Moreover we have the following complement to Lemma 4.2.

<u>Lemma 5.1</u>. Let be $1 \le p_0 \le q_0$, $1 \le p_1 \le q_1$, $p_0 < p < p_1$, $q_0 \le q \le q_1$, $q_0 \neq q_1$ and assume

$$\frac{q_0}{p_0}\,\frac{p-p_0}{q-q_0} = \frac{q_1}{p_1}\,\frac{p-p_1}{q-q_1} = \frac{1}{v} \tag{5.1}$$

((5.1) is equivalent to $(1/p_0 - 1/p)/(1/p - 1/p_1) = (1/q_0 - 1/q)/(1/q - 1/q_1)$ and is satisfied for $p = p_t$, $q = q_t$ with the same $t \in (0, 1)$.) Then calling

$$|||(g^\lambda, h_\lambda)|||_v = \max\{(q - q_0)\left[\int_0^\infty \lambda^{q-q_0-1} \|g^{\lambda^v}\|_{p_0}^{q_0}\, d\lambda\right]^{p_0/q_0} \tag{5.2}$$

$$(q_1 - q)\left[\int_0^\infty \lambda^{q-q_1-1} \|h_{\lambda^v}\|_{p_1}^{q_1}\, d\right]^{p_1/q_1}\}$$

we have

$$|||(f^\lambda, f_\lambda)|||_v \le \int_{\mathscr{X}} |f|^p\, d\mu = \|f\|_p^p \tag{5.3}$$

Remark 5.1. For $p_0 = q_0$, $p_1 = q_1$, Lemma 5.1 reduces to one half of Lemma 4.2, that is to (4.4), (4.4a) with $\geq$ replacing $=$.

Proof. Let us consider the expression

$$\left[\int_0^\infty \lambda^{q-q_0-1} \|f^{\lambda^v}\|_{p_0}^{q_0}\, d\lambda\right]^{p_0/q_0}$$

$$= \left[\int_0^\infty \left(\int_{\mathscr{X}} |f^{\lambda^v}(x)|^{p_0}\, d\mu\right)^{q_0/p_0} \lambda^{q-q_0-1}\, d\lambda\right]^{p_0/q_0} \tag{5.4}$$

$$= \left[\int_0^\infty \left(\int_{\mathscr{X}} F(\lambda, x)\, d\mu\right)^r d\nu\right]^{1/r}$$

where $F(\lambda, x) = |f^{\lambda^v}(x)|^{p_0}$, $r = q_0/p_0$, $d\nu = \lambda^{q-q_0-1}\, d\lambda$.

Applying the Minkowski integral inequality to (5.4) it becomes bounded by

$$\int_{\mathscr{X}} \Big(\int_0^\infty |F(\lambda, x)|^r d\nu\Big)^{1/r} d\mu$$

$$\int_{\mathscr{X}} \Big(\int_0^\infty |f^{\lambda^v}(x)|^{q_0} \lambda^{q-q_0-1} d\lambda\Big)^{p_0/q_0} d\mu \tag{5.5}$$

As for fixed x,

$$f^{\lambda^v}(x) = 0 \quad \text{if} \quad \lambda^v > |f(x)| \quad \text{and} \quad |f^{\lambda^v}(x)| = |f(x)| \quad \text{if} \quad \lambda^v \le |f(x)|$$

(5.5) becomes

$$= \int_{\mathscr{X}} \Big(\int_0^{|f|^{1/v}} \lambda^{q-q_0-1} |f(x)|^{q_0} d\lambda\Big)^{p_0/q_0} d\mu$$

$$= \int_{\mathscr{X}} |f(x)|^{p_0} (q - q_0)^{-p_0/q_0} |f(x)|^{\frac{1}{v}(q-q_0)p_0/q_0} d\mu$$

Thus

$$(q - q_0)^{p_0/q_0} \Big[\int_0^\infty \lambda^{q-q_0-1} \|f^{\lambda^v}\|_{p_0}^{q_0} d\lambda\Big]^{p_0/q_0} \le \int |f(x)|^p d\mu$$

by virtue of (5.1) and from this follows the thesis. ∇

Theorem 5.2 (The Marcinkiewicz interpolation theorem: general case). Let T be, as in Theorem 4.1, a sublinear operator in S and let

$$1 \le p_i \le q_i \le \infty, \quad i = 0, 1, \quad q_0 \ne q_1 \tag{5.6}$$

$$0 < t < 1, \quad 1/p_t = (1 - t)/p_0 + t\text{-}p_1, \quad 1/q_t = (1 - t)/q_0 + t/q_1 \tag{1.4}$$

Assume that T is of weak type (p_0, q_0) with constant M_0 and of weak type (p_1, q_1) with constant M_1, i.e.,

$$(Tf)_*(\alpha) \leq (M_i \|f\|_{p_i} / \alpha)^{q_i}, \quad i = 0, 1 \tag{5.7}$$

Then

(a) T is of type (p, q), for $p = p_t$, $q = q_t$, all $t \in (0, 1)$, i.e.,

$$\|Tf\|_q \leq M_t \|f\|_p, \quad \text{for all} \quad f \in L^p \tag{5.8}$$

and

(b) $M_t \leq K M_0^{1-t} M_1^t$

where $K = K(p_0, q_0, p_1, q_1, t)$ is bounded for $0 < \varepsilon < t < 1 - \varepsilon$ but tends to infinity when $t \to 0$ or $t \to 1$.

Remark 5.2. Conditions (5.6) on the numbers p_i, q_i mean that the theorem is valid in the lower triangle of the type square, with the restriction that the segment from $(1/p_0, 1/q_0)$ to $(1/p_1, 1/q_1)$ must not be horizontal.

Remark 5.3. As it was already noted in Remark 4.1, for the diagonal case, this theorem is partially stronger than the M. Riesz-Thorin theorem but does not include it, because of the factor K depending on t that appears in (b). (See [16].) For a general survey on the subject see [17].

Remark 5.4. The following simple example (due to R. Panzone) shows that the condition $q_0 \neq q_1$ in (5.6) is necessary to the validity of the theorem. Let $f : [0, 1] \to \mathbb{C}$ be an integrable function and let T be defined by $Tf(x) = F(x) = x^{-1/2} \int_0^t f(t)dt$. T is linear and $F : [0,1] \to \mathbb{C}$ is also integrable. For every $\alpha > 0$, $E_\alpha = E_\alpha(Tf) = \{x \in [0,1] : |Tf(x)| > \alpha\} = \{x : \alpha x^{1/2} < J\}$, where $J = \int_0^1 f \leq \|f\|_1$. If $J < 0$ then $E_\alpha = \phi$. If $0 \leq J < \alpha$, $x \in E_\alpha$ implies $x \leq (J/\alpha)^2$ and $E_\alpha \subset [0, (\|f\|_1/\alpha)^2]$, so $(Tf)_*(\alpha) \leq (\|f\|_1/\alpha)^2$. If $J > \alpha$ then every $x \leq 1$ satisfies $|Tf(x)| > \alpha$ and $E_\alpha = [0,1]$, so $(Tf)_*(\alpha) = 1 < J/\alpha < (J/\alpha)^2 \leq (\|f\|_1/\alpha)^2$. Thus, in every case, $(Tf)_*(\alpha) \leq (\|f\|_1/\alpha)^2$ and, furthermore, since $\|f\|_1 \leq \|f\|_p$ for all $p \geq 1$, $(Tf)_*(\alpha) \leq (\|f\|_p/\alpha)^2$ i.e., T is of weak type $(p, 2)$ for all $p \geq 1$. If Theorem 5.2 were valid for $q_0 = q_1$, we could apply it to T at $(1, 2)$ and $(2, 2)$ and T would be of type $(p, 2)$ for all $1 < p < 2$. But T <u>cannot be</u> of type $(p, 2)$ for any $p \geq 1$ because this would mean that $\|Tf\|_2 \leq M_p \|f\|_p$ and that $|Tf|^2$ is integrable whenever $|f|^p$ is integrable, which is impossible as seen by taking $f \in L^p$ such that $J \neq 0$, since then $|Tf(x)|^2 = J^2/x$ which is not integrable in $[0,1]$.

<u>Remark 5.5</u>. Theorem 5.2 is still valid for quasi-linear operators, as defined in Section 1.

<u>Remark 5.6</u>. Using abstract interpolation methods it has been proved [18] that Theorem 5.2 is still valid if only one of the end-points $(1/p_0, 1/q_0)$, $(1/p_1, 1/q_1)$ is in the lower triangle. But a counterexample given by R. Hunt [19] shows that $(1/p_t, 1/q_t)$ must lie on the lower triangle, i.e., $p_t \leq q_t$, in order to have (5.8).

Before giving the proof of Theorem 5.2 we shall indicate some of its applications.

The following example shows that certain (strong) type of two operators each at a point of the type square entail the weak type of its sum on the segment in between.

Proposition 5.3. If T is an operator that for each $\alpha > 0$ can be decomposed as $T = T_1 + T_2$ where T_1 is of type (p, ∞) with constant $\alpha/2$ and T_2 is of type (p, p) with constant $c\alpha^{1-q/p}$, then T is of weak type (p, q).

Proof. It suffices to consider $\|f\|_p = 1$. As $(Tf)_*(\alpha) \le (T_1f)_*(\alpha/2) + (T_2f)_*(\alpha/2)$, and $\|T_1f\|_\infty \le \alpha/2$ by hypothesis, for each $\alpha > 0$, then $(T_1f)_*(\alpha) = 0$ and $(Tf)_*(\alpha) \le (Tf_2)_*(\alpha/2) \le 2^p\alpha^{-p}c^p\alpha^{p-q}$. Thus $\sup_{\alpha>0} \alpha((Tf)_*(\alpha))^{1/q} = [Tf]_q \le (2c)^{p/q} = (2c)^{p/q}\|f\|_p$. ∇

In Section 3 of Chapter 3, while introducing some problems dealing with harmonic functions in $\mathbb{R}^n$, we stated that the fundamental solution of Δ is given by the Newtonian potential $U^f = \int f(t)|x - t|^{2-n}dt$ (see (3.13) of Chapter 3). More generally, it is important to consider the Riesz potentials.

Definition 5.1. Let $0 < \gamma < n$. The Riesz potential or order γ, denoted by I_γ, is defined as

$$I_\gamma f(x) = c_\gamma \int_{\mathbb{R}^n} f(t)|x - t|^{\gamma-n}dt = c_\gamma k_\gamma * f(x) \tag{5.9}$$

where $k_\gamma(x) = |x|^{\gamma-n}$ and $c_\gamma = \pi^{n/2}2^\alpha\Gamma(\alpha/2)/\Gamma(n/2 - \alpha/2)$.

Proposition 5.3 enables us to prove that this operator I_γ is of weak type on a segment parallel to the main diagonal of the type square and, through the Marcinkiewicz theorem, that it is of type there. More precisely, we have

Corollary 5.4. The operator I_γ, $0 < \gamma < n$, is of weak type (p, q) for $1/p - 1/q = \gamma/n$, $1 \le p$, $q < \infty$. (See Figure 6.)

Proof. Let $\lambda > 0$ be fixed and let $k'(x) = k_\gamma(x).\chi_\lambda(x)$, where χ_λ is the characteristic function of the ball centered at 0 and of radius λ, $k''(x) = k_\gamma(x) - k'(x)$, $T'f = k' * f$, $T''f = k'' * f$. Then it is immediate that $k' \in L^1$, with $\|k'\|_1 \le c_1\lambda$ and so T' is of type (p, q) with constant $c_1\lambda^\gamma$, and that $k'' \in L^{p'}$, with $\|k''\|_{p'} \le c_2\lambda^{\gamma-n/p}$ and so T'' is of type (p, ∞) with constant $c_2\lambda^{\gamma-n/p}$.

For each $\alpha > 0$ we can chose λ so that the hypotheses of Proposition 5.3 are satisfied for $I_\gamma = T' + T''$, so that I_γ will be of weak type (p, q), taking $c_2\lambda^{\gamma-n/p} = \alpha/2$ and $c_1\lambda^\gamma = c\alpha^{1-q/p}$ and this is possible under the hypothesis on p, q and γ. ∇

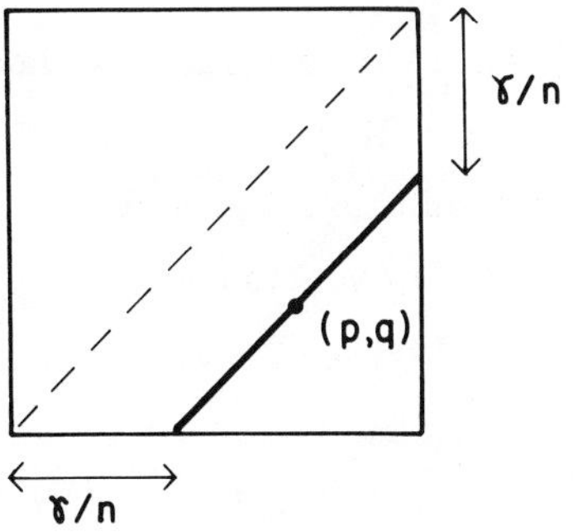

Figure 6

From this result, interpolation leads to

Theorem 5.5 (The Hardy-Littlewood-Sobolev theorem on Riesz potentials). The operator I_γ, $0 < \gamma < n$, is of type (p, q) for $1/p - 1/q = \gamma/n$, i.e., I_γ is of type on the open segment parallel to the main diagonal of the type square at distance γ/n. At the endpoint $(1, n/(n - \gamma))$ I_γ is of weak type.

We introduce now another family of interesting operators, whose weak type (p, p) is a direct consequence of Theorem 5.5.

Definition 5.2. Let be $0 < \gamma < n$ and define

$$H_\gamma : f \to |x|^{-\gamma} I_\gamma(f) \tag{5.10}$$

i.e.,

$$H_\gamma f(x) = c_\gamma |x|^{-\gamma}(f * |x|^{\gamma-n})$$

$$= c_\gamma |x|^{-\gamma}\int_{\mathbb{R}^n} f(t)|x - t|^{\gamma-n}dt \tag{5.10a}$$

Proposition 5.6. The operator H_γ, $0 < \gamma < n$, is of type (p, p) for all $1 < p < n/\gamma$ and of weak type $(1, 1)$.

Proof. We shall prove, as a corollary to Theorem 5.5, that H_γ is of weak type (p, p) for all $1 \le p < n/\gamma$. This fact together with the Marcinkiewicz interpolation theorem yields the thesis.

Given $f \in L^p$, $1 \le p < n/\gamma$, consider the nonincreasing rearrangement of $H_\gamma f$, that by Proposition 3.4(d) is bounded by

$$(H_\gamma f)^*(t) = (|x|^{-\gamma}I_\gamma f)^*(t) \le (|x|^{-\gamma})^*(t/2)(I_\gamma f)^*(t/2) \tag{5.11}$$

If $g(x) = |x|^{-\gamma}$, $0 < \gamma < n$, we have

$$g_*(\alpha) = |\{x \in \mathbb{R}^n : |x|^{-\gamma} > \alpha\}| = |\{x : |x|^\gamma < 1/\alpha\}| = \Omega_n \alpha^{-n/\gamma}$$

for all $\alpha > 0$, and so, $g^*(t) = \Omega_n^{\gamma/n} t^{-\gamma/n} = ct^{-\gamma/n}$. Replacing in (5.11),

$$(H_\gamma f)^*(t) \le c(t/2)^{-\gamma/n}(I_\gamma f)^*(t/2)$$

and then

$$t^{1/p}(H_\gamma f)^*(t) \le c2^{\gamma/n}t^{1/p-\gamma/n}(I_\gamma f)^*(t/2) \tag{5.12}$$

Since $1/p - \gamma/n = 1/q$ for q as in Theorem 5.5, and I_γ is of weak type (p, q) for $p \geq 1$ and such q, i.e., $t^{1/q}(I_\gamma f)^*(t) \leq A_{pq}\|f\|_p$ by (3.10b), inequality (5.12) becomes

$$t^{1/p}(H_\gamma f)^*(t) \leq c_{\gamma np} t^{1/q}(I_\gamma f)^*(t/2)$$
$$\leq A_{\gamma np}\|f\|_p$$

which is the weak type (p, p) inequality for H_γ and holds for every p such that $1/p - \gamma/n > 0$. ∇

<u>Proof of the Marcinkiewicz theorem (lower triangle)</u>. We shall give the proof for the case $p_0 \neq p_1$, $q_0 < q_1 < \infty$. The case $p_0 \neq p_1$, $q_1 < q_0 < \infty$ is completely similar. The cases $p_0 = p_1$ and $p_0 \neq p_1$ with $q_j = \infty$ for $j = 0$ or 1 are left as exercises to the reader.

As before, given $f \in L^p$, we consider the decomposition $f = f^\beta + f_\beta$, $f^\beta \in L^{p_0} \cap L^p$, $f_\beta \in L^{p_1} \cap L^p$, but where $\beta = \beta(\alpha)$ is a function that will be chosen later. Thus,

$$\|Tf\|_q^q = q\int_0^\infty \alpha^{q-1}(Tf)_*(\alpha)d\alpha$$

$$\leq q\int_0^\infty \alpha^{q-1}((Tf^\beta)_*(\alpha/2) + (Tf_\beta)_*(\alpha/2))d\alpha$$

$$\leq q(2M_0)^{q_0}\int_0^\infty \alpha^{q-q_0-1}\|f^\beta\|_{p_0}^{q_0}d\alpha$$

$$+ q(2M_1)^{q_1}\int_0^\infty \alpha^{q-q_1-1}\|f_\beta\|_{p_1}^{p_1}d\alpha$$

(continued)

$$= q(2M_0)^{q_0} \int_0^\infty \alpha^{q-q_0-1} \left(\int_{\mathscr{X}} |f^\beta|^{p_0} d\mu \right)^{q_0/p_0} d\alpha$$

$$+ q(2M_1)^{q_1} \int_0^\infty \alpha^{q-q_1-1} \left(\int_{\mathscr{X}} |f_\beta|^{p_1} d\mu \right)^{q_1/p_1} d\alpha$$

Now, by the hypothesis, we can choose v as to have

$$p_i + \frac{q-q_i}{v} \frac{p_i}{q_i} = p_i + p - p_i = p \quad \text{for} \quad i = 0,1$$

in fact, these two equalities are equivalent to

$$v = \frac{p_0}{q_0} \frac{q-q_0}{p-p_0} \quad \text{and} \quad v = \frac{p_1}{q_1} \frac{q-q_1}{p-p_1}$$

and this holds, since it is the same as

$$(1/p_0 - 1/p)/(1/p - 1/p_1) = (1/q_0 - 1/q)/(1/q - 1/q_1)$$

which is true whenever $p = p_t$ and $q = q_t$ for the same t, i.e., whenever the point $(1/p, 1/q)$ lies on the segment joining $(1/p_0, 1/q_0)$ with $(1/p_1, 1/q_1)$. Therefore we may consider $\beta = \beta(\alpha) = \alpha^v$, for such v, and by Lemma 5.1,

$$\|Tf\|_q^q \leq \frac{q(2M_0)^{q_0}}{q-q_0} \left(\int_{\mathscr{X}} |f|^p d\mu \right)^{q_0/p_0}$$

$$+ \frac{q(2M_1)^{q_1}}{q_1-q} \left(\int_{\mathscr{X}} |f|^p d\mu \right)^{q_1/p_1} \tag{5.13}$$

If $\|f\|_p = 1$, this can be written as

$$\|Tf\|_q \le C\|f\|_p \tag{5.14}$$

with

$$C^q = \frac{q}{q-q_0}(2M_0)^{q_0} + \frac{q}{q_1-q}(2M_1)^{q_1} \tag{5.15}$$

Since both sides of (5.11) are homogeneous functions of f, this inequality for f, with $\|f\|_p = 1$, remains true for any $f \in L^p$. This proves part (a) of the thesis.

If $M_0 = M_1 = 1$, (5.14) can be rewritten as

$$\|Tf\|_q \le KM_0^{1-t}M_1^t\|f\|_p$$

for $K = (q/(q - q_0) + q/(q_1 - q))^{1/q}$, which is part (b) of the thesis.

Since T is sublinear, $|T(cf)| \le |c||Tf|$. Letting $T_1 = c_1T$ and $d\nu_1 = c_2 d\nu$ we have

$$\nu_1(\{|T_1f| > \alpha\}) = c_2\nu\{|c_1Tf| > \alpha\}) = c_2\nu(\{|Tf| > \alpha/c_1\})$$

$$\le c_2(M_i\|f\|_{p_i}c_1/\alpha)^{q_i}$$

$$= (c_2^{1/q_i}c_2M_i\|f\|_{p_i}/\alpha)^{q_i}, \qquad i = 0,1$$

and we can take c_1, c_2 as to have $c_1c_2^{1/q_0}M_0 = c_1c_2^{1/q_1}M_1 = 1$. Hence T_1 is of weak type (p_i, q_i) with constants $M_i = 1$, $i = 0, 1$, and by what has been already proved, $\|T_1f\|_q^1 \le K\|f\|_p$, where the norm $\|\cdot\|_q^1$ is taken with respect to the measure ν_1. An easy computation shows that going back to T and ν we obtain an inequality of the form (b). This proves the theorem under the above assumptions on p_i, q_i. As was already pointed out, if $1 \leq q_1 < q_0 < \infty$, the proof will be the same changing v to $-v$.

6. THE CONDITIONS OF KOLMOGOROFF AND ZYGMUND

Let $(\mathscr{X}, \mu)$ and $(\mathscr{Y}, \nu)$ be two measure spaces and let T be a (sublinear) operator that transforms measurable functions in $(\mathscr{X}, \mu)$ into measurable functions in $(\mathscr{Y}, \nu)$. If T is of weak type (p, q), $1 \leq p, q < \infty$ we cannot assure that $Tf \in L^q(\mathscr{Y}, \nu)$ if $f \in L^p(\mathscr{X}, \mu)$, nor even that $|Tf(y)|^q$ is locally integrable. Nevertheless, the following proposition shows that $|T(y)|^r$ is locally integrable for all $r < q$.

<u>Theorem 6.1 (Kolmogoroff's condition)</u>. (a) If T is an operator of weak type (p, q), $1 \leq p, q < \infty$, with constant M, then for all $0 < r < q$, $|Tf|^r$ is locally integrable for each $f \in L^p$ and, furthermore, the Kolmogoroff inequality

$$\left(\int_K |Tf(y)|^r d\nu\right)^{1/r} \leq M\left(\frac{q}{q-r}\right)^{1/r} \nu(K)^{1/r-1/q} \|f\|_p \tag{6.1}$$

holds, where $K \subset \mathscr{Y}$ is any set of finite ν-measure.

(b) Conversely, if there exist an $0 < r < q$ and a positive constant M_1 such that for all $K \subset \mathscr{Y}$,

$$\left(\int_K |Tf(y)|^r d\nu\right)^{1/r} \leq M_1 \nu(K)^{1/r-1/q} \|f\|_p \tag{6.1a}$$

holds for all $f \in L^p$, then T is of weak type (p, q) with constant $M \leq M_1$.

(c) These results hold also for $q = \infty$ with the convention that $q/(q - r) = 1$.

Remark 6.1. (a) says in particular that if $(\mathcal{Y}, \nu)$ is a space of finite measure, then T being of weak type (p, q) implies that T is of type (p, r) for every $r < q$.

Proof. (a) Let $(Tf)_*^K(\alpha) = \nu\{y \in K : |Tf(y)| > \alpha\} = \nu(E_\alpha(Tf) \cap K)$. Thus $(Tf)_*^K(\alpha) \leq \nu(K)$ and $(Tf)_*^K(\alpha) \leq (Tf)_*(\alpha) \leq (M\alpha^{-1}\|f\|_p)^q$. Then

$$\int_K |Tf|^r d\nu = r\int_0^\infty \alpha^{r-1}(Tf)_*^K(\alpha)d\alpha = r\int_0^N + r\int_N^\infty$$

$$\leq r\int_0^N \alpha^{r-1}\nu(K)d\alpha + r\int_N^\infty \alpha^{r-1}(M\alpha^{-1}\|f\|_p)^q d\alpha$$

$$= \nu(K)N^r + (M\|f\|_p)^q(r/(q-r))N^{r-q}$$

The last sum is minimized by $N = M\|f\|_p \nu(K)^{-1/q}$ and this value of N gives (6.1).

(b) Let (6.1a) hold. If $K = E_\alpha(Tf)$ then, by Chebyshev's inequality,

$$\alpha^r \nu(K) \leq \int_K |Tf(y)|^r d\nu \leq M_1^r \nu(K)^{1-r/q}\|f\|_p^r$$

and

$$\nu(K) \leq (M_1\alpha^{-1}\|f\|_p)^q \quad \text{for all} \quad \alpha > 0 \tag{6.2}$$

that is, T is of weak type (p, q).

The proof of (c) is left as an exercise to the reader. ∇

If K is of finite measure, then $L^q(K) \subset L^p(K)$ if $q > p$. Now, the logarithm grows more slowly than any power, so if for some $\varepsilon > 0$, $\int_K |f|^{p+\varepsilon}$ converges, then $\int_K |f|^p \log^+|f|$ will also converge. That is, the condition $\int_K |f|^p + \int_K |f|^p \log^+|f| < \infty$ is stronger than the condition $f \in L^p(K)$ but weaker than $f \in L^{p+\varepsilon}(K)$, any $\varepsilon > 0$. Therefore, for $\varepsilon > 0$ and K of finite measure we have that

$$L^{p+\varepsilon}(K) \subset L^p \log^+ L(K) \subset L^p(K)$$

for $1 \leq p < \infty$. (For the definition of the Zygmund class $L^p \log^+ L$, see Chapter 0, Section 1.)

<u>Theorem 6.2 (Zygmund's condition)</u>. Let T be an operator of weak type (p, p) and of weak type (q, q), $1 \leq p < q < \infty$. Then for every $f \in L^p \log^+ L$, $Tf \in L^p(K)$, for all K with $\nu(K) < \infty$, and

$$\int_K |Tf(y)|^p d\nu \leq M(\nu(K) + \int_{\mathscr{X}} |f(x)|^p (1 + \log^+ |f(x)|) d\mu) \tag{6.3}$$

<u>Proof</u>. Let $(Tf)_*^K$ be as in the proof of Theorem 6.1. Then, $(Tf)_*^K \leq \nu(K)$ and $(Tf)_*^K(\alpha) \leq (Tf)_*(\alpha)$. Since $(Tf)_*(\alpha)$ is less than or equal to $(M_1 \|f\|_p/\alpha)^p$ and $(M_2 \|f\|_q/\alpha)^q$, then,

$$\begin{aligned}
\int_K |Tf|^p d\nu &= p\int_0^\infty \alpha^{p-1}(Tf)_*^K(\alpha)d\alpha = 2^p p\int_0^\infty \alpha^{p-1}(Tf)_*^K(2\alpha)d\alpha \\
&\leq 2^p p\int_0^1 \alpha^{p-1}\nu(K)d\alpha + 2^p p\int_1^\infty \alpha^{p-1}(Tf^\alpha)_*^K(\alpha)d\alpha \\
&\quad + 2^p p\int_1^\infty \alpha^{p-1}(Tf_\alpha)_*^K(\alpha)d\alpha \\
&\leq 2^p \nu(K) + 2^p p\int_1^\infty \alpha^{p-1}(M_1/\alpha)^p(\int_{\mathscr{X}} |f^\alpha(x)|^p d\mu)d\alpha \\
&\quad + 2^p p\int_1^\infty \alpha^{p-1}(M_2/\alpha)^q(\int_{\mathscr{X}} |f_\alpha(x)|^q d\mu)d\alpha \\
&= 2^p p\nu(K) + 2^p pM_1^p \int_{\mathscr{X}} |f(x)|^p (\int_1^{|f|} \alpha^{-1} d\alpha)d\mu \\
&\quad + 2^p pM_2^q \int_{\mathscr{X}} |f(x)|^q (\int_{|f|}^\infty \alpha^{p-q-1} d\alpha)d\mu
\end{aligned} \tag{6.4}$$

Since $\int_1^{|f|} \alpha^{-1} d\alpha = 0$ if $|f| < 1$ and $= \log|f|$ if $|f| \geq 1$, then $\int_1^{|f|} \alpha^{-1} d\alpha = \log^+|f|$, and (6.4) becomes

$$\int_K |Tf|^p d\nu \leq 2^p p\nu(K) + 2^p p M_1^p \int_{\mathscr{X}} |f|^p \log^+|f| d\mu$$

$$+ 2^p p/(q - p) M_2^q \int_{\mathscr{X}} |f|^p d\mu$$

which yields (6.3) on taking $M = 2^p p(1 + M_1^p + M_2^q/(q - p))$. ∇

Exercise 6.1. Prove the analog of Theorem 6.2 for $q = \infty$, that is, that (6.3) holds for $M \leq 2^p p(M_1^p + M_\infty)$ where M_1 is the constant for weak type (p, p), as before, and M_∞ is the constant for (strong) type (∞, ∞).

REFERENCES

1. M. Riesz, Acta Math., 49:465 (1926).
2. G. Thorin, Kungl. Fysiogr. Sällsk i Lund Förh., 8:166 (1938).
3. G. Thorin, Med. Lunds Univ. Mat. Sem., 9:1 (1948).
4. J. D. Tamarkin and A. Zygmund, Bull. A.M.S., 50:279 (1944).
5. A. P. Calderón and A. Zygmund, Amer. J. Math., 78:282 (1956).
6. E. M. Stein and G. Weiss, Trans. A.M.S., 87:159 (1958).
7. M. A. Krasnoselskii, P. P. Zabreyko, E. I. Pustelnik and P. E. Arbolevskii, Integral operators in spaces of summable functions, Nauka, Moscow, 1966 (in Russian).
8. E. M. Stein, Trans. A.M.S., 83:482 (1956).
9. P. Duren, Theory of H^p spaces, Academic Press, London-New York, 1970.
10. J. Bergh and J. Löfstrom, Interpolation Spaces, Springer-Verlag, Berlin-Heidelberg-New York, 1976.

11. E. T. Oklander, Interpolación, espacios de Lorentz, y el teorema de Marcinkiewicz, Cursos y Seminarios de Matemáticas, Univ. Buenos Aires, fasc. 20, Buenos Aires, 1965.

12. R. Hunt, L'Enseign. Math., 12:249 (1966).

13. J. Marcinkiewicz, C. R. Acad. Sci. Paris, A, 208:1272 (1939).

14. A. Zygmund, J. Math. Pures Appl., 35:223 (1956).

15. J. L. Lions and J. Peetre, Inst. Hautes Etudes Sci. Publ. Math., 19:5 (1964).

16. M. Cotlar and M. L. Bruschi, Rev. Univ. La Plata, 3:162 (1956).

17. A. Zygmund, Trigonometric Series, 2nd edition, Cambridge Univ. Press, Cambridge, 1959.

18. C. A. Berenstein, M. Cotlar, N. Kerzman, P. Krée, Studia Math., 29:79 (1967).

19. R. Hunt, Bull. A.M.S., 70:803 (1964).

Chapter 5

MAXIMAL THEORY AND THE SPACE BMO

1. THE HARDY-LITTLEWOOD MAXIMAL THEOREM

When $\phi(x)$ is a continuous function defined on $\mathbb{R}$ and $\Phi(x) = \int^x \phi(t)dt$ is its indefinite integral, it is a fact from elementary calculus that $\phi(x) = \Phi'(x) = \lim_{r\to 0} (\Phi(x + r) - \Phi(x))/r = \lim_{r\to 0} (1/r)\int_x^{x+r} \phi(t)dt = \lim_{r\to 0} (1/2r)\int_{x-r}^{x+r} \phi(t)dt$. The same holds in several dimensions, namely if ϕ is a continuous function defined in $\mathbb{R}^n$, $n \geq 1$, and $Q(x, r)$ is a cube of sides parallel to the axes, centered at x and of side $r > 0$, then

$$\phi(x) = \lim_{r\to 0} \frac{1}{|Q(x, r)|} \int_{Q(x, r)} \phi(t)dt \tag{1.1}$$

where $|Q(x, r)| = r^n$ is the volume (n-dimensional measure) of $Q(x, r)$. More generally, if for every $r > 0$ we fix a cube Q_r of sides parallel to the axes of length r, such that Q_r contains (but is not necessarily centered at) the point x, then

$$\phi(x) = \lim_{r\to 0} \frac{1}{|Q_r|} \int_{Q_r} \phi(t)dt \tag{1.2}$$

Let $\rho(x)$ be the characteristic function of the cube $Q(0, 1) \subset \mathbb{R}^n$ and $\rho_r(x) = r^{-n}\rho(x/r)$, r^{-n} times the characteristic function of the cube $Q(0, r)$. Set

$$L_r\phi = \phi * \rho_r \tag{1.3}$$

Then, for each $r > 0$, $L_r\phi(x)$ is a function of x such that

$$L_r\phi(x) = \frac{1}{|Q(x, r)|}\int_{Q(x, r)} \phi(t)dt \tag{1.4}$$

Thus (1.1) can be rewritten as

$$\phi(x) = \lim_{r \to 0} L_r\phi(x) \tag{1.1a}$$

The classical Lebesgue theorem on the differentiation of the integral, already mentioned in Chapter 1, generalizes (1.1) to a wider class of functions.

<u>Theorem 1.1 (The Lebesgue differentiation theorem)</u>.

(a) If $f \in L^p(\mathbb{R}^n)$, $1 \leq p \leq \infty$, or more generally, if $f \in L_{loc}(\mathbb{R}^n)$, then

$$\lim_{r \to 0} L_r f(x) = \lim_{r \to 0} \frac{1}{|Q(x, r)|}\int_{Q(x, r)} f(t)dt = f(x) \quad \text{a. e.} \tag{1.5}$$

(b) If $f \in L^p(\mathbb{R}^n)$, $1 \leq p < \infty$, then

$$\lim_{r \to 0} \|L_r f - f\|_p = 0 \tag{1.6}$$

Since, as in (1.3), $L_r f = f * \rho_r$ and $\{\rho_r\}$ is a convolution unit, part (b) of Theorem 1.1 is an immediate consequence of Theorem 1.4 of Chapter 1 (as well as (1.1) for continuous functions

ϕ.). To prove part (a) of Theorem 1.1, we cannot equally rely on Theorem 2.1 of Chapter 1 since in that theorem the pointwise convergence was proved precisely on the points where (1.5) holds. We shall therefore introduce the method of maximal functions. For this purpose we define the Hardy-Littlewood maximal function, which plays an important role in real variable theory and in harmonic analysis.

Definition 1.1. Given $f : \mathbb{R}^n \to \mathbb{C}$, a measurable function, its Hardy-Littlewood maximal function $\Lambda f(x)$ is defined by

$$\Lambda f(x) = \sup_{r>0} \frac{1}{|Q(x,r)|} \int_{Q(x,r)} |f(t)|dt \tag{1.7}$$

where $Q(x, r)$ is any cube centered at x of sides parallel to the axes of length r.

For every $f \in L_{loc}(\mathbb{R}^n)$, Λf is also a measurable function, since for each $r > 0$ the integral is continuous, and the supremum of continuous functions, being semicontinuous, is measurable.

The operator, denoted by Λ, such that $\Lambda : f \to \Lambda f$ is called the Hardy-Littlewood maximal operator. It majorizes many important operators, such as the Poisson integral operator (see Sections 2 and 4 below and Section 3 of Chapter 6), thus sharing with them its continuity properties.

Remark 1.1. If $S(x, r)$ is a sphere of radius r and centered in x,

$$\Lambda_1 f(x) = \sup_{r>0} |S(x,r)|^{-1} \int_{S(x,r)} |f(t)|dt =$$

$$= \sup_{r>0} (\omega_n r^n)^{-1} \int_{|x-t|\le r} |f(t)|dt = \sup_{r>0} (\omega_n r^n)^{-1} \int_{|t|\le r} |f(x-t)|dt$$

differs from $\Lambda f(x)$ only in a (multiplicative) constant and is called the "spherical" maximal function, as different from the "cubical" maximal function $\Lambda f(x)$. The study of Λf and of $\Lambda_1 f$ is essentially equivalent.

Let $\{f_r\}$ be a sequence of functions, $f_r(x) \to f(x)$ a.e. Then $\int |f_r(x) - f(x)|^p dx \to 0$ whenever all the $|f_r(x)|^p$ are dominated by an integrable function, i.e., whenever $\sup_r |f_r(x)| = M(x) \in L^p$. $M(x)$ is called the maximal function of the sequence and thus it is useful to know that certain sequences have their maximal functions in L^p. In the case $f_r = L_r f$, the maximal function is Λf. Therefore, in classical analysis, the complement, due to Hardy and Littlewood [1], to the Lebesgue differentiation theorem became very important, since it asserts that, for $f \in L^p$, $1 < p \leq \infty$, the maximal function of the sequence $f_r = L_r f$ is in L^p. This is not true for $p = 1$, but in this case the maximal function lies in L^1_*.

The Hardy-Littlewood maximal operator Λ is a sublinear operator since it satisfies the conditions

(i) $\Lambda(f + g)(x) \leq \Lambda f(x) + \Lambda g(x)$ for all $f, g \in L_{loc}$

(ii) $\Lambda(\alpha f)(x) = |\alpha| \Lambda f(x)$ for all $\alpha \in \mathbb{C}$, $f \in L_{loc}$

Λ is also a positive operator; moreover, for all $f \in L_{loc}$ it is $\Lambda f(x) \geq 0$ for all $x \in \mathbb{R}^n$.

From Definition 1.1, it is obvious that Λ transforms bounded functions into bounded functions and that

$$\|\Lambda f\|_\infty \leq \|f\|_\infty \tag{1.8}$$

i.e., Λ is of type (∞, ∞) with constant less than or equal to 1. On the other hand, $f \in L^1$ does not imply $\Lambda f \in L^1$. Consider, for example, $f(x) = \chi_{(0,1)}(x)$. Then, $\Lambda f(x) = \sup_{r>0} (2r)^{-1} \int_x^{x+r} |f(t)| dt \geq (2x)^{-1} \int_0^{2x} \chi_{(0,1)}(t) dt = (2x)^{-1}$ for $x > 1/2$ and $\Lambda f \notin L^1(\mathbb{R})$. Therefore, Λ is not an operator of type $(1, 1)$. But we shall prove that it is an operator of weak type $(1, 1)$. This is the kernel of the Hardy-Littlewood maximal theorem, and the Lebesgue theorem follows as a corollary.

Theorem 1.2 (The Hardy-Littlewood maximal theorem). Let f be a measurable function defined in $\mathbb{R}^n$, Λf its maximal function, as given in (1.7), and $\Lambda : f \to \Lambda f$ the Hardy-Littlewood maximal operator. Then Λ is of type (p, p), for $1 < p \leq \infty$, and of weak type (1, 1).

Remark 1.2. As seen in (1.8), the result of type (∞, ∞) for Λ is immediate. Thus the proof of Theorem 1.2 consist essentially in proving the weak type (1, 1), for Λ, since the type (p, p), $1 < p < \infty$, will then be a consequence of the Marcinkiewicz interpolation theorem. To prove the weak type (1, 1) for Λ, we are going to give a covering lemma of the Vitali type. This lemma is not stated in its more general form, but only as it will be used here. For refinements and related questions see [2].

Lemma 1.3. Let $E \subset \mathbb{R}^n$ be a measurable set and $\{Q_\alpha\}_{\alpha \in A}$ be a family of cubes of sides parallel to the axes such that $\sup_\alpha |Q_\alpha| < \infty$ and $\{Q_\alpha\}_\alpha$ covers E. It is possible to choose from the family a disjoint sequence (eventually finite) of cubes $\{Q_k\}_{k \in \mathbb{N}}$ such that

$$\Sigma_k |Q_k| \geq C|E| \tag{1.9}$$

where C is a constant depending only on the dimension n.

Remark 1.3. In this statement the family of cubes can be changed to a family of spheres without alteration. (Cfr. Remark 1.1.)

Proof. By assumption, $\sup_\alpha |Q_\alpha| < \infty$ or, equivalently, $\sup_\alpha$ (sidelength $Q_\alpha) < \infty$, hence we may choose a Q_1 such that its side is longer or equal to $1/2 \sup_\alpha$ (sidelength Q_α). Among the cubes disjoints from Q_1, we choose Q_2 such that

$$\text{side } Q_2 \geq \frac{1}{2} \sup_{Q_\alpha \cap Q_1 = \phi} (\text{side } Q_\alpha)$$

Again, among the cubes disjoints from Q_1 and Q_2 we choose Q_3 such that

$$\text{side } Q_3 \geq \frac{1}{2} \sup_{Q_\alpha \cap (Q_1 \cup Q_2) = \phi} (\text{side } Q_\alpha)$$

and so on. There are two possibilities:

(a) that once $Q_1, \ldots, Q_k$ are already chosen, one can always pick a Q_{k+1} and get an infinite sequence $\{Q_k\}$, or

(b) that at a certain step no cube can be chosen and the procedure gives a finite sequence $Q_1, \ldots, Q_k$.

In both cases, the chosen cubes are mutually disjoint by construction. If $\Sigma_k |Q_k| = +\infty$, (1.9) follows immediately. Let us suppose then that $\Sigma_k |Q_k| < \infty$; for each k we consider Q_k^* the cube that has the same center as Q_k and 5 times its side, thus $|Q_k^*| = 5^n |Q_k|$. Our aim is to show that $\{Q_k^*\}$ cover E, i.e., $\bigcup_k Q_k^* \supset E$. For this purpose it is enough to show that for each $\alpha \in A$, $Q_\alpha \subset \bigcup_k Q_k^*$.

Let Q_α be a cube of the initial family. If there is a $k \in \mathbb{N}$ such that $Q_\alpha = Q_k$ then, a fortiori, $Q_\alpha \subset Q_k^*$ and there is nothing more to prove. We may assume then that $Q_\alpha \neq Q_k$ for all k. Since the Q_k's are disjoint and $\Sigma_k |Q_k| < \infty$, there is a largest $k(\alpha)$ such that

$$\text{side } Q_{k(\alpha)} \geq \frac{1}{2} \text{ side } Q_\alpha \tag{1.10}$$

(that there is such a $k(\alpha)$ follows from the construction of the Q_k's). Being $k(\alpha)$ the largest index for which (1.10) holds,

$$\text{side } Q_{k(\alpha)+m} < \frac{1}{2} \text{ side } Q_\alpha, \quad \forall m \geq 1 \tag{1.11}$$

This implies the existence of an integer $r \in \{1, \ldots, k(\alpha)\}$ such that $Q_r \cap Q_\alpha \neq \phi$, because if $Q_r \cap Q_\alpha = \phi$ for every $r \in \{1, \ldots, k(\alpha)\}$ then Q_α would have been chosen instead of $Q_{k(\alpha)+1}$, since by (1.11) its side is longer than twice the side of $Q_{k(\alpha)+1}$. Let $r(\alpha)$ be the least such integer r for which $Q_r \cap Q_\alpha \neq \phi$. Since $Q_{r(\alpha)} \cap Q_\alpha \neq \phi$ and, by construction, side $Q_{r(\alpha)} \geq 1/2$ side Q_α, $Q^*_{r(\alpha)} \supset Q_\alpha$ as we wanted to prove.

Then, $E \subset \bigcup_k Q^*_k$, and

$$|E| \leq |\cup_k Q^*_k| \leq \Sigma_k \; |Q^*_k| = 5^n \; \Sigma_k \; |Q_k|$$

and (1.9) is proved with $C = 5^{-n}$. ∇

Remark 1.4. If $d(x, y)$ is the Euclidean distance in $\mathbb{R}^n$, then $S(x, r) = \{y : d(x, y) \leq r\}$, and the Lebesgue measure μ on $\mathbb{R}^n$ has the following property:

$$\mu(S(x, 2r)) \leq A\mu(S(x, r)), \quad A = 2^n, \quad \text{a fixed constant} \qquad (1.12)$$

This property was essential in the proof of the lemma. More generally, if X is any set, $d(x, y)$ is a pseudo-distance in X (i.e., $d(x, y) \leq K(d(x, z) + d(y, z))$, K a fixed constant), and μ is a measure in X satisfying (1.12) for the sphere corresponding to d, then it can be proved that the preceding lemma holds for (X, d, μ). Such spaces (X, d, μ) are called homogeneous spaces and the Hardy-Littlewood maximal theorem, as well as other results given in the next sections, can be extended to such spaces and, in particular, to general groups. For these generalizations we refer to [3].

Proof of Theorem 1.2. For a given $f \in L^1$, let $\alpha > 0$ be fixed and $E = E_\alpha(\Lambda f) = \{x : \Lambda f(x) > \alpha\}$. By definition of $\Lambda f(x)$, for each $x \in E$, there is a cube Q_x centered at x and of side $r = r(x)$ such that

$$\frac{1}{|Q_x|}\int_{Q_x} |f(t)|\,dt > \alpha \tag{1.13}$$

i.e., for each $x \in E$ there is a cube Q_x with

$$|Q_x| = \alpha^{-1}\int_{Q_x} |f(t)|\,dt \leq \alpha^{-1}\|f\|_1 < \infty \tag{1.13a}$$

Thus the family of cubes $\{Q_x\}_{x\in E}$ fulfills the conditions of Lemma 1.3, since $E \subset \bigcup_x Q_x$, and there is a sequence $\{Q_k\}$ of disjoint cubes such that $\Sigma_k |Q_k| \geq |E| = C(\Lambda f)_*(\alpha)$. Therefore, by (1.13)

$$\begin{aligned}(\Lambda f)_*(\alpha) &\leq C^{-1}\sum_k |Q_k| \leq (C\alpha)^{-1}\sum_k \int_{Q_k} |f(t)|\,dt \\ &= (C\alpha)^{-1}\int_{\bigcup Q_k} |f(t)|\,dt \\ &\leq (C\alpha)^{-1}\int_{\mathbb{R}^n} |f(t)|\,dt\end{aligned}$$

and $\sup_\alpha \alpha(\Lambda f)_*(\alpha) \leq C^{-1}\|f\|_1$, which is the weak type (1, 1) condition for Λ, with constant $C^{-1} = 5^n$. ∇

We now prove Theorem 1.1 as a corollary of Theorem 1.2, by a method of subordination of operators that will be generalized in Section 4.

Proof of Theorem 1.1(a). Let $f \in L^p$, $1 \leq p \leq \infty$, and $L_r f(x)$ be as in (1.5). The thesis is that $L_r f(x) \to f(x)$ a.e. as $r \to 0$. The problem is local and we may assume that f is just locally integrable.

(Thus we shall obtain the convergence seeked for a.e. point in a sphere of radius R, and letting $R \to \infty$, the convergence a.e. in $\mathbb{R}^n$.)

Observe that, for all positive r,

$$|L_r f(x)| \leq \Lambda f(x) \tag{1.14}$$

For $f \in L^1_{loc}$ and each $x \in \mathbb{R}^n$, let

$$\Omega f(x) = \limsup_{r \to 0} L_r f(x) - \liminf_{r \to 0} L_r f(x) \tag{1.15}$$

Thus

$$|\Omega f(x)| \leq 2 \sup_r |L_r f(x)| \leq 2\Lambda f(x) \tag{1.16}$$

If g is a continuous function with compact support then by elementary calculus, $\lim_{r \to 0} L_r g(x) = g(x)$ for all $x \in \mathbb{R}^n$ and $\Omega g = 0$. As every $f \in L^1$ can be written as $f = g + h$ for such a g and $\|h\|_1 < \eta$, $\eta > 0$ arbitrary,

$$\Omega f(x) \leq \Omega g(x) + \Omega h(x) = \Omega h(x) \tag{1.17}$$

By Proposition 3.1(b) of Chapter 4, (1.16) and the weak type (1, 1) of Λ,

$$(\Omega f)_*(\alpha) \leq (\Omega h)_*(\alpha) \leq (\Lambda h)_*(\alpha/2) \leq 2A\alpha^{-1} \|h\|_1 < 2A\alpha^{-1}\eta$$

and choosing η very small with respect to α we get $\Omega f = 0$ a.e. So $\lim_{r \to 0} L_r f(x)$ exists a.e. and it remains to show that it is equal to $f(x)$. Since

$$\|L_r f - f\|_1 = \int_{\mathbb{R}^n} \left| \frac{1}{|Q|} \int_{Q(x,r)} f(t)dt - f(x) \right| dx$$

$$= \int_{\mathbb{R}^n} \frac{1}{|Q|} \left| \int_{Q(0,r)} (f(x-t) - f(x))dt \right| dx$$

$$\leq \frac{1}{|Q|} \int_{Q(0,r)} \left(\int_{\mathbb{R}^n} |f(x-t) - f(x)| dx \right) dt$$

$$\to 0 \quad \text{when} \quad r \to 0$$

by the continuity of the L^1 norm, there exists a subsequence $\{L_{r_k} f\}$ converging to f a.e. in x and the thesis (a) follows. ∇

<u>Corollary 1.4</u>. If $f \in L^p(\mathbb{R}^n)$, $1 \leq p \leq \infty$, then $\Lambda f(x) < \infty$ a.e.

<u>Proof</u>. Since $f \in L^p$, by Theorem 1.1(a), for almost all $x \in \mathbb{R}^n$ we have that $I_r = |Q(x, r)|^{-1} \int_{Q(x,r)} |f(t)|dt \to |f(x)|$ as $r \to 0$. Therefore I_r is bounded for all $r < r_0$. By Hölder's inequality, $|I_r| \leq \|f\|_p |Q(x, r)|^{-1/p}$ which tends to zero as $r \to \infty$. So I_r is also bounded for $r \geq r_0$ and the thesis is proved. ∇

<u>Definition 1.2</u>. Let $\Lambda' : f \to \Lambda' f$ be defined by

$$\Lambda' f(x) = \sup \frac{1}{|Q_x|} \int_{Q_x} |f(t)| dt \tag{1.18}$$

where the supremum is taken over all the cubes Q_x of sides parallel to the axes that contain x.

Since $Q(x, r)$ is a Q_x and $Q_x \subset Q(x, 4\ell)$, if side $Q_x = \ell$, $|Q_x| = 4^{-n}|Q(x, 4\ell)|$, and

$$\frac{1}{|Q_x|}\int_{Q_x}|f|dt \le \frac{4^n}{|Q(x, 4\ell)|}\int_{Q(x, 4\ell)}|f|dt \le 4^n\Lambda f(x) \qquad (1.19)$$

We obtain thus

$$\Lambda f(x) \le \Lambda' f(x) \le 4^n \Lambda f(x) \qquad (1.19a)$$

so Λ' shares the same type and weak type properties of Λ. In particular we get the following consequence of the Hardy-Littlewood maximal theorem and its corollary, the Lebesgue theorem,

Corollary 1.5. If $1 < p \le \infty$ then

$$f \in L^p \iff \Lambda f \in L^p \iff \Lambda' f \in L^p$$

Proof. Since Λ (and therefore, by (1.19a), Λ') is of type (p, p), $p > 1$, $f \in L^p$ implies $\Lambda f \in L^p$ ($\Lambda' f \in L^p$). But, by (1.5) and (1.14), $|f(x)| = \lim L_r|f(x)| \le \Lambda f(x)$ a.e. in x, so $\int|f(x)|^p dx \le \int(\Lambda f(x))^p dx$, and $\Lambda f \in L^p$ implies $f \in L^p$. ∇

We have already seen the importance of the subordination of the "differentiation operators" L_r by the Hardy-Littlewood maximal operator Λ. In Section 2 we show that Λ also majorizes the operators assigning to each function its Poisson integrals.

2. APPLICATIONS TO POISSON INTEGRALS

In Chapter 3 we considered harmonic functions defined in $\mathbb{R}^{n+1}_+ = \mathbb{R}^n \times \mathbb{R}^+$ and studied them as Poisson integrals of L^p functions. The Hardy-Littlewood maximal function provides useful estimates for such harmonic functions.

<u>Proposition 2.1.</u> Let $f \in L^p(\mathbb{R}^n)$, $1 \leq p \leq \infty$, and let

$$u(x, t) = P_t * f(x), \quad t > 0$$

be its Poisson integral, where $P_t(x) = P(x, t) = C_n t(|x|^2 + t^2)^{-(n+1)/2}$ $C_n = \Gamma((n+1)/2)\pi^{-(n+1)/2}$. Then

$$\sup_{t>0} |u(x, t)| \leq A_n \Lambda f(x) \tag{2.1}$$

where A_n is a constant depending only on n.

<u>Proof.</u> For $y = ry'$, $y' \in \Sigma$, the unit sphere in $\mathbb{R}^n$, let

$$g(r) = g_x(r) = r^{n-1} \int_\Sigma f(x - ry')dy'$$

and

$$G(r) = G_x(r) = \int_0^r g(s)ds = \int_{|y|\leq r} f(x - y)dy \tag{2.2}$$

By integration by parts we get

$$u(x, t) = C_n t \int_{\mathbb{R}^n} \frac{f(x-y)}{(|y|^2+t^2)^{(n+1)/2}} dy = C_n t \int_0^\infty \frac{g(r)}{(r^2+t^2)^{(n+1)/2}} dr$$

$$= (n + 1)C_n t \int_0^\infty \frac{rG(r)}{(r^2+t^2)^{(n+3)/2}} dr$$

By definition of Λf and from (1.14), for γ_n a constant,

$$|G(r)| \leq \gamma_n r^n \Lambda f(x) \tag{2.3}$$

Thus,

$$|u(x,t)| \leq (n+1)C_n\gamma_n t\Lambda f(x) \int_0^\infty (r^2+t^2)^{-(n+3)/2} r^{n+1}\, dr$$

$$= ((n+1)C_n\gamma_n \int_0^\infty (s^2+1)^{-(n+3)/2} s^{n+1}\, ds)\Lambda f(x) \quad (2.4)$$

and the thesis follows since the integral in the last bracket is finite. ∇

Remark 2.1. A similar result holds for the Weierstrass integrals $W_t * f$ introduced in Section 1 of Chapter 3, and for any convolutions with other "bell shaped" kernels, i.e., positive, radial kernels that are decreasing as functions of $|x|$. More generally, a proof entirely similar to the above one yields the following result, involving the least decreasing radial majorant of a convolution unit as defined in (2.1) of Chapter 1.

Exercise 2.1. Let $\{\phi_\varepsilon\}$ be a convolution unit given by $\phi \in L^1(R^n)$, $\int \phi = 1$, $\phi_\varepsilon(x) = \varepsilon^{-n}\phi(x/\varepsilon)$ and let its least decreasing radial majorant $\psi(x) = \sup_{|y|\geq|x|} |\phi(y)| \in L^1(\mathbb{R}^n)$. Prove that

$$\sup_{\varepsilon>0} |\phi_\varepsilon * f(x)| \leq A_n C\Lambda f(x) \quad (2.5)$$

for every $f \in L^p(\mathbb{R}^n)$, $1 \leq p \leq \infty$, where $C = \int \psi$.

A converse of Proposition 2.1 holds for positive functions.

Proposition 2.2. Let $f(x) \geq 0$ have a Poisson integral $u(x,t)$ defined for each $t > 0$ (e.g.: let $f \in L^p$, $1 \leq p \leq \infty$), then

$$\Lambda f(x) \leq B_n \sup_{t>0} u(x,t) \quad (2.6)$$

where B_n is constant depending only on n.

Proof. For each $r > 0$,

$$\sup_{t>0} u(x,t) \geq u(x,r) = \int_{\mathbb{R}^n} f(x-y)P(y,r)dy$$

$$\geq \int_{|y|\leq 2r} f(x-y)P(y,r)dy$$

$$= C_n \int_{|y|\leq 2r} f(x-y)r(|y|^2 + r^2)^{-(n+1)/2}dy$$

$$\geq C_n r(5r^2)^{-(n+1)/2} \int_{|y|\leq 2r} f(x-y)dy$$

$$= 5^{-(n+1)/2} C_n r^{-n} \int_{|y|\leq 2r} f(x-y)dy$$

$$= B_n^{-1} \frac{1}{|Q(x,r)|} \int_{Q(x,r)} f(y)dy$$

and (2.6) follows. ∇

If $u(x,t)$ is the Poisson integral of an L^p function, then Theorem 4.4A of Chapter 3 insures the existence, for almost every $x_0 \in \mathbb{R}^n$, of the boundary value $\lim u(x,t)$ whenever (x,t) tends to $(x_0, 0)$ "vertically", i.e., on the line $x = x_0$.

More generally such limit values exist a.e., even when (x,t) tends to $(x_0, 0)$ in a less restricted way.

To be precise in our statements we introduce some definitions.

Definition 2.1. Given $x_0 \in \mathbb{R}^n$ and $\alpha > 0$, the cone $\Gamma_\alpha(x_0)$ in $\mathbb{R}^{n+1}_+$ of vertex $(x_0, 0)$ and aperture α is the region defined by

$$\Gamma_\alpha(x_0) = \{(x,t) \in \mathbb{R}^{n+1}_+ : |x - x_0| < \alpha t\} \tag{2.7}$$

Definition 2.2. A function $u(x, t)$ defined in $\mathbb{R}_+^{n+1}$ is nontangentially bounded at $x_0 \in \mathbb{R}^n$ if, for every $\alpha > 0$, $u(x, t)$ is bounded in $\Gamma_\alpha(x_0) \cap \{(x, t) \in \mathbb{R}_+^{n+1} : t \leq C\}$ for some constant C.

Remark 2.2. The height of the truncation of the cone $\Gamma_\alpha(x_0)$ is irrelevant when dealing with harmonic functions, since these functions are continuous and therefore if bounded for a finite C they will also be bounded for another one.

Definition 2.3. A function $u(x, t)$ defined in $\mathbb{R}_+^{n+1}$ has nontangential limit at $x_0 \in \mathbb{R}^n$ if, for every $\alpha > 0$, $\lim u(x, t) = \ell$ exists whenever (x, t) tends to $(x_0, 0)$ inside the cone $\Gamma_\alpha(x_0)$.

Proposition 2.3. Let $f \in L^p(\mathbb{R}^n)$, $1 \leq p \leq \infty$, and let

$$u(x, t) = P_t * f(x), \qquad t > 0$$

be its Poisson integral. Then for all $\alpha > 0$

$$\sup_{(x,t)\in\Gamma_\alpha(x_0)} |u(x, t)| \leq C_{\alpha n} \Lambda f(x_0) \tag{2.8}$$

where $C_{\alpha n}$ is a constant depending only on α and on n.

Proof. This result is an immediate consequence of Proposition 2.1. Let $\Gamma_\alpha(x_0)$ be fixed. As for any x such that $|x - x_0| < \alpha t$,

$$|x_0|^2 + t^2 \leq \max(1 + 2\alpha^2, 2)(|x|^2 + t^2) \tag{2.9}$$

and $P_t(x) = P(x, t) = C_n t(|x|^2 + t^2)^{-(n+1)/2}$, then

$$P(x, t) \leq C_\alpha P(x_0, t) \tag{2.10}$$

for every $(x, t) \in \Gamma_\alpha(x_0)$, where $C_\alpha = \max(1 + 2\alpha^2, 2)^{(n+1)/2}$.

From (2.10),

$$\sup_{(x,t)\in\Gamma_\alpha(x_0)} |u(x, t)| \le C_\alpha \sup_{t>0} |u(x_0, t)|$$

and (2.8) follows from (2.1), with $C_{\alpha n} = C_\alpha A_n$. ∇

The following result for Poisson integrals also holds.

<u>Proposition 2.4</u>. Let $f \in L^p(\mathbb{R}^n)$, $1 \le p \le \infty$, and let $u(x, t)$ be its Poisson integral. Then the nontangential limit of $u(x, t)$ exists for almost all $x \in \mathbb{R}^n$ and is equal to $f(x)$.

<u>Proof</u>. Consider a fixed point $x_0 \in \mathbb{R}^n$ and, for $\alpha > 0$, $(x, t) \in \Gamma_\alpha(x_0)$. Then

$$u(x, t) - f(x_0) = \int_{\mathbb{R}^n} |f(y) - f(x_0)| P(x - y, t)dy$$

Since by (2.10), $P(x, t) \le C_\alpha P(x_0, t)$ for every $(x, t) \in \Gamma_\alpha(x_0)$,

$$|u(x, t) - f(x_0)| \le C_\alpha \int_{\mathbb{R}^n} |f(y) - f(x_0)| P(x_0 - y, t)dy \tag{2.11}$$

But for the Poisson kernel, $P(x, t) = P_t(x) = t^{-n}P_1(x/t)$, for all $t > 0$, so $\int |f(y) - f(x_0)| P(x_0 - y, t)dy = \int |f(y) - f(x_0)| P_t(x_0 - y)dy$ and it was proved in Theorem 2.1 of Chapter 1 that this last integral tends to zero as $t \to 0$, whenever $x_0 \in \mathscr{L}_f$, the Lebesgue set of f. For $f \in L^p$, $1 \le p \le \infty$, $|\mathbb{R}^n - \mathscr{L}_f| = 0$, hence $u(x, t) \to f(x_0)$ a.e. whenever $(x, t) \to (x_0, 0)$ inside $\Gamma_\alpha(x_0)$. ∇

Proposition 2.4 implies partially Proposition 2.3, in the sense that every function with nontangential limit at a certain point $x_0 \in \mathbb{R}^n$ is nontangentially bounded at x_0, but (2.8) is a more precise statement than boundedness. Nevertheless, the nontangential bound for Poisson integrals has the interest of being easily established through comparison with the Hardy-Littlewood maximal functions. Like for Poisson integrals, the nontangential bound of a function at a point is more easily established than the existence of the nontangential limit at the point, but an important result for harmonic functions in $\mathbb{R}^{n+1}_+$ is that both are equivalent. The corresponding theorem requires a proof as strong as the result itself and we do not give it here. (For the proof see [4], p. 64 or [5], p. 201.)

3. MAXIMAL OPERATORS AND THE SPACE BMO

We indicate $Q \subset \mathbb{R}^n$ any cube of sides parallel to the axes and by $|Q|$ its Lebesgue measure. For every locally integrable f, let $f_Q = |Q|^{-1} \int_Q f(t)dt$ be the mean value of f in Q. If f is a constant C, then $f_Q = C$ for all Q.

<u>Definition 3.1</u>. For $f \in L_{loc}$, let $f^{\#}_Q$ denote the <u>mean oscillation</u> of f in Q,

$$f^{\#}_Q = |Q|^{-1} \int_Q |f(t) - f_Q| dt \tag{3.1}$$

<u>Definition 3.2</u>. For $f \in L_{loc}$, let

$$\Lambda^{\#} f(x) = f^{\#}(x) = \sup_{r>0} f^{\#}_{Q(x,r)} \tag{3.2}$$

where $Q(x, r)$ is a cube of side r centered at x. The operator $\Lambda^{\#} : f \to \Lambda^{\#} f$ will be called the <u>sharp maximal operator</u>.

<u>Definition 3.3</u>. Similarly, for every $f \in L_{loc}$, let

$$\Lambda^{\#}{}'f(x) = \sup_{Q_x} f^{\#}_{Q_x} \tag{3.2a}$$

where Q_x is a cube containing x.

The operator $\Lambda^{\#}{}' : f \to \Lambda^{\#}{}'f$ is equivalent to $\Lambda^{\#}$. In fact, we have

<u>Lemma 3.1.</u> For every $f \in L_{loc}$, and every $x \in \mathbb{R}^n$,

$$\Lambda^{\#} f(x) \leq \Lambda^{\#}{}' f(x) \leq 2^{n+1}\Lambda^{\#} f(x) \tag{3.3}$$

<u>Proof.</u> The first inequality is evident from (3.2) and (3.2a). Let now $x \in Q$ of side ℓ, and let be $Q^* = Q(x, 2\ell)$, so that $Q \subset Q^*$ and $|Q^*| = 2^n|Q|$. Then, for every Q that contains x,

$$|f_Q - f_{Q^*}| = \frac{1}{|Q|}\int_Q (f(t) - f_{Q^*})dt| \leq \frac{1}{|Q|}\int_{Q^*} |f(t) - f_{Q^*}|dt$$

$$= \frac{2^n}{|Q^*|}\int_{Q^*} |f(t) - f_{Q^*}|dt \leq 2^n\Lambda^{\#} f(x)$$

and

$$f^{\#}_Q = \frac{1}{|Q|}\int_Q |f(t) - f_Q|dt \leq \frac{1}{|Q|}\int_Q |f(t) - f_{Q^*}|dt + |f_{Q^*} - f_Q|$$

$$\leq \frac{2^n}{|Q^*|}\int_{Q^*} |f(t) - f_{Q^*}|dt + 2^n\Lambda^{\#}f(x)$$

$$\leq 2^n f^{\#}_{Q^*}(x) + 2^n\Lambda^{\#} f(x) = 2^{n+1}\Lambda^{\#} f(x) \qquad \nabla$$

<u>Lemma 3.2.</u> For every $f \in L_{loc}$ and every $x \in \mathbb{R}^n$,

$$\Lambda^{\#}f(x) \leq 2\Lambda f(x) \quad \text{and} \quad \Lambda^{\#}{}'f(x) \leq 2\Lambda 'f(x) \tag{3.4}$$

Proof.

$$f_Q^{\#} = \frac{1}{|Q|} \int_Q |f(t) - f_Q| dt \leq \frac{1}{|Q|} \int_Q |f(t)| dt + |f_Q|$$

and this is bounded by $2\Lambda f(x)$ or $2\Lambda' f(x)$ respectively if Q is centered at x or merely contains x. ∇

From Lemmas 3.1 and 3.2 we conclude that $\Lambda^{\#}$ and $\Lambda^{\#'}$ share the same type and weak type properties, and that those of Λ (or Λ') imply the same for $\Lambda^{\#}$ (or $\Lambda^{\#'}$).

Recall that if $|f(x)| \leq M$ a.e. in x then $\Lambda f(x) \leq M$ for all x, since $|Q|^{-1}\int_Q |f(t)| dt \leq M$ for all Q. Conversely, if $\Lambda f(x) \leq M$ for all x, then $|f(x)| \leq M$ a.e. in x, since by the Lebesgue differentiation theorem,

$$|f(x)| = \left|\lim_{r \to 0} \frac{1}{|Q(x, r)|} \int_{Q(x, r)} f(t) dt\right|$$

$$\leq \sup_r \frac{1}{|Q(x, r)|} \int_{Q(x, r)} |f(t)| dt \leq M$$

Thus,

$$f \in L^{\infty} \iff \Lambda f \in L^{\infty} \quad \text{and} \quad \|f\|_{\infty} = \|\Lambda f\|_{\infty} \tag{3.5}$$

From (3.4) and (3.5) it follows that $f \in L^{\infty}$ implies $\Lambda^{\#} f \in L^{\infty}$, but the converse may be false (a counterexample is given by the function $\log|x| \in \text{BMO} - L^{\infty}$. See [6] for the proof).

Definition 3.4. A function $f \in L_{loc}$ has bounded mean oscillation if $\Lambda^{\#} f \in L^{\infty}$ and we set

$$\|f\|_{\text{BMO}} = \|\Lambda^{\#} f\|_{\infty} \tag{3.6}$$

Exercise 3.1. Show that $\|\cdot\|_{BMO}$ is a seminorm in L_{loc}. (Hint: $\|f\|_{BMO} = 0$ if and only if $f(t) = C$ a.e.)

Definition 3.4. We define the equivalence relation

$$f_1 \sim f_2 \Longleftrightarrow f_1 - f_2 = \text{constant a.e.}$$

The space BMO is the quotient of the class $\{f \in L_{loc} : \|f\|_{BMO} < \infty\}$ under this equivalence relation, normed by $\|\cdot\|_{BMO}$.

Exercise 3.2. Prove that BMO is a complete (thus, Banach) space. (For reference, see [7]).

By the remark preceding Definition 3.4, $L^\infty \subset$ BMO but $L^\infty \nsupseteq$ BMO. By (3.3), $f \in \text{BMO} \Longleftrightarrow \Lambda^{\#}{}' f \in L^\infty$, i.e., if there exists a finite constant M such that, for every cube Q,

$$|Q|^{-1} \int_Q |f(t) - f_Q|\,dt \leq M \tag{3.7}$$

Lemma 3.3. If $f \in L_{loc}$, then $f \in$ BMO if and only if there exist a finite constant M and a constant C_Q for every Q, such that

$$|Q|^{-1} \int_Q |f(t) - C_Q|\,dt \leq M \tag{3.7a}$$

Proof. It is obvious that (3.7) implies (3.7a). If (3.7a) holds, then $|f_Q - C_Q| = \left||Q|^{-1}\int_Q (f(t) - C_Q)dt\right| \leq M$, so

$$|Q|^{-1}\int_Q |f(t) - f_Q|\,dt \leq |Q|^{-1}\int_Q |f(t) - C_Q|\,dt + |C_Q - f_Q| \leq 2M \quad \nabla$$

For the study of the operator $\Lambda^{\#}$ and the space BMO we need a lemma due to Calderón and Zygmund that extends to $\mathbb{R}^n$, $n > 1$, an earlier result of F. Riesz, valid for $n = 1$. This lemma was proved in [8] to deal with singular integrals, but has become in recent years a useful tool in analysis and will be used in different contexts in this and the following chapter.

Lemma 3.4 (The Calderón-Zygmund lemma). Let $f : \mathbb{R}^n \to \mathbb{C}$ be a positive locally integrable function and $\alpha > 0$, a fixed constant. The space $\mathbb{R}^n$ admits a decomposition $\mathbb{R}^n = P \cup Q$, $P \cap Q = \phi$ such that

(i) $Q = \bigcup_{k=1}^{\infty} Q_k$, where Q_k is a cube, whose interior $\mathring{Q}_k$ satisfies $\mathring{Q}_k \cap \mathring{Q}_j = \phi$ for $k \neq j$;

(ii) $f(x) \leq \alpha$ a.e. $x \in P$;

(iii) $\alpha < |Q_k|^{-1} \int_{Q_k} f(x)dx < 2^n \alpha$ for every Q_k, $k = 1, 2, \ldots$.

Proof. As $f \in L_{loc}$, there is always an ℓ such that $|Q|^{-1} \int_Q f(x)dx < \alpha$ for all cubes Q of side ℓ.

We divide $\mathbb{R}^n$ in a mesh of cubes of sides of length ℓ parallel to the axes. Let Q^0 be such a cube. We next divide each side of Q^0 in two, thus obtaining 2^n cubes contained in Q^0 and of side $\ell/2$. Let Q' be such a cube. There are two possibilities:

$$(1) \quad |Q'|^{-1} \int_{Q'} f \leq \alpha \quad \text{or} \quad (2) \quad |Q'|^{-1} \int_{Q'} f > \alpha$$

In the case (2) we stop the subdivision: Q', conveniently numbered, will belong to the sequence Q_k since it satisfies condition (iii). In fact, if (2),

$$\alpha < \frac{1}{|Q'|} \int_{Q'} f \leq \frac{2^n}{|Q^0|} \int_{Q^0} f \leq 2^n \alpha$$

In the case (1) we go on subdividing Q' into 2^n cubes of sides $\ell/4$, etc. Let Q be the union of all cubes Q_k that are in case (2) in each step of the subdivision process. By construction the Q_k's satisfy both conditions (i) and (iii).

It remains to show that condition (ii) holds for a.e. $x \in P = \mathbb{R}^n - Q$, i.e., $x \notin Q_k$, for all k. This is to say that x belongs to a sequence of cubes, of sidelengths tending to zero, that are all in case (1). By the Lebesgue differentiation Theorem 1.1(a), $f(x) = \lim |Q|^{-1} \int_Q f(t)dt$ a.e. and so, $f(x) \le \alpha$ a.e. ∇

The decomposition $\{P, Q_1, \dots, Q_k, \dots\}$ of $\mathbb{R}^n$ is called the Calderón-Zygmund (C-Z) decomposition of the space, corresponding to a given f and α.

We can now consider, for a fixed $f \ge 0$, the C-Z decomposition as a function of $\alpha > 0$, as in the following two results.

Lemma 3.5. Under the same hypothesis of Lemma 3.4, for a fixed integrable $f \ge 0$, let $\{Q_k^\alpha\}$ be the sequence of cubes corresponding to a given $\alpha > 0$. Then, if $t(\alpha) = \Sigma_{k=1}^{\infty} |Q_k^\alpha|$,

(a) $t(\alpha) \le \|f\|_1/\alpha$;

(b) if Λf is the Hardy-Littlewood maximal function f, its distribution function satisfies $(\Lambda f)_*(5^n\alpha) \le 2^n t(\alpha)$.

Proof.

(a) follows immediately from (iii) of Lemma 3.4.

(b) Let for each $k = 1, 2, \dots,$ Q_k^* be the cube Q_k^α expanded twice its side.

We claim that if $x \notin \bigcup_k Q_k^*$ then for every Q centered in x

$$\frac{1}{|Q|} \int_Q f(t)dt \le 5^n \alpha \tag{3.8}$$

and thus $\Lambda f(x) \leq 5^n\alpha$. Therefore, (3.8) implies

$$(\Lambda f)_*(5^n\alpha) = |\{x : \Lambda f(x) > 5^n\alpha\}| \leq |\bigcup_k Q_k^*| = 2^n \sum_k |Q_k^\alpha| = 2^n t(\alpha)$$

In order to prove (3.8) observe first that it holds obviously for any $Q \subset P$ since then $f \leq \alpha$ in Q and $|Q|^{-1} \int_Q f \leq \alpha < 5^n\alpha$. So it remains to consider the case when there is a k such that $Q_k^\alpha \cap Q \neq \phi$. As the center x of Q is outside $\bigcup_k Q_k^*$ and Q intersects Q_k^α we must have that $Q_k^\alpha \subset Q^*$, where Q^* is Q expanded twice its side, thus $\bigcup\{Q_k^\alpha : Q_k^\alpha \cap Q \neq \phi\} \subset Q^*$. By (ii) and (iii) of Lemma 3.4,

$$\int_Q f = \int_{Q\cap P} f + \sum_{Q_k^\alpha \cap Q \neq \phi} \int_{Q_k^\alpha} f$$

$$\leq \alpha|Q| + \sum_{Q_k^\alpha \cap Q \neq \phi} 2^n\alpha|Q_k^\alpha|$$

$$\leq \alpha|Q| + 2^n\alpha|Q^*|$$

and therefore,

$$|Q|^{-1} \int_Q f \leq \alpha + 2^n\alpha . 2^n < 5^n\alpha \qquad \nabla$$

<u>Remark 3.1</u>. Given a fixed $f \geq 0$ and $0 < \alpha' < \alpha$, we may start the C - Z decomposition of $\mathbb{R}^n$ for both α and α' by the same mesh of cubes where the mean value f_Q of f will be less than or equal to $\alpha' < \alpha$. When subdivided, the mean value of f on one of the smaller cubes, say Q', may be larger than α' but less than or equal to α; then Q' will not be further subdivided for the α'

decomposition, but will be subdivided for the α decomposition. When subdivided, a $Q'' \subset Q'$ can become a member of the α decomposition but will not be one of the α' decomposition. Thus, given $\alpha' < \alpha$ we can choose $\{Q_k^\alpha\}$ and $\{Q_j^{\alpha'}\}$ such that every Q_k^α be contained in a $Q_j^{\alpha'}$.

Lemma 3.6. Let $f \geq 0$ be locally integrable, $\alpha > 0$ and $\beta > 0$ and $t(\alpha)$ as in Lemma 3.5. Then

$$t(\alpha) \leq (\Lambda^{\#\prime} f)_*(\alpha\beta/2) + \beta t(2^{-n-1}\alpha) \tag{3.9}$$

Proof. Let $\alpha' = 2^{-n-1}\alpha < \alpha$. As shown in Remark 3.1, we may choose two C - Z decompositions for α and α' such that every cube of the α decomposition be contained in a cube of the α' decomposition. Let I be the family of cubes $Q_j^{\alpha'}$ of the α' decomposition such that

$$Q_j^{\alpha'} \subset \{x : \Lambda^{\#\prime} f(x) > \alpha\beta/2\} \tag{3.10}$$

and let II be the family of the remaining cubes of the α' decomposition. Let $Q' \in$ II: there is $x \in Q'$ such that $\Lambda^{\#\prime} f(x) \leq \alpha\beta/2$ and, by Definition 3.2,

$$f_{Q'}^{\#} = |Q'|^{-1} \int_{Q'} |f(t) - f_{Q'}|\,dt \leq \frac{\alpha\beta}{2} \tag{3.10a}$$

and, by Lemma 3.4,

$$|f_{Q'}| \leq 2^n \alpha' = \alpha/2 \tag{3.10b}$$

For the Q_k^α's corresponding to the α decomposition, such that $Q_k^\alpha \subset Q'$, by (3.10a) and (3.10b) it holds that

$$\alpha \Sigma |Q_k^\alpha| \leq \Sigma \int_{Q_k^\alpha} |f - f_{Q'}| + |f_{Q'}| \Sigma |Q_k^\alpha|$$

$$\leq \int_{Q'} |f - f_{Q'}| + |f_{Q'}| \Sigma |Q_k^\alpha|$$

$$< \frac{\alpha\beta}{2} |Q'| + \frac{\alpha}{2} \Sigma |Q_k^\alpha|$$

or

$$\Sigma |Q_k^\alpha| < \beta |Q'|$$

where the sum is taken over all $Q_k^\alpha \subset Q'$. Thus,

$$\sum_{Q' \in II} \left(\sum_{Q_k^\alpha \subset Q'} |Q_k^\alpha| \right) < \beta \Sigma |Q'| = \beta t(\alpha') = \beta t(2^{-n-1}\alpha)$$

On the other hand, by (3.10),

$$\sum_{Q' \in I} \left(\sum_{Q_k^\alpha \subset Q'} |Q_k^\alpha| \right) \leq \sum_{Q' \in I} |Q'| \leq (\Lambda^{\#\prime} f)_*(\alpha\beta/2)$$

Thus

$$t(\alpha) = \left(\sum_{Q' \in I} + \sum_{Q' \in II} \right) \left(\sum_{Q_k^\alpha \subset Q'} |Q_k^\alpha| \right)$$

$$\leq \beta t(2^{-n-1}\alpha) + (\Lambda^{\#\prime} f)_*(\alpha\beta/2) \qquad \nabla$$

We state now the maximal theorem for $\Lambda^\#$. This analog to Theorem 1.2, due to C. Fefferman and E. M. Stein [9], will enable us to interpolate between L^p and BMO spaces.

<u>Theorem 3.7 (Sharp maximal theorem).</u> Let $1 \le p_0 \le p$, $1 < p < \infty$. For every $f \in L^{p_0}$ there exists a constant C_p, independent of f, depending only on n and p, such that

$$\|\Lambda^{\#} f\|_p \le C_p \|f\|_p \quad \text{and} \quad \|f\|_p \le C_p \|\Lambda^{\#} f\|_p \tag{3.11}$$

and thus

$$f \in L^p \Longleftrightarrow \Lambda^{\#} f \in L^p \tag{3.11a}$$

<u>Proof.</u> Since by (3. 4) $\Lambda^{\#} f \le 2\Lambda f$, the Hardy-Littlewood maximal theorem 1. 2 implies that we only have to prove the second inequality in (3.11). Thus, as $\|f\|_p \le \|\Lambda f\|_p$ by Corollary 1. 5, by (3. 3) it will be sufficient to show that, for $1 < p < \infty$,

$$\|\Lambda f\|_p \le B_p \|\Lambda^{\#'} f\|_p \tag{3.12}$$

By Lemma 3. 5(b) (that can be applied since $f \in L^{p_0}$, $p_0 < \infty$, and is therefore locally integrable),

$$\int (\Lambda f(x))^p dx = p \int_0^\infty \alpha^{p-1} (\Lambda f)_*(\alpha) d\alpha$$

$$\le 5^{np} 2^n p \int_0^\infty \alpha^{p-1} t(\alpha) d\alpha \tag{3.13}$$

and by Lemma 3. 6,

$$\int_0^\infty \alpha^{p-1} t(\alpha) d\alpha \le \int_0^\infty \alpha^{p-1} (\Lambda^{\#'} f)_*(\alpha\beta/2) d\alpha + \beta \int_0^\infty \alpha^{p-1} t(2^{-n-1}\alpha) d\alpha$$

$$= (2/\beta)^p \int_0^\infty \alpha^{p-1}(\Lambda^{\#}{}'f)_*(\alpha)d\alpha + 2^{(n+1)p}\beta \int_0^\infty \alpha^{p-1}t(\alpha)d\alpha \tag{3.14}$$

Choose $\beta = 2^{-(n+1)p-1}$ and substract the last term of the sum in (3.14) to the term at the left,

$$\int_0^\infty \alpha^{p-1}t(\alpha)d\alpha \leq A_p \int_0^\infty \alpha^{p-1}(\Lambda^{\#}{}'f)_*(\alpha)d\alpha$$

$$= p^{-1}A_p \int_{\mathbb{R}^n} (\Lambda^{\#}{}'f(x))^p dx \tag{3.15}$$

Now (3.12) follows from (3.13) and (3.15). ∇

Many classical operators T, like the Hilbert transform, which will be treated in Chapter 6, are bounded from L^p to L^p for $1 < p < \infty$, but not for $p = 1$ or $p = \infty$. Since the substitute result for $p = 1$ is $T : L^1 \to L^1_*$, and $L^\infty_* = L^\infty$ by definition, one may ask for a substitute result for $p = \infty$. We shall see that this will be $T : L^\infty \to BMO$, since BMO plays with respect to L^∞ the same role of L^p_* with respect to L^p, for $1 \leq p < \infty$. The next theorem of C. Fefferman and E. M. Stein is, thus, an extension of the Marcinkiewicz interpolation theorem for the endpoint (∞, ∞).

Theorem 3.8 (Interpolation theorem between L^p and BMO). Let $1 < p < \infty$ and T be a linear operator, continuous from L^p into itself and from L^∞ into BMO, i.e.,

$$T : L^p \to L^p \quad \text{and} \quad T : L^\infty \to BMO \quad \text{continuously}$$

Then $T : L^q \to L^q$ continuously for all $p < q < \infty$.

Proof: Let T_1 be defined by $T_1 f = \Lambda^{\#}(Tf)$. If $f \in L^\infty$ then, by hypothesis, $Tf \in BMO$ and by (3.6), $\Lambda^{\#}(Tf) \in L^\infty$. Thus, $T_1 : L^\infty \to L^\infty$. Also $T_1 : L^p \to L^p$, since $f \in L^p$ implies $Tf \in L^p$ and $\Lambda^{\#}(Tf) \in L^p$, $1 < p < \infty$. By the M. Riesz-Thorin interpolation theorem, $T_1 : L^q \to L^q$ for all q such that $p < q < \infty$, i.e., if $f \in L^q$ then $\Lambda^{\#}(Tf) \in L^q$ and, by (3.11a), this implies that $Tf \in L^q$ and $T : L^q \to L^q$ boundedly. ∇

In the remainder of the section we give some basic properties of the functions in BMO. For more details see [10].

Lemma 3.9. If $f \in BMO$ then $f(x)(1 + |x|^{n+1})^{-1} \in L^1(\mathbb{R}^n)$ and, more precisely, if $Q = Q(0, 1)$, then

$$\int_{\mathbb{R}^n} \frac{|f(x)-f_Q|}{1+|x|^{n+1}}\, dx \leq A\,\|f\|_{BMO} \tag{3.16}$$

Thus the Poisson integral of f, $P_t * f(x)$, exists for every $(x, t) \in \mathbb{R}^{n+1}_+$.

Proof. Let I be the integral in (3.16) and let $I = I_0 + \Sigma_{k=1}^{\infty} I_k$, where

$$I_k = \int_{S_k} |f(x) - f_Q|(1 + |x|^{n+1})^{-1} dx, \quad S_k = Q_k - Q_{k-1}$$

$$Q_k = Q(0, 2^k)$$

As

$$I_0 = \int_Q |f(x) - f_Q|(1 + |x|^{n+1})^{-1} dx \leq C\,\|f\|_{BMO}$$

it will be enough to prove that $I_k \leq C_k \|f\|_{BMO}$ with $\Sigma_k C_k < \infty$.

Since $x \in S_k$ implies $|x| > 2^{k-1}$ and $1 + |x|^{n+1} > 1 + 2^{(k-2)(n+1)} >$ $> 4^{-n-1}2^{k(n+1)}$, we have

$$I_k \leq 4^{n+1}2^{-(n+1)k} \int_{Q_k} |f(x) - f_Q|\,dx \tag{3.17}$$

But

$$\begin{aligned}\int_{Q_k} |f(x) - f_Q|\,dx &\leq \int_{Q_k} (|f(x) - f_{Q_k}| + |f_{Q_k} - f_Q|)\,dx \\ &\leq |Q_k|\,\|f\|_{BMO} + |Q_k|\,|f_{Q_k} - f_Q| \\ &= 2^{kn}(\|f\|_{BMO} + |f_{Q_k} - f_Q|) \qquad (3.18)\end{aligned}$$

Since

$$\begin{aligned}|f_{Q_k} - f_{Q_{k-1}}| &= \left| |Q_{k-1}|^{-1} \int_{Q_{k-1}} f(x)\,dx - f_{Q_k} \right| \\ &= \frac{2^n}{2^{kn}} \left| \int_{Q_{k-1}} (f(x) - f_{Q_k})\,dx \right| \\ &\leq 2^n |Q_k|^{-1} \int_{Q_k} |f(x) - f_{Q_k}|\,dx \\ &\leq 2^n \|f\|_{BMO}\end{aligned}$$

it follows that

$$|f_{Q_k} - f_Q| \leq |f_{Q_k} - f_{Q_{k-1}}| + |f_{Q_{k-1}} - f_{Q_{k-2}}| + \ldots \leq k2^n \|f\|_{BMO}$$

and from (3.17) and (3.18) we obtain that

$$I_k \le 4^{n+1} 2^{-(n+1)k} 2^{nk} (1 + k2^n) \|f\|_{BMO}$$

$$\le C_k \|f\|_{BMO}$$

where $C_k = c_n 2^{-k}$ and $\Sigma_k C_k < \infty$. ∇

Let now be, if $|A|$ is the Lebesgue measure of the set A,

$$\mu_Q(\alpha) = |\{x \in Q : |f(x) - f_Q| > \alpha\}| \tag{3.19}$$

Lemma 3.10. If there exist two constant B, b such that, for all cubes Q,

$$\mu_Q(\alpha) \le B|Q|e^{-b\alpha} \tag{3.20}$$

then $f \in BMO$.

Proof. If (3.20) holds, thus for every Q, $\int_Q |f(x) - f_Q| dx \le \int_0^\infty \mu_Q(\alpha) d\alpha \le Bb^{-1}|Q|$ and $f \in BMO$. ∇

In fact, property (3.20) characterizes BMO, as shown in the following result, due to F. John and L. Nirenberg, who proved it in [6], where the BMO space was first introduced. The proof we are going to give is due to Calderón and is taken from [7].

Theorem 3.11 (John-Nirenberg). There exist two constants B and b, depending only on the dimension n, such that for every $f \in BMO$, and every $Q \subset \mathbb{R}^n$ and $\alpha > 0$,

$$\mu_Q(\alpha) \le B|Q|\exp(-b\alpha/\|f\|_{BMO}) \tag{3.20a}$$

where $\mu_Q(\alpha)$ is given by (3.19).

Proof. If $\alpha < \|f\|_{BMO}$ then (3.20a) holds, taking $B = e$, $b = 1$, since $\mu_Q(\alpha) \leq |Q|$ always. If $\alpha > \|f\|_{BMO}$, we consider a fixed cube Q_0 and assume that $f_{Q_0} = 0$ (if this is not the case we deal with $g = f - f_{Q_0}$, that has $g_{Q_0} = 0$ and $\|g\|_{BMO} = \|f\|_{BMO}$). Let be $\mu(\alpha) = |\{x \in Q_0 : |f(x)| > \alpha\}| = |E_\alpha|$. As $f \in L^1(Q_0)$, for each $\lambda > \|f\|_{BMO}$, Lemma 3.4 provides a λ C-Z decomposition of Q_0 given by $\{P^\lambda, Q_1^\lambda, \ldots, Q_k^\lambda, \ldots\}$, such that

$$\lambda < |Q_k^\lambda|^{-1} \int_{Q_k^\lambda} |f|dx \leq 2^n\lambda$$

Since $f \in BMO$ and $\|f\|_{BMO} < \lambda$ we may obtain a better estimate, namely, that for every Q_k^λ,

$$\lambda < |Q_k^\lambda|^{-1} \int_{Q_k} |f|dx \leq 2^n \|f\|_{BMO} + \lambda \tag{3.21}$$

In fact, Q_k^λ is a component of the λ decomposition of Q_0 that comes from subdividing a Q such that $m = |Q|^{-1}\int_Q |f|dx \leq \lambda$, so

$$\begin{aligned} |Q_k^\lambda|^{-1} \int_{Q_k^\lambda} |f|dx &\leq |Q_k^\lambda|^{-1} \int_{Q_k^\lambda} |f - m|dx + m \\ &\leq 2^n |Q|^{-1} \int_Q |f - m|dx + \lambda \\ &\leq 2^n \|f\|_{BMO} + \lambda \end{aligned}$$

It was shown in Remark 3.1 that if $\lambda < \nu$, we can choose the λ and ν decompositions of Q_0 as to insure that for each j there is a k such that $Q_j^\nu \subset Q_k^\lambda$.

Assume $\|f\|_{BMO} = 1$ and take $\nu = \lambda + 2^{n+1}\|f\|_{BMO} = \lambda + 2^{n+1} > \lambda$. We want to estimate $\Sigma_j |Q_j^\nu|$ in terms of $\Sigma_k |Q_k^\lambda|$. For every Q_j^ν there is a $Q_k^\lambda \supset Q_j^\nu$ and $m_\lambda = |Q_k^\lambda|^{-1}\int_{Q_k^\lambda} |f|\,dx \le \lambda + 2^n\|f\|_{BMO} < \nu$, by (3.21). Furthermore,

$$\nu < |Q_j^\nu|\int_{Q_j^\nu} |f|\,dx \le |Q_j^\nu|^{-1}\int_{Q_j^\nu} |f(x) - m_\lambda|\,dx + m_\lambda$$

$$\le |Q_k^\lambda|\,|Q_j^\nu|^{-1}(|Q_k^\lambda|^{-1}\int_{Q_k^\lambda} |f(x) - m_\lambda|\,dx) + m_\lambda$$

$$\le |Q_k^\lambda|\,|Q_j^\nu|^{-1}\|f\|_{BMO} + \lambda + 2^n \tag{3.22}$$

As $\nu = \lambda + 2^{n+1}$, (3.22) yields $|Q_j^\nu| \le 2^{-n}|Q_k^\lambda|$.

Adding up the cubes in the ν decomposition this gives

$$\sum_j |Q_j^\nu| \le 2^{-n}\sum_k |Q_k^\lambda| \tag{3.23}$$

We started with $\alpha > \|f\|_{BMO} = 1$. Let $r = [(\alpha - 1)2^{-n-1}]$, where $[t]$ denotes the integral part of t. Thus if $\nu = 1 + 2^{n+1}r$, then $1 \le \nu \le \alpha$ and then $E_\alpha \subset E_\nu \subset Q^\nu \cup Z$, where $|Z| = 0$ and $Q^\nu = \bigcup_j Q_j^\nu$. Thus $\mu(\alpha) = |E_\alpha| \le |Q^\nu|$ for $\nu = 1 + 2^{n+1}r$. Applying (3.23) r times this gives

$$\mu(\alpha) \le 2^{-nr}\sum_k |Q_k^\lambda| \le 2^{-nr}|Q_0|$$

We take $B = 2^{(n/2^{n+1})+n}$ and $b = (\log 2)^{n2^{-n-1}}$.

Then

$$B|Q_0|\exp(-b\alpha) = 2^n|Q_0|2^{n(1-\alpha)2^{-n-1}}$$

$$\geq 2^n|Q_0|2^{-n(r+1)} = 2^{-nr}|Q_0|$$

$$\geq \mu(\alpha)$$

and (3.20a) is proved. ∇

Corollary 3.12. If for each p, $1 < p < \infty$,

$$\|f\|_{BMOp} = \sup_Q(|Q|^{-1}\int_Q |f(x) - f_Q|^p dx)^{1/p} \tag{3.24}$$

then

$$\|f\|_{BMOp} \leq C\|f\|_{BMO} \leq C\ \|f\|_{BMOp} \tag{3.25}$$

Proof. The second inequality in (3.25) is an immediate consequence of Hölder's inequality, since

$$\int_Q |f - f_Q|dx \leq (\int_Q |f - f_Q|^p dx)^{1/p}|Q|^{1/p'}$$

and thus

$$f_Q^{\#} = |Q|^{-1}\int_Q |f - f_Q|dx \leq |Q|^{-1/p}(\int_Q |f - f|^p dx)^{1/p}$$

On the other hand, by Theorem 3.11,

$$f_Q^{\#} = |Q|^{-1}\int_Q |f - f_Q|^p dx = |Q|^{-1}p\int_0^\infty \alpha^{p-1}\mu(\alpha)d\alpha$$

$$\leq pB\int_0^\infty \alpha^{p-1}\exp(-b\alpha/\|f\|_{BMO})d\alpha$$

$$= C_p(\|f\|_{BMO})^p$$

and the first inequality in (3.25) follows. ∇

Remark 3.2. We know that if $f \in L^\infty \cap L^1$ then $f \in L^p$, $1 < p < \infty$. A similar result holds for $BMO \supset L^\infty$, namely if $f \in BMO \cap L^1$ then $f \in L^p$, $1 < p < \infty$, and furthermore $\|f\|_p \leq C_p \|f\|_1^{1/p} \|f\|_{BMO}^{1/p'}$. This fact can be proved using (a refinement of) Theorem 3.11.

4. THE METHOD OF MAXIMAL FUNCTIONS

In Section 1 we dealt with a sequence of operators L_r, $r > 0$, with the following properties: (a) for each $r > 0$, $L_r f$ is defined for every $f \in L^p$, for all $1 \leq p \leq \infty$ (since $L_r f = f * \rho_r$ and $\rho_r \in L^1$); (b) if D is the class of all continuous functions with compact support then, for each $\phi \in D$, $\lim_{r\to 0} L_r \phi(x) = T\phi(x)$ exists (here $T\phi = \phi$). The Hardy-Littlewood maximal theorem asserts, in particular, that the maximal operator Λ of the family $\{L_r\}$ is of weak type (p, p) for all p, $1 \leq p \leq \infty$. From this property of Λ follows Theorem 1.1, using (a) and (b). Many theories in analysis deal with sequences of linear operators $\{T_r\}_{r\in R}$ where R is a (continuous) family of indexes, such that the operators T_r satisfy properties (a) and (b). Here we introduce the maximal operator M of the sequence $\{T_r\}_{r\in R}$, defined by $M : f \to Mf$, where

$$Mf(x) = \sup_{r\in R} |T_r f(x)| \tag{4.1}$$

is the maximal function of the sequence $\{T_r f(x)\}$.

In the above mentioned theories one of the main goals is to prove the following propositions.

Proposition I. For every $f \in L^p$, $1 \leq p < \infty$, the sequence of functions $\{T_r f(x)\}$ converges as $r \to 0$ pointwise a.e. to a finite limit $Tf(x)$.

Proposition II. For every $f \in L^p$, $1 < p < \infty$, the sequence of functions $\{T_r f(x)\}$ converges as $r \to 0$ to Tf in the L^p norm, i.e., $\|T_r f - Tf\|_p \to 0$ as $r \to 0$.

Proposition III. The limit operator T, $T : f \to Tf$, is of type (p, p), $1 < p < \infty$, and of weak type $(1, 1)$.

Proposition IV. If M is the maximal operator of $\{T_r\}$, as defined in (4.1), then M is of type (p, p), $1 < p < \infty$, and of weak type $(1, 1)$.

Proposition IIa (Substitute for Proposition II for $p = 1$). For every $f \in L^1$, $T_r f \to Tf$ in the norm of $L^s(K)$, for $0 < s < 1$ and K a set of finite measure (Kolmogoroff's property). For every $f \in L \log^+ L$, $T_r f \to Tf$ in the norm of $L^1(K)$, K a set of finite measure (Zygmund's property).

Remark 4.1. The validity of Proposition I implies that for every $f \in L^p$, $1 \leq p < \infty$, both $Tf(x)$ and $Mf(x)$ are finite a.e., and so the statements of Propositions III and IV make sense.

In the summability theory of Fourier series and integrals, in the theory of differentiation of integrals, in the theory of conjugate functions and Hilbert transforms, in ergodic theory, sequences of operators $\{T_r\}$ appear, and a goal of these theories is to establish the validity of Propositions I-IV for the different cases. Let us observe that in ergodic theory we deal with $\{T_r\}$ for $r \to \infty$ and the theory holds unchanged for $r \to \infty$ as well as for R a discrete family of indexes.

In the case of the sequence $\{T_r = L_r\}$ of Section 1, we deduced the properties corresponding to Propositions I and II (Theorem 1.1) from the properties of Λ in the Hardy-Littlewood theorem 1.2, that is, from the properties corresponding to Propositions III and

IV. This idea applies to the general case, and the "method of maximal functions" that follows, greatly simplifies the proofs of Propositions I, II and IIa.

Theorem 4.1. Let $\{T_r\}$ be a family of linear operators that satisfies the following conditions:

(A) The maximal operator M of the family is of weak type (p, p) for all p, $1 \leq p < \infty$;

(B) there exists a q, $1 \leq q < \infty$, and a dense subset $D \subset L^q$ such that $\{T_r \phi(x)\}$ converges to $T\phi(x)$ a.e. for every $\phi \in D$.

Then Propositions I, II, IIa, III and IV hold for the family $\{T_r\}$.

Remark 4.2. The proof involves several steps: first to prove Proposition IV, using (A) and the Marcinkiewicz interpolation theorem, and then prove Proposition I. Then we use Propositions IV and I to conclude the proofs of Propositions II, IIa and III.

Remark 4.3. The proof of Proposition I (Step 2 of the proof below) is the same as the one given for Theorem 1.1.

Proof. (Step 1) By the Marcinkiewicz interpolation theorem, condition (A) for $p = 1$ and any $p_0 < \infty$ implies the type (p, p) of M for all p, $1 < p < \infty$. So Proposition IV holds.

(Step 2) Let us first prove Proposition I for L^q, the space of condition (B). If this is done, as $D_1 = L^q \cap L^p$ is a dense subset of every L^p, $1 \leq p < \infty$, by the same argument, Proposition I will be valid for every L^p, $1 \leq p < \infty$.

Given $f \in L^q$, let

$$R(f, x) = \lim_{r_1, r_2} \sup |T_{r_1} f(x) - T_{r_2} f(x)| \tag{4.2}$$

We claim that $R(f;x) = 0$ a.e. in x. From (4.2) it follows that $R(f;x) \leq 2Mf(x)$. Since D is dense in L^q, there exists a $\phi \in D$ such that $\|f - \phi\|_q < \eta$, for any arbitrary $\eta > 0$. Thus, as from (4.2), $R(f,x) \leq R(\phi;x) + R(f - \phi;x)$ and, as from condition (B), $R(\phi;x) = 0$ a.e. in x, it is

$$\begin{aligned} R(f;x) &\leq R(f - \phi;x) \\ &\leq 2M(f - \phi)(x) \quad \text{a.e.} \end{aligned} \tag{4.3}$$

Since by condition (A) M is an operator of weak type (p, p) for all $p \geq 1$, M is, in particular, of weak type (q, q). Thus, as $q < \infty$,

$$(Mf)_*(\alpha) \leq C\alpha^{-q}\|f\|_q^q, \quad \alpha > 0 \tag{4.4}$$

for every $f \in L^q$. By (4.3) the measure of the set $\{x : R(f;x) > \alpha\}$ will be bounded by

$$\begin{aligned} (M(f - \phi))_*(\alpha/2) &\leq C2^q\alpha^{-q}\|f - \phi\|_q^q \\ &< C2^q\alpha^{-q}\eta^q \end{aligned}$$

by (4.4), for every $\eta > 0$. Therefore $R(f;x) = 0$ a.e. in x. So Proposition I is proved.

(Step 3) Let now be $p > 1$ and we shall prove Proposition II. By Proposition I, for every $f \in L^p$,

$$T_r f(x) \to Tf(x) \quad \text{a.e.}$$

or equivalently,

$$|T_r f(x) - Tf(x)|^p \to 0 \quad \text{a.e.}$$

and, since we want to prove that

$$\int |T_r f(x) - Tf(x)|^p dx \to 0$$

it will be enough to take limits under the integral sign.

This will be possible, by the Lebesgue dominated convergence theorem, if the sequence $\{|T_r f(x) - Tf(x)|^p\}_r$ is bounded by an integrable function. But this is precisely the case, since $|T_r f(x)| \leq Mf(x)$, for all r, and $|Tf(x)| \leq Mf(x)$. Thus

$$|T_r f(x) - Tf(x)|^p \leq (2Mf(x))^p$$

and, since by Proposition I, M is of type (p, p) for $p > 1$, $(2Mf)^p \in L^1$ for $f \in L^p$ and Proposition II is proved.

(Step 4) The same reasoning as in Step 3 plus the fact that, under the hypotheses, M satisfies Kolmogoroff's and Zygmund's conditions (cfr. Section 6 of Chapter 4) and thus provides a majorant for $\{T_r f\}$ in $L^s(K)$ and $L^1(K)$ respectively, proves Proposition IIa.

(Step 5) From their definitions follows that $|Tf(x)| \leq Mf(x)$ for every $f \in L^p$, $1 \leq p < \infty$, and almost every x. Thus, T inherits the type and weak type properties from M, and Proposition III follows from Proposition IV. ∇

<u>Remark 4.4.</u> Proposition I still holds if hypothesis (A) is changed into

(A') M is of weak type (q, s), for q as in (B), and $1 \leq s < \infty$.

The proof of Step 2 goes unchanged.

Remark 4.5. If hypothesis (A) is changed to

(A'') M is of weak type (p, p) for all p, $1 \le p \le p_0$, $1 < p_0 < \infty$, then, without altering the proof of Theorem 4.1, Propositions I to IV are valid for every L^p, $1 < p \le p_0$.

Remark 4.6. In the cases when in addition to (A), M is of type (∞, ∞), as happens for $M = \Lambda$, the Hardy-Littlewood maximal operator, then Propositions I to IV are valid for every L^p, $1 < p \le \infty$.

The differentiability theory developed in Section 1 is an application of the method of maximal functions. In that case, $T_r = L_r$ for $r \in \mathbb{R}^+$, $M = \Lambda$ the Hardy-Littlewood maximal operator, $D = C_0$ and $q = 1$. An application to ergodic theory is developed in the next section.

*Remark 4.7. In the above formulation of the method of maximal functions we have followed [11], where a more precise theory is developed. Let us fix p, $1 \le p < \infty$. If T is of weak type (p, q) for some q, $1 \le q < \infty$, then $\alpha^q (Tf)_*(\alpha) \le (M\|f\|_p)^q$, and therefore, given a sequence $\{f_k\}$,

$$\|f_k\|_p \to 0 \Longrightarrow (Tf_k)_*(\alpha) \to 0 \quad \text{for all} \quad \alpha > 0 \tag{4.5}$$

We shall say that a sublinear operator T is of meager type p if (4.5) is satisfied for every sequence $\{f_r\}$ in L^p.

*Exercise 4.1. If T is an operator of meager type p and $f \in L^p$ prove that $\lim_{\alpha \to \infty} (Tf)_*(\alpha) = 0$.

*Exercise 4.2. If T is an operator of meager type p then the set

$$N_{\alpha\varepsilon} = \{f \in L^p : (Tf)_*(\alpha) \le \varepsilon\} \tag{4.6}$$

is closed in L^p, for every $\varepsilon > 0$, $\alpha > 0$ fixed.

*Exercise 4.3. Given two measure spaces $(\mathscr{X}, \mu)$ and $(\mathscr{Y}, \nu)$ with $\nu(\mathscr{Y}) < \infty$, and T acting from μ-measurable functions defined in $\mathscr{X}$ into ν-measurable functions defined in $\mathscr{Y}$, prove that $\lim_{\alpha\to\infty} (Tf)_*(\alpha) = 0$ is equivalent to $|Tf(y)| < \infty$ a.e. In the general case $\lim_{\alpha\to\infty} (Tf)_*(\alpha) = 0$ implies $|Tf(y)| < \infty$ a.e.

*Proposition 4.2. The operator T is of meager type p if and only if the two following conditions are satisfied: (a) if $f \in L^p$ then $(Tf)_*(\alpha) \to 0$ as $\alpha \to \infty$, (b) the set $N_{\alpha\varepsilon}$ defined in (4.6) is closed in L^p, for every α and $\varepsilon > 0$.

Proof. By Exercises 4.1 and 4.2 it remains only to prove that (a) and (b) imply (4.5). Let us fix $\varepsilon > 0$. By (a), for every $f \in L^p$ there is a λ such that $(Tf)_*(\lambda) < \varepsilon$, so that $f \in N_{\lambda\varepsilon}$ and $L^p = \bigcup_{\lambda=1}^{\infty} N_{\lambda\varepsilon}$. By (b) and Baire's theorem, there is a $N_{\lambda_0\varepsilon}$ which contains a ball of center f_0 and radius $\rho_0 > 0$. Thus, for every g, $\|g\|_p \le 1$, we have $(T(f_0 + \rho_0 g))_*(\lambda_0) \le \varepsilon$ and $(Tf_0)_*(\lambda_0) \le \varepsilon$. Since $(T(\rho_0 g))_*(2\lambda_0) \le (T(f_0 + \rho g_0))_*(\lambda_0) + (Tf_0)_*(\lambda_0)$ it follows that there is a λ_1 such that $(Tg)_*(\lambda_1) \le 2\varepsilon$ for all g, $\|g\|_p \le 1$ and hence, for such g, $(Tg)_*(\lambda) \le 2$ for all $\lambda > \lambda_1$. From this and $(T(\varepsilon g))_*(\alpha) = (Tg)_*(\alpha/\varepsilon)$, it follows that $\|g_k\|_p \to 0$ implies $(Tg_k)_*(\alpha) < \varepsilon$ for every $\alpha > 0$ and for large enough k. This proves (4.5). ∇

*Exercise 4.4. Let $\{T_k\}$ be a sequence of operators such that $T_1 f(x) \le T_2 f(x) \le \dots \uparrow Tf(x)$ for each $f \in L^p$. If each T_k is of meager type p and if T satisfies condition (a) of Proposition 4.2, prove that T is also of meager type p.

Now let $\{T_\varepsilon\}_{\varepsilon>0}$ be a sequence of operators such that: (1) for each $\varepsilon > 0$, T_ε is of weak type (p, p) for every p, $1 \leq p < \infty$; (2) for $\phi \in D$, where D is a dense subset of every L^p, $1 \leq p < \infty$, $T_\varepsilon \phi(x) \to T\phi(x)$ a. e. Let M be the maximal operator of the sequence $\{T_\varepsilon\}$.

*Proposition 4.3. Let $\{T_\varepsilon\}$ satisfy (1) and (2) and M be its maximal operator.

(i) If M is of meager type p, $1 \leq p < \infty$, then $T_\varepsilon f(x) \to Tf(x)$ a. e. for every $f \in L^p$.

(ii) If M is of type (p, p), $1 < p < \infty$, then $T_\varepsilon f \to Tf$ in the L^p norm.

Proof. (ii) was already proved as part of Theorem 4.1. (i) is proved in a way similar to Proposition I and we leave it to the reader as an exercise. ∇

*Exercise 4.5. Prove that the maximal operator M is of meager type p if and only if $\lim_{\alpha\to\infty} (Mf)_*(\alpha) = 0$ for all $f \in L^p$.

5. ERGODIC THEOREMS

As an important example of the theory of convergence of operators introduced in the preceding section we present, in a context similar to the one leading to the Lebesgue theorem on differentiation of integrals, several results on ergodic theory, following the unified approach developed by M. Cotlar in [11] and [12].

In $\mathbb{R}^1$ the differential operator L_r can be written as

$$L_r f(x) = \frac{1}{r}\int_0^r f(x - t)dt = \frac{1}{r}\int_0^r f(\tau_t x)dt \tag{5.1}$$

where τ_t is the translation operator in $t \in \mathbb{R}^1$.

The translation is a transform of $\mathbb{R}^1$ into $\mathbb{R}^1$ such that every interval--and therefore every measurable set--is transformed into another one of same measure. Furthermore, the translations form a group:

$$\tau_{t+s}x = s + (t + s) = (x + t) + s = \tau_s(\tau_t x) = \tau_s\tau_t x$$

This situation is a particular case of the following more general one: let (X, μ) be a general measure space and let $\{\sigma_t\}$, $\sigma_t : X \to X$, $t \in \mathbb{R}$, be a collection of operators such that

(1) $\{\sigma_t\}$ is a <u>group</u>: $\sigma_{t+s}x = \sigma_t(\sigma_s x)$, $\sigma_0 x = x$, $\forall x \in X$;

(2) each σ_t is <u>isomeasurable</u>:
if $E \subset X$ and $E_t = \{y = \sigma_t x : x \in E\}$ then E_t is measurable and $\mu(E_t) = \mu(E)$ for all $t \in \mathbb{R}$;

(3) $\{\sigma_t\}$ is a <u>measurable group</u>:
if $f(x)$ is a measurable function of x (in X) then $f(x, t) = f(\sigma_t x)$ is a measurable function of (x, t) (in $X \times \mathbb{R}$).

(More generally, a parameter $t \in \mathbb{R}^n$ can be considered.)

Each function $f(x)$ gives rise two a 2 variables function $f(x, t) = f(\sigma_t x)$; fixing x, it can be considered as a function of t, $f_x(t)$. In particular, if $f(x) = \chi_E(x)$ is the characteristic function of a set $E \subset X$, then $f(\sigma_{-t}x) = \chi_{\sigma_t E}(x)$, that is, $\chi_E(\sigma_{-t}x) = \chi_{\sigma_t E}(x)$ (in fact, $x \in \sigma_t E$ implies that $\sigma_{-t}x \in E$.). From this we can conclude that

(2a) each σ_t is <u>isointegrable</u>:
if $f(x)$ is an integrable function of x then $f(x, t) = f(\sigma_t x)$ is an integrable function of (x, t) and

$$\int_X f_t(x)d\mu = \int_X f(\sigma_t x)d\mu = \int_X f(x)d\mu \tag{5.2}$$

i.e., the integral is invariant with respect to μ.

Proof. If $f(x) = \chi_E(x)$, E measurable, then by (2),

$$\int_X f(\sigma_t x)d\mu = \int_X \chi_{\sigma_{-t}E}(x)d\mu = \mu(\sigma_{-t}E) = \mu(E) = \int_X \chi_E(x)d\mu$$

So (5. 2) holds for characteristic functions and, therefore, for simple functions and for integrable functions. ∇

The same proof lets us conclude that, for each t,

$$\int_E f(\sigma_t x)d\mu = \int_{\sigma_t E} f(x)d\mu \tag{5.2a}$$

We can now introduce the operators $\{T_r\}_{r>0}$ defined by

$$T_r f(x) = \frac{1}{r}\int_0^r f(\sigma_t x)dt \tag{5.3}$$

Definition 5.1. Given $x \in X$ fixed, the path of x is the set $\{\sigma_t x : t \in \mathbb{R}\}$.

Ergodic theory studies the properties of paths.

A basic problem is to determine the frequency with which the path of a point x enters a given set E. Since $\sigma_t x \in E$ if and only if $\chi_E(\sigma_t x) = 1$, if we consider m instants $t_1, \ldots, t_m$ and the corresponding points in the path, $\sigma_{t_1} x, \ldots, \sigma_{t_m} x$, the sum $\Sigma_{k=1}^m \chi_E(\sigma_{t_k} x)$ gives the number of times $\sigma_{t_k} x \in E$. By analogy it is said that the integral

$$\int_0^r \chi_E(\sigma_t x)dt$$

represents the "number" of times the path of x enters E in the interval $(0, r)$. Therefore, the mean frequency of permanence of the path of x in E during $(0, r)$ is defined by

$$T_r \chi_E(x) = \frac{1}{r} \int_0^r \chi_E(\sigma_t x)dt \qquad (5.4)$$

Substituting the characteristic function by a general f, we define the <u>mean frequency with respect to f</u> as $T_r f(x)$.

For formula (5.3) to make sense we must prove the integrability of $f(\sigma_t x)$ with respect to t.

<u>Lemma 5.1.</u> Let $f(x)$ be an integrable function in (X, μ). Then $f(x, t) = f(\sigma_t x) = f_x(t)$ is, for almost every x, an integrable function of t in every finite interval, i.e.,

$$F(x) = \int_0^r f_x(t)dt < \infty \qquad (5.5)$$

<u>Proof.</u> It is enough to consider the case $f \geq 0$. By condition (3), $f(x, t)$ is a measurable function of (x, t) and, being positive, $F(x)$ is defined a.e. in x. We still have to show that $F(x)$ is finite, but from the isomeasurability, we have that

$$\int_X F(x)d\mu = \int_X \left(\int_0^r f(\sigma_t x)dt\right)d\mu(x) = \int_0^r \left(\int_X f(\sigma_t x)d\mu(x)\right)dt$$

$$= \int_0^r \left(\int_X f(x)d\mu(x)\right)dt = r\left(\int_X f d\mu\right) < \infty$$

therefore $F(x) < \infty$ a.e. ▽

The operators $\{T_r\}_{r>0}$, that are thus well defined on integrable functions, are called the <u>ergodic operators</u>.

When $X = \mathbb{R}$, $\sigma_t = \tau_t$, we get $T_r = L_r$, the differentiation operators of Lebesgue theory. Furthermore, T_r relates to L_r in the following way, that will be useful later.

Lemma 5.2. For a function $g : \mathbb{R} \to \mathbb{R}$ and a constant $A > 0$, define $g^{(A)}(t) = g(t)\chi_{(0, A)}(t)$. Then for $f \in L^1(X)$ and $0 \leq r,\ s \leq N$,

$$(T_r f)_x(s) = (L_r f_x^{(2N)})(s) \tag{5.6}$$

where $f_x(s) = f(\sigma_s x)$ for a fixed $x \in X$.

Proof.

$$(T_r f)_x(s) = (T_r f)(\sigma_s x) = \frac{1}{r}\int_0^r f(\sigma_t(\sigma_s x))dt = \frac{1}{r}\int_0^r f(\sigma_{t+s} x)dt$$

$$= \frac{1}{r}\int_0^r f_x(t + s)dt = \frac{1}{r}\int_0^r f_x^{(2N)}(t + s)dt$$

(since $0 \leq t + s \leq 2N$)

$$= \frac{1}{r}\int_0^{s+r} f_x^{(2N)}(t)dt = (L_r f_x^{(2N)})(s) \qquad \nabla$$

Let M be the maximal operator of the family $\{T_r\}$,

$$Mf(x) = \sup_{r>0} |T_r f(x)| \tag{5.7}$$

for each $f \in L^1(X)$, $x \in X$.

Theorem 5.3. Propositions I, II, III and IV of Section 4 hold for the ergodic operators $\{T_r\}$ as $r \to \infty$.

Remark 5.1. Proposition I corresponds in this case to the Birkhoff-Khinchin ergodic theorem (that gives the pointwise convergence a.e.). Proposition II corresponds to the mean ergodic theorem of von Neumann. Proposition IV corresponds to the dominated ergodic theorem of Wiener. These theorems give information on the frequency of the permanence of the path over an infinite period.

Proof. By Theorem 4.1, the thesis will follow if $\{T_r\}$ satisfy the conditions (A) and (B) of that theorem:

(A) M is of type (∞, ∞): This follows from the definition, and the constant of type is again less than or equal to 1. Furthermore, M is of weak type (1, 1): We claim that there exists $C > 0$ such that $(Mf)_*(\alpha) \leq C\alpha^{-1}\|f\|_1$ for all $\alpha > 0$.

If $M_N f(x) = \sup_{0<r<N} |T_r f(x)|$, then $M_N f \uparrow Mf$, and by Proposition 3.1(c) of Chapter 4, $(M_N f)_* \uparrow (Mf)_*$. Therefore it will be enough to show that for all $N > 0$,

$$(M_N f)_*(\alpha) \leq C\alpha^{-1} \int_X |f(x)|\, d\mu. \tag{5.8}$$

If $\psi(x)$ is the characteristic function of the set $\{x : M_N f(x) > \alpha\}$, (5.8) can be written as

$$\int_X \psi(x)\, d\mu \leq C\alpha^{-1} \int_X |f(x)|\, d\mu \tag{5.8a}$$

Let $\psi(x, s) = \psi(\sigma_s x) = \psi_x(s)$. For fixed x, $\psi_x(s)$ is the characteristic function of the set $\{s : M_N f(\sigma_s x) > \alpha\}$.

Let

$$\Lambda_N f(x) = \sup_{0<r<N} \frac{1}{r} \int_0^r |f(x - t)|\, dt, \quad \text{with} \quad \Lambda_N f \uparrow \Lambda f$$

By Lemma 5.2, for every s, $0 \leq s \leq N$,

$$(M_N f)_x(s) \leq (\Lambda_N f_x^{(2N)})(s) \leq (\Lambda f_x^{(2N)})(s)$$

Writing $\phi_x(s)$ as the characteristic function of the set $\{s : \Lambda f_x^{(2N)}(s) > \alpha\}$ then,

$$\int_0^N \psi(x, s)ds \leq \int_0^N \phi_x(s)ds \leq \int_0^\infty \phi_x(s)ds = (\Lambda f_x^{(2N)})_*(\alpha) \tag{5.9}$$

and, as Λ is an operator of weak type (1, 1),

$$\int_0^N \psi(x, s)ds \leq \frac{A}{\alpha} \|f_x^{(2N)}\|_1 = \frac{A}{\alpha} \int_0^\infty |f_x^{(2N)}(s)|ds = \frac{A}{\alpha} \int_0^{2N} |f_x(s)|ds \tag{5.10}$$

Since (5.10) holds for every x, we can integrate over X,

$$\int_X (\int_0^N \psi(x, s)ds)d\mu(x) \leq \frac{A}{\alpha} \int_X^{2N} (\int_0^{2N} |f(x, s)|ds)d\mu(x) \tag{5.11}$$

Using Fubini's theorem and the isointegrability condition (2a), (5.11) becomes

$$\int_X \psi(x)d\mu \leq \frac{2A}{\alpha} \int_X |f(x)|d\mu$$

which is (5.8) with $C = 2A$. Thus, by the Marcinkiewicz interpolation theorem, M is of type (p, p) for all $1 < p \leq \infty$ and, a fortiori, of weak type (p, p), $1 \leq p \leq \infty$, and condition (A) is satisfied.

(B) We want to prove that there exists a set $D \subset L^2$, dense in L^2, such that for every $g \in D$,

$$\lim_{r\to\infty} T_r g(x) = \lim_{r\to\infty} \frac{1}{r}\int_0^r g(\sigma_t x)dt = Tg(x) \quad \text{a. e.} \tag{5.12}$$

We consider two classes of particular functions in L^2:

(i) functions invariant with respect to the group $\{\sigma_t\}$, i.e., $A = \{\phi(x) \in L^2 : \phi(\sigma_t x) = \phi(x) \text{ a. e.}, \ \forall t \in \mathbb{R}\}$

(the null set can vary with t),

and

(ii) $B = \{\psi(x) \in L^2 : \exists\ \gamma(x) \in L^2 \cap L^\infty$ and $s \in \mathbb{R}$ with

$$\psi(x) = \gamma(x) - \gamma(\sigma_s x) = \gamma(x) - \gamma_s(x)\}$$

The limit in (5.12) exists for every $\phi \in A$:

$$T_r\phi(x) = \frac{1}{r}\int_0^r \phi(\sigma_t x)dt = \frac{1}{r}\int_0^r \phi(x)dt = \phi(x)$$

for every r, thus $\lim_{r\to\infty} T_r\phi(x) = \phi(x)$.

Similarly, the limit in (5.12) exists for every $\psi \in B$: let $\psi = \gamma - \gamma_s$ and $C \geq |\gamma(x)|$, then for every r,

$$T_r\psi(x) = \frac{1}{r}\int_0^r (\gamma(\sigma_t x) - \gamma(\sigma_t\sigma_s x))dt$$

$$= \frac{1}{r}\int_0^r \gamma_r(t)dt - \frac{1}{r}\int_0^r \gamma_x(t+s)dt$$

$$= \frac{1}{r}\int_0^r \gamma_x(t)dt - \frac{1}{r}\int_s^{s+r} \gamma_x(t)dt$$

$$= \frac{1}{r}\int_0^s \gamma_x(t)dt - \frac{1}{r}\int_r^{r+s} \gamma_x(t)dt$$

and

$$|T_r\psi(x)| \leq 2Cr^{-1}s$$

Since s is fixed and $r \to \infty$, $\lim_{r\to\infty} T_r\psi(x) = 0$.

Now consider $D = \{g \in L^2 : g = \phi + \psi,\ \phi \in A,\ \psi \in B\}$. It remains to prove that such D is a dense set of L^2. Since L^2 is a Hilbert space, we need only show that $A = B^\perp$ (see Chapter 0, Corollary 4.3), i.e., that $h \in B^\perp$ if and only if $h = \phi \in A$.

Let $h \in B^\perp$ and $\{h_n\} \subset L^2 \cap L^\infty$ be such that $\|h_n - h\|_2 \to 0$. Thus $\psi(x) = h_n(x) - h_n(\sigma_s x) \in B$ and

$$(h, h_n - h_{ns}) = 0 = \int_X h(x)\overline{(h_n(x) - h_n(\sigma_s x))}\,d\mu$$

$$\to \int_X h(x)\overline{(h(x) - h(\sigma_s x))}\,d\mu$$

$$= (h, h - h_s) \quad \text{as} \quad n \to \infty$$

Thus, $\|h - h_s\|_2^2 = (h - h_s, h - h_s) = (h, h) - (h_s, h_s) + (h, h - h_s) - (h_s, h - h_s) = 0 + 0 - (h_s, h - h_s)$. Furthermore,

$$(h_s, h - h_s) = \int_X h(\sigma_s x)(h(x) - h(\sigma_s x))\,d\mu =$$

$$= \int_X h(x)(h(\sigma_{-s} x) - h(x))\,d\mu = (h, h_{-s} - h) = 0$$

Collecting these results, $\|h - h_s\|_2 = 0$, for every s, which implies that $h_s = h$ a.e. for every s, i.e., $h \in A$. ∇

Properties of the limit operator. Since $\lim_{r\to\infty} T_r f(x)$ exists a.e. in x, the operator $T : f \to Tf = \lim T_r f$ is defined for every $f \in L^p$, $p \geq 1$.

Lemma 5.4. If $f \in L^1(X)$ then the functions $\{T_r f\}_{r>0}$ are equi-continuous (in the Vitali sense), i.e., for every $\varepsilon > 0$ there exists a $\delta > 0$ such that

$$\left| \int_E T_r f(x) d\mu \right| < \varepsilon \quad \text{for every} \quad r \tag{5.13}$$

if $\mu(E) < \delta$.

Proof. As $f \in L^1$, for every $\varepsilon > 0$, there exists a $\delta > 0$ such that if $\mu(E) < \delta$ then $\left| \int_E f d\mu \right| < \varepsilon$. Then,

$$\int_E T_r f(x) d\mu = \int_E \left(\frac{1}{r} \int_0^r f(\sigma_t(x) dt\right) d\mu(x)$$

$$= \frac{1}{r} \int_0^r \left(\int_E f(\sigma_t x) d\mu(x)\right) dt$$

$$= \frac{1}{r} \int_0^r \left(\int_{\sigma_t E} f(x) d\mu\right) dt$$

And, as $\mu(\sigma_t E) = \mu(E) < \delta$,

$$\left| \int_E T_r f(x) d\mu \right| \leq \frac{1}{r} \int_0^r \varepsilon \, dt = \varepsilon \qquad \nabla$$

Vitali's theorem on integrals of equicontinuous functions enables us to take limits under the integral sign if the integration is over a finite domain (see [13], p. 150). More precisely we have:

Corollary 5.5. If $E \subset X$, $\mu(E) < \infty$, then

$$\lim_{r \to \infty} \int_E T_r f(x) d\mu = \int_E Tf(x) d\mu \tag{5.14}$$

Definition 5.2. A set $H \subset X$ is invariant with respect to the group $\{\sigma_t\}$ if $\sigma_t H = H$ for every t.

If $\sigma_t H = H$ for almost every $x \in H$ and almost every t, H is invariant in the broad sense.

A function $f(x)$ is invariant if $f(\sigma_t x) = f(x)$ for every t and almost every x.

If f is an invariant function then $\{x : f(x) > \alpha\}$ is an invariant set (in the broad sense). In the proof of Lemma 5.4, it was shown that

$$\int_H T_r f(x) d\mu = \frac{1}{r} \int_0^r \left(\int_{\sigma_t H} f d\mu \right) dt$$

If H is invariant (in the broad sense),

$$\int_H T_r f d\mu = \int_H f d\mu \tag{5.15}$$

Theorem 5.6. Let $f \in L^1(X)$ and let $\{T_r\}_{r>0}$ be the ergodic operators.

(a) If $\lim_{r \to \infty} T_r f(x) = Tf(x)$ exists for $x \in X$, then the limit exists for every point $\sigma_s x$, $\forall s$, and $Tf(\sigma_s x) = Tf(x)$.

Thus, for every integrable function f, Tf is an invariant function.

(b) For every invariant set H, $\mu(H) < \infty$,

$$\int_H Tf(x) d\mu = \int_H f(x) d\mu \tag{5.16}$$

(c) (5.16) does not hold if $\mu(H) = \infty$. But $Tf \in L^1(X)$ and

$$\int_X |Tf(x)|\,d\mu \le \int_X |f(x)|\,d\mu \tag{5.16a}$$

(d) T is a <u>linear</u>, <u>positive</u> operator of <u>type (p, p)</u> for every $p \ge 1$.

<u>Proof.</u>

(a) For every r and every s it follows from the definitions that $T_r f(\sigma_s x) = T_{r+s} f(x) + \frac{s}{r}(T_{r+s} f(x) - T_s f(x).)$ Keeping s fixed and $r \to \infty$, $\lim T_r f(\sigma_s x) = \lim T_r f(x)$ and both exist simultaneously.

(b) is a consequence of Corollary 5.5 applied to (5.15).

(c) We may assume that $f \ge 0$. By (a), Tf is an invariant function and then for $\varepsilon > 0$, $H_\varepsilon = \{x : Tf(x) > \varepsilon\}$ is an invariant set (in the broad sense). As M, the maximal operator of $\{T_r\}$, is of weak type (1, 1) by Theorem 5.3, T is also of weak type (1, 1) and so $\mu(H_\varepsilon) < \infty$. Thus, by (b),

$$\int_{H_\varepsilon} Tf\,d\mu = \int_{H_\varepsilon} f\,d\mu \le \int_X f\,d\mu$$

and making $\varepsilon \to 0$, (5.16a) is proved.

(d) It is immediate that T is a linear and positive operator.

Since $|Tf(x)| \le Mf(x)$ and M is of type (p, p) for all $p > 1$, T is also of type (p, p) for $p > 1$. Furthermore, (c) asserts that T is of type (1, 1). ∇

<u>Definition 5.3.</u> $\{\sigma_t\}$ is a <u>metrically transitive or ergodic group</u> if every invariant set H satisfies

$$\mu(H) \times \mu(X - H) = 0 \tag{5.17}$$

that is to say, $\{\sigma_t\}$ is ergodic if X and ϕ are (but for a null set) the only sets invariant with respect to it.

If $\{\sigma_t\}$ is an ergodic group then every invariant function f is equal to a constant a.e.: assume $f(x) \neq c$ a.e.; then there exist $a < b$ such that the sets $H_1 = \{x : f(x) < a\}$ and $H_2 = \{x : f(x) > b\}$ are of positive measure. But if f is invariant, H_1 and H_2 are invariant sets and $H_2 \subset X - H$, and the group $\{\sigma_t\}$ cannot be ergodic.

By Theorem 5.6(a), Tf is an invariant function for any integrable f and, if $\{\sigma_t\}$ is ergodic, it must be $Tf(x) = C$ a.e. Furthermore, by Theorem 5.6(b), if $\mu(X) < \infty$, then

$$\int_X f d\mu = \int_X Tf d\mu = C\mu(X)$$

and $C = (\int_X f d\mu)/\mu(X)$. We thus have

<u>Corollary 5.7</u>. Let $\{\sigma_t\}$ be an ergodic group. Then, for every integrable f, Tf is a constant a.e. Furthermore, if the space X is of finite measure, then

$$Tf(x) = \frac{\int_X f d\mu}{\mu(X)} \quad \text{a.e.} \tag{5.18}$$

In particular, for $f(x) = \chi_E(x)$, then $Tf(x) = \mu(E)/\mu(X)$.

This expresses the <u>uniformization principle</u>: "The paths of almost all points are uniformly distributed in X."

<u>Corollary 5.8. (Poincare's return theorem)</u>. Let be $E \subset X$, with $\mu(X) < \infty$ and $\mu(E) > 0$. For almost every $x \in E$ the path of x returns an infinite number of times to E.

If $\{\sigma_t\}$ is an ergodic group the thesis holds not only a.e. in E but a.e. in X.

Proof. Let $f(x) = \chi_E(x)$, then $T_r f(x)$ indicates the frequency with which the path of x enters E in the interval (0, r). Thus, if for every $x \in E$ it is $Tf(x) > 0$, then for infinitely many r's it is $T_r f(x) > 0$, i.e., the path of x will enter infinitely many times into E. But this is the case, since if $H = \{x : Tf(x) = 0\}$ then, Tf being invariant makes H invariant, and by Theorem 5.6(b),

$$\mu(H \cap E) = \int_H f(x)d\mu = \int_H Tf(x)d\mu = 0$$

As for every $x \in X - H$, $Tf(x) > 0$, then $Tf(x) > 0$ a.e. $x \in E$. If $\{\sigma_t\}$ is an ergodic group then $Tf(x) = C$ a.e. and the thesis holds a.e. in X. ∇

In the last proof it was shown that for every set $E \subset X$, $\mu(E) > 0$, it corresponds a set K, $\mu(K) = 0$ such that for every $x \in X - K$ the path of x enters E infinitely many times. This exceptional set varies in general with E, but given a (countable) sequence $\{E_k\}$ it is possible to choose a K common to all the sets of the sequence. In particular, if $X = \mathbb{R}^n$, it is possible to choose a countable number of open sets E_k in such a way that any open set contains at least one E_k. Therefore as the path of any $x \in X - K$ enters infinitely many times in each E_k, it will enter infinitely many times in each open set.

Corollary 5.9 (Boltzman's ergodic principle). Let $X = \mathbb{R}^n$, $\mu(X) < \infty$ and $\{\sigma_t\}$ be an ergodic group. Then almost every point $x \in X$ has the property that for every open set $E \subset X$ the path of x enters infinitely many times in E.

This corollary says that the path of almost every point enters every arbitrarily small neighborhood of any other point, that is, that almost every path enters "everywhere."

REFERENCES

1. G. H. Hardy and J. E. Littlewood, Acta Math., 54:81 (1930).
2. Miguel de Guzmán, Differentiation of integrals in $\mathbb{R}^n$, Lecture Notes in Math. # 481, Springer-Verlag, New York-Heidelberg-Berlin, 1975.
3. R. R. Coifman and G. Weiss, Analyse Harmonique Non Commutative sur certains Espaces Homogènes, Lecture Notes in Math. # 242, Springer-Verlag, New York-Heidelberg-Berlin, 1971.
4. E. M. Stein and G. Weiss, Introduction to Fourier Analysis on Euclidean Spaces, Princeton Univ. Press, Princeton, 1971.
5. E. M. Stein, Singular Integrals and Differentiability Properties of Functions, Princeton Univ. Press, Princeton, 1970.
6. F. John and L. Nirenberg, Comm. Pure Appl. Math., 14:415 (1961).
7. U. Neri, Studia Math., 61:63 (1977).
8. A. P. Calderón and A. Zygmund, Acta Math., 88:85 (1952).
9. C. Fefferman and E. M. Stein, Acta Math., 129:137 (1972).
10. H. M. Reimann and T. Rychener, Funktionen beschränkter mittlerer Oszillation, Lecture Notes in Math. # 487, Springer-Verlag, New York-Heidelberg-Berlin, 1975.
11. M. Cotlar, Rev. Mat. Cuyana, 1:105 (1955).
12. M. Cotlar, Rev. U. M. A., 15 (1956).
13. N. Dunford and J. T. Schwartz, Linear Operators, vol. I, Interscience, 1958.

Chapter 6

SINGULAR INTEGRALS

1. THE HILBERT TRANSFORM IN L^2

The following extension of the (Lebesgue) integral is important for the problems of this chapter.

Let $f(x)$ be measurable in $\mathbb{R}^n$, and for some $x_0 \in \mathbb{R}^n$, absolutely integrable over each set $|x - x_0| > \varepsilon > 0$. We say then that $f(x)$ is integrable over $\mathbb{R}^n$ in the principal value sense if

$$\lim_{\varepsilon \to +0} \int_{|x-x_0|>\varepsilon} f(x)dx \tag{1.1}$$

exists and is finite. The value of this limit will be denoted by

$$\text{P.V.} \int_{\mathbb{R}^n} f(x)dx \tag{1.2}$$

When $n = 1$, clearly,

$$\text{P.V.} \int_{\mathbb{R}^n} f(x)dx = \lim_{\varepsilon \to +0} \left(\int_{-\infty}^{x_0-\varepsilon} + \int_{x_0+\varepsilon}^{\infty} \right) f(x)dx$$

Similar definitions are easily applicable to cases when f is defined in subsets of $\mathbb{R}^n$. For example, if f(x) is defined in an interval $(a, b) \subset \mathbb{R}^1$ and x_0 is interior to (a, b), we may set

$$\text{P.V.} \int_a^b f(x)dx = \lim_{\varepsilon \to +0} \left(\int_a^{x_0-\varepsilon} + \int_{x_0+\varepsilon}^b \right) f(x)dx \quad (1.3)$$

The case that interests us initially is when n = 1 and f(x) = $g(x)/(x - x_0)$. By definition,

$$\text{P.V.} \int_{-\infty}^{\infty} \frac{g(x)}{x-x_0} dx = \lim_{\varepsilon \to +0} \left(\int_{-\infty}^{x_0-\varepsilon} + \int_{x_0+\varepsilon}^{\infty} \right) \frac{g(x)}{x-x_0} dx \quad (1.4)$$

The two integrals on the right of (1.4) exist, in the Lebesgue sense, if $g \in L^1(-\infty, \infty)$ or, more generally (as can be seen using Hölder's inequality) if $g \in L^p(-\infty, \infty)$, $1 \le p < \infty$. It is clear that in this case the existence of the limit in (1.4) is equivalent to the existence of

$$\lim_{\varepsilon \to +0} \int_{\varepsilon < |x-x_0| \le \eta} \frac{g(x)}{x-x_0} dx \quad (1.5)$$

for any fixed $\eta > 0$. Since the integral in (1.5) equals

$$\int_{\varepsilon < |x-x_0| \le \eta} \frac{g(x)-g(x_0)}{x-x_0} dx + g(x_0) \int_{\varepsilon < |x-x_0| \le \eta} \frac{dx}{x-x_0}$$

$$= \int_{\varepsilon < |x-x_0| \le \eta} \frac{g(x)-g(x_0)}{x-x_0} dx$$

it is immediate that the limit in (1.5) exists if g satisfies at x_0 the condition Lip 1 (i.e., $|g(x) - g(x_0)| \le C|x - x_0|$ for some C), and in particular, if $g'(x_0)$ exists and is finite.

To sum up, the integral (1.4) exists if $g \in L^p(-\infty, \infty)$, $1 \le p < \infty$, and is Lip 1 at x_0. The integral (1.4) is called the Hilbert transform of g at the point x_0. We shall use the notation

$$H_\varepsilon g(x_0) = \frac{1}{\pi} \int_{|x-x_0|>\varepsilon} \frac{g(x)}{x-x_0}\, dx \tag{1.6}$$

and

$$Hg(x_0) = \lim_{\varepsilon \to +0} H_\varepsilon g(x_0) \tag{1.6a}$$

The purpose of this chapter is to establish the existence and properties of Hg and its extensions to higher dimensions under very general conditions.

Hilbert transforms occur naturally in the theory of analytic functions. For suppose that $g(x) \in L^p(-\infty, \infty)$, $1 \le p < \infty$, and consider the integral

$$\frac{1}{2\pi i} \int_{-\infty}^{\infty} \frac{g(t)}{t-z}\, dt \tag{1.7}$$

where $z = x + iy$ is a complex number, say, in the upper half-plane $\mathbb{R}^2_+$; thus $y > 0$. Without loss of generality, we may assume, whenever it is convenient, that $g(x)$ is real valued.

It is clear that, for each $N > 0$, the function

$$F_N(z) = \frac{1}{2\pi i} \int_{-N}^{N} \frac{g(t)}{t-z}\, dt$$

is holomorphic in $\mathbb{R}^2_+$ and, for z belonging to any half-plane $y \geq y_0 > 0$, tends uniformly, as $N \to \infty$, to

$$F(z) = \frac{1}{2\pi i} \int_{-\infty}^{\infty} \frac{g(t)}{t-z}\, dt \tag{1.8}$$

which therefore is holomorphic in $\mathbb{R}^2_+$. In what follows we assume that $z \in \mathbb{R}^2_+$.

Decomposing $1/i(t - z)$ into its real and imaginary parts, we can write

$$F(z) = \frac{1}{2\pi} \int_{-\infty}^{\infty} g(t) \frac{y}{(x-t)^2+y^2}\, dt + \frac{i}{2\pi} \int_{-\infty}^{\infty} g(t) \frac{x-t}{(x-t)^2+y^2}\, dt$$

$$= \frac{1}{2}(g * P_y)(x) + \frac{i}{2}(g * Q_y)(x) \tag{1.9}$$

where $P_y(x) = \frac{1}{\pi} \frac{y}{x^2+y^2}$ is the Poisson kernel in $\mathbb{R}^2_+$ (see Chapter 3, (1.11)), and

$$Q_y(x) = \frac{1}{\pi} \frac{x}{x^2+y^2} \tag{1.10}$$

is called the <u>conjugate Poisson kernel</u> in $\mathbb{R}^2_+$.

The integral

$$u(x, y) = (g * P_y)(x) = \frac{1}{\pi} \int_{-\infty}^{\infty} \frac{y}{(x-t)^2+y^2}\, dt$$

is the Poisson integral of g and, as we know (see Chapter 3, Theorem 1.8 and Chapter 5, Proposition 2.4), it tends to $g(x)$ for almost every x, as $z = x + iy$ tends nontangentially to x, even if $g \in L^{\infty}$. Under the hypothesis that g is real valued,

$$u(x, y) = \operatorname{Re} F(z)$$

We shall now consider the integral

$$v(x, y) = \operatorname{Im} F(z) = (g * Q_y)(x) = \frac{1}{\pi}\int_{-\infty}^{\infty} g(t) \frac{x-t}{(x-t)^2+y^2} dt \qquad (1.11)$$

which is called the <u>conjugate Poisson integral</u> of g, and prove the following

<u>Theorem 1.1</u>. For any $g \in L^p(-\infty, \infty)$, $1 \leq p < \infty$, and almost every x, the conjugate Poisson integral of g tends, as $z \to x$ nontangentially, to a finite limit.

<u>Proof</u>. Decomposing g into its positive and negative parts, it is enough to assume that $g \leq 0$. Thus $u \leq 0$, and $G(z) = \exp(u+iv)$ is regular and bounded by 1 in absolute value. By Theorem 4.7 of Chapter 3 and Proposition 2.4 of Chapter 5), $G(z)$ has a nontangential limit a.e. as $z \to x$. This limit cannot be zero in a set of positive measure because then u would tend nontangentially to $-\infty$ in a set of positive measure, impossible since u is the Poisson integral of a function in L^p. Therefore, the limit of $v(x, y)$ exists nontangentially and is finite a.e. ∇

From this result we deduce the fundamental

<u>Theorem 1.2</u>. For any $g \in L^p(-\infty, \infty)$, $1 \leq p < \infty$, the Hilbert transform $Hg(x)$ exists and is finite a.e. Moreover, it is equal to the limit of the conjugate Poisson integral of g at every point of the Lebesgue set $\mathscr{L}_g$ of g, i.e.,

$$\lim_{y \to +0} \left(\int_{-\infty}^{\infty} g(t) \frac{x-t}{(x-t)^2+y^2} dt - \int_{|x-t|>y} \frac{g(t)}{x-t} dt\right) = 0 \qquad (1.12)$$

if $x \in \mathscr{L}_g$.

Proof. Observe that the difference of the two integrals in (1.12) can be rewritten as

$$\int_{-\infty}^{\infty} g(x - t)\Phi_y(t)dt \tag{1.13}$$

where $\Phi_y(t) = y^{-1}\Phi(y^{-1}t)$ and

$$\Phi(t) = \begin{cases} \dfrac{t}{t^2+1} - \dfrac{1}{t} & \text{if } |t| > 1 \\ \dfrac{t}{t^2+1} & \text{if } |t| \le 1 \end{cases}$$

Since

$$\Psi(x) = \sup_{|t| \ge |x|} |\Phi(t)| = \begin{cases} \dfrac{1}{|x|(1+x^2)} & \text{if } |x| > 1 \\ \dfrac{1}{2} & \text{if } |x| \le 1 \end{cases}$$

Ψ is integrable in $(-\infty, \infty)$. Therefore, by the remark following Theorem 2.1 of Chapter 1, the integral (1.13) tends, at every $x \in \mathscr{L}_g$, to $g(x).(\int_{-\infty}^{\infty} \Phi(t)dt)$. As $\int_{-\infty}^{\infty} \Phi(t)dt = 0$, since Φ is odd, the result is proved. ▽

Remark 1.1. The reader will observe that in Theorems 1.1 and 1.2 we exclude the case $p = \infty$. The example $g(x) \equiv 1$ shows that Hg need not exist if g is merely bounded. If, however, we assume that g is bounded and that

$$\int_{|x|>1} |g(x)/x|dx$$

is finite, then it is easily seen that the $Hg(x)$ exists and is finite a.e.

Thus we may define the Hilbert transform operator $H : g \to Hg$ for $g \in L^p(\mathbb{R}^1)$, $1 \leq p < \infty$. Later on we shall prove that this operator (and its generalizations to $\mathbb{R}^n$, $n > 1$) is bounded in L^p for $1 < p < \infty$, but since the proof is simpler in the important case $p = 2$, we give it here in a more complete form.

Theorem 1.3. If $g \in L^2(-\infty, \infty)$ then

$$(Hg)\hat{}(x) = (-i \operatorname{sgn} x)\hat{g}(x) \tag{1.14}$$

where $\operatorname{sgn} x$ denotes the signum of x. In particular,

$$\|Hg\|_2 = \|g\|_2, \quad \forall g \in L^2(-\infty, \infty) \tag{1.15}$$

Proof. The argument is based on the formula

$$\hat{Q}_y(x) = (-i \operatorname{sgn} x)e^{-2\pi|yx|} \tag{1.16}$$

that we shall prove later. In fact, (1.16) gives

$$(Q_y * g)\hat{}(x) = (-i \operatorname{sgn} x)e^{-2\pi|yx|}\hat{g}(x) \tag{1.17}$$

and by Plancherel's theorem, (1.17) implies that $Q_y * g$ tends, as $y \to +0$, in L^2, to an L^2 function whose Fourier transform is $(-i \operatorname{sgn} x)\hat{g}(x)$. But by Theorem 1.2, $(Q_y * g)(x)$ tends to $Hg(x)$ a.e. as $y \to +0$, thus (1.14) holds.

In order to establish (1.16), let us recall that, if $z = x + iy$,

$$\frac{1}{2}(P_y(x) + iQ_y(x)) = I(z) = \frac{-1}{2\pi iz}$$

hence,

$$P_y(x) = 2 \operatorname{Re} I(z)$$

and (1.18)

$$Q_y(x) = 2 \operatorname{Im} I(z)$$

Since for $y > 0$,

$$I(z) = \int_0^\infty e^{2\pi izt}\,dt = \int_0^\infty e^{2\pi ixt}e^{-2\pi yt}\,dt$$

we may rewrite (1.18) as

$$P_y(x) = \int_{-\infty}^\infty e^{2\pi ixt}e^{-2\pi|yt|}(\chi_+(t) + \chi_-(t))dt$$

and (1.19)

$$Q_y(x) = -i\int_{-\infty}^\infty e^{2\pi ixt}e^{-2\pi|yt|}(\chi_+(t) - \chi_-(t))dt$$

where χ_+ and χ_- are the characteristic functions of $[0, \infty)$ and $(-\infty, 0)$, respectively. Since $\chi_+(t) + \chi_-(t) \equiv 1$ and $\chi_+(t) - \chi_-(t) = \operatorname{sgn} t$, for $t \neq 0$, (1.19) yields both the already known formula

$$\hat{P}_y(t) = e^{-2\pi|yt|}$$

(see Chapter 3, Section 1) and

$$\hat{Q}_y(t) = (-i \operatorname{sgn} t)e^{-2\pi|yt|}$$

and the theorem is proved. ∇

This enables us to see that the conjugate Poisson integral of an L^2 function coincides a.e. with the Poisson integral of its Hilbert transform.

Corollary 1.4. If $g \in L^2(-\infty, \infty)$ then for $y > 0$,

$$(g * Q_y)(x) = (Hg * P_y)(x) \quad \text{a.e.} \tag{1.20}$$

Proof. Formula (1.20) means that

$$\frac{1}{\pi}\int_{-\infty}^{\infty} g(x-t)\frac{t}{t^2+y^2}\,dt = \frac{1}{\pi}\int_{-\infty}^{\infty} Hg(x-t)\frac{y}{t^2+y^2}\,dt \tag{1.20a}$$

holds a.e. Check that both sides of (1.20a) have the same Fourier transform, equal to $(-i \operatorname{sgn} x)e^{-2\pi|yx|}\hat{g}(x)$. ∇

Note that if $H^2 g = H(Hg)$ then, by (1.14), $H^2 = -I$, where I is the identity operator in L^2. This fact, together with (1.15), implies that H is a unitary operator in $L^2(\mathbb{R}^1)$.

Remark 1.2. To each real-valued function $g \in L^p(\mathbb{R}^1)$, $1 \leq p < \infty$, we have associated an analytic function $F(z)$, defined in $\mathbb{R}^2_+$ by (1.8), such that, for $z = x + iy$ tending nontangentially to $x \in \mathbb{R}^1$, $F(z)$ tends to

$$F(x) = \frac{1}{2}g(x) + \frac{i}{2}Hg(x)$$

Furthermore, (1.14) yields, for $g \in L^2$,

$$\hat{g}(x) + i(Hg)^{\wedge}(x) = 0 \tag{1.21}$$

whenever $x < 0$.

In the preceding considerations we discussed the case of functions defined on $\mathbb{R}^1$ and the next step will be to extend our results to functions in $\mathbb{R}^n$. But before that, we shall briefly discuss the case when the functions are defined in other domains, the simplest of which is a circular domain. We limit ourselves here to dimension one and functions defined on the circumference of the unit circle. We will not aim for completeness of results or proofs but shall indicate the analogies between this case and that of $\mathbb{R}^1$. The reader should have no serious difficulty in completing the arguments.

In what follows we consider points z in the disc $U = \{z : |z| < 1\}$ and functions $f(z)$ defined in U, as well as functions $g(t) = f(e^{it})$, defined on the circumference $\mathbb{T} = \{z : |z| = 1\}$. Let $f(z) = f(re^{it})$ be analytic in U. For $0 < r < R < 1$ we have the Cauchy formula

$$f(z) = f(re^{it}) = \frac{1}{2\pi i} \int_{|\zeta|=R} \frac{f(\zeta)d\zeta}{\zeta - z}$$

$$= \frac{1}{2\pi} \int_0^{2\pi} f(Re^{i\theta}) \frac{Re^{i\theta}d\theta}{Re^{i\theta} - re^{it}}$$

and if we can justify the passage to the limit when $R \to 1$ (which can be done under very general hypothesis), and write $g(\theta) = \lim_{R\to 1} f(Re^{i\theta})$, we obtain

$$f(z) = \frac{1}{2\pi} \int_0^{2\pi} g(\theta) \frac{e^{i\theta}d\theta}{e^{i\theta} - re^{it}}$$

$$= \frac{1}{2\pi} \int_0^{2\pi} g(\theta) \frac{d\theta}{1 - re^{i(t-\theta)}}$$

$$= \frac{1}{2\pi} \int_0^{2\pi} g(\theta) C_r(t - \theta)d\theta$$

where

$$C_r(t) = \frac{1}{1-re^{i\theta}} = \sum_{n=0}^{\infty} r^n e^{int}$$

is the Cauchy kernel.

Consider the expression

$$\sum_{n=-\infty}^{\infty} r^{|n|} e^{int} = (C_r(r) + \overline{C_r(t)}) - 1$$

$$= \mathrm{Re}\left(\frac{1+re^{it}}{1-re^{it}}\right)$$

$$= \frac{1-r^2}{1-2r\cos t+r^2}$$

This last function is usually denoted by $P_r(t) = P(r, t)$, and called the Poisson kernel <u>for the circle</u> (see Chapter 3, Section 3). Its harmonic conjugate is

$$Q_r(t) = Q(r, t) = \mathrm{Im}\left(\frac{1+re^{it}}{1-re^{it}}\right)$$

$$= \frac{2r \sin t}{1-2r\cos t+t^2} \tag{1.22}$$

and is called the conjugate Poisson kernel for the circle. Kernels $P_r(t)$ and $Q_r(t)$ have properties very similar to those of their analogues for $\mathbb{R}^2_+$. The latter were denoted by $P_y(x)$ and $Q_y(x)$ (see (1.10)) and we immediately see the ressemblance if we set $x = r\cos t$, $y = r\sin t$.

For any function $g(t)$, $-\infty < t < \infty$, periodic of period 2π, and integrable over a period, the expressions $\frac{1}{2\pi}(g * C_r)(t)$, $\frac{1}{2\pi}(g * P_r)(t)$ and $\frac{1}{2\pi}(g * Q_r)(t)$ will be called the Cauchy integral, the Poisson integral and the conjugate Poisson integral of g, respectively. For real-valued g, an easy computation yields that the Poisson integral and the conjugate Poisson integral of g are, respectively, the real and imaginary parts of the Cauchy integral of g.

We observed in Section 3 of Chapter 3 that for any $g \in L^p(0, 2\pi)$, $1 \leq p \leq \infty$, the Poisson integral of g tends a.e. to g as $r \to 1$. An argument parallel to the proof of Theorem 1.1 shows that we have for the case of the circle a situation similar to that of the line namely, for any $g \in L^p(0, 2\pi)$, $1 \leq p < \infty$, the conjugate Poisson integral of g tends a.e. nontangentially to a finite limit. Since

$$\lim_{r \to 1} Q_r(t) = \frac{\sin t}{1 - \cos t} = \cotan t/2$$

it is not difficult to check that the value of this limit is equal a.e. to

$$Hg(t) = \tilde{g}(y) = \frac{1}{2\pi}\ \text{P.V.} \int_0^{2\pi} g(\theta) \cotan \frac{t-\theta}{2}\, d\theta \tag{1.23}$$

which is called the <u>conjugate function</u> of g.

Consider the Fourier series development of real-valued $g \in L^2$,

$$g(t) \sim \sum_{-\infty}^{\infty} \hat{g}(n) e^{int}$$

It follows that

$$(P_r * g)(t) \sim \sum_{-\infty}^{\infty} \hat{g}(n) r^{|n|} e^{int}$$

and

$$(Q_r * g)(t) \sim \sum_{-\infty}^{\infty} (-i \operatorname{sgn} n)\hat{g}(n)r^{|n|}e^{int}$$

since $(P_r + iQ_r) * g$ must be an analytic function. From this follows, as in Theorem 1.3, that

$$(\tilde{g})^\wedge(n) = (-i \operatorname{sgn})\hat{g}(n) \tag{1.14a}$$

and

$$(g + i\tilde{g})^\wedge(n) = 0 \quad \text{for} \quad n < 0 \tag{1.21a}$$

We shall say that an integrable periodic function $h(t)$ belongs to the Hardy space $H^1(\mathbb{T})$ if the Fourier coefficients $\hat{h}(n)$ vanish for $n < 0$, or equivalently, if it is the boundary function of an analytic function $F(z) = F(re^{it})$, defined in the disc, $h(t) = F(e^{it})$. Formula (1.21a) tells us, then, that $g + i\tilde{g} \in H^1(\mathbb{T})$. In Section 6 we shall use this fact to further study the continuity of the Hilbert operator $H : g \to \tilde{g}$, as acting in the circle.

Let us consider now the n-dimensional case. Since in $\mathbb{R}^1$ the function $1/x$ may be written as $x/|x|^2 = x/|x|^{n+1}$, where n is the dimension of $\mathbb{R}^1$, in $\mathbb{R}^n$, $n > 1$, we have the following n analogs of the function $1/x$:

$$k_j(x) = \frac{x_j}{|x|^{n+1}}, \qquad j = 1, \ldots, n \tag{1.24}$$

There are therefore n analogs of the Hilbert transform in $\mathbb{R}^n$, namely,

$$R_j f(x) = c_n \text{ P.V.} \int_{\mathbb{R}^n} f(x - t) \frac{t_j}{|t|^{n+1}}\, dt \tag{1.25}$$

$j = 1, \ldots, n$, called the Riesz transforms of f.

Since in $\mathbb{R}^1$ it is

$$\frac{1}{x} = \frac{\operatorname{sgn} x}{|x|} = \frac{\Omega(x)}{|x|}$$

where $\Omega(x)$ is a homogeneous function (i.e., $\Omega(\lambda x) = \Omega(x)$ for all $\lambda > 0$), such that

$$\Omega(x) + \Omega(-x) = 0 \tag{1.26}$$

this suggests another generalization of the Hilbert transform to $\mathbb{R}^n$. Observe that condition (1.26) is essential to the existence of the P.V. of the integral of $1/x$ and it can be rewritten formally as

$$\int_\Sigma \Omega(x)dx = 0 \tag{1.26a}$$

where $\Sigma = \Sigma_1 = \{-1, 1\}$ is the "unit sphere" in $\mathbb{R}^1$.

So we consider in $\mathbb{R}^n$, $n > 1$, kernels k of the type

$$k(x) = \frac{\Omega(x)}{|x|^n} \tag{1.27}$$

with two special properties on Ω,

(i) Ω is homogeneous (of degree zero):

$$\Omega(\lambda x) = \Omega(x) \quad \text{for all} \quad \lambda > 0 \tag{1.27a}$$

(ii) Ω has mean value zero on the unit sphere:

$$\int_\Sigma \Omega(x)dx = 0 \tag{1.27b}$$

These kernels k are called Calderón-Zygmund kernels and will be denoted throughout this chapter as C-Z kernels. Since $\Omega(x) = \Omega(x')$ for $x' \in \Sigma$, Ω can be considered as a function defined on Σ.

For each C-Z kernel k we consider the singular integral

$$Kf = f * \text{P.V. } k(x) = \text{P.V.} \int_{\mathbb{R}^n} f(x - t) \frac{\Omega(t)}{|t|^n} dt$$

$$= \lim_{\varepsilon \to +0} \int_{|x-t|>\varepsilon} f(x - t) \frac{\Omega(t)}{|t|^n} dt \qquad (1.28)$$

and the operator $K : f \to Kf$ is called a C-Z operator.

For $\Omega_j(x) = x_j/|x|$ we obtain the Riesz operator R_j.

In order to insure the existence of (1.28) and to obtain for it properties similar to those of the Hilbert transform, we still have to ask a smoothness condition on Ω, for instance that Ω belongs to a Lipschitz class or to C^1. Since $\Omega \in \text{Lip } \alpha$, $0 < \alpha \leq 1$, means that $\omega(\rho) \leq C\rho^\alpha$, where

$$\omega(\rho) = \sup\{|\Omega(x') - \Omega(y')| : x', y' \in \Sigma, |x' - y'| < \rho\} \qquad (1.29)$$

is the modulus of continuity of Ω on Σ, and $\Omega \in C^1(\Sigma)$ entails that $\omega(\rho) \leq C\rho$, both conditions imply

$$\int_0^1 (\omega(\rho)/\rho)d\rho < \infty \qquad (1.30)$$

That Ω satisfies condition (1.30), usually known as a "Dini type" condition, is thus more general than $\Omega \in \text{Lip } \alpha$ or $\Omega \in C^1$ and is preferred for the treatment of the C-Z kernels. Observe that if Ω satisfies (1.30), then $\Omega \in L^\infty(\Sigma)$.

In Section 2 we shall see that the above results concerning the boundedness of the Hilbert operator H in $L^2(\mathbb{T})$ and $L^2(\mathbb{R}^1)$ extend to the Riesz operators R_j, $j = 1, \ldots, n$, and also to the general C-Z operators K.

The chapter deals with the fundamental properties of the singular integral operators given by convolution with kernels with singularities at the origin and at infinity. The theory for such singular integrals was developed by Calderón and Zygmund in [1] and extended by them to the more general case of variable kernels (singular integrals that are not given by convolution) in [2]. For references see the survey papers [3], [4] and [5].

2. SINGULAR INTEGRALS: THE L^2 THEORY

To study the C-Z operators acting on $L^2(\mathbb{R}^n)$, we first establish some essential properties of the multiplier operators.

Definition 2.1. An operator $T : L^2(\mathbb{R}^n) \to L^2(\mathbb{R}^n)$, $n \geq 1$, is called a multiplier operator with symbol $\sigma(x) \in L^\infty(\mathbb{R}^n)$ if

$$(Tf)^\wedge(x) = \sigma(x)\hat{f}(x) \tag{2.1}$$

for all $f \in L^2(\mathbb{R}^n)$.

Remark 2.1. If E_N is an N-dimensional Hilbert space and if $e_1, \ldots, e_N$ is a basis for E_N, then every operator T in E_N is given by a matrix (t_{jk}) in this basis. If $t_{jk} = \sigma_j \delta_{jk}$, $\delta_{jj} = 1$, $\delta_{jk} = 0$ for $j \neq k$, then T is a diagonal operator in this basis and, for every $x = \Sigma\, c_j e_j \in E_N$, we have that $T = \Sigma \sigma_j c_j e_j$, that is the coordinates of Tx are obtained from those of x by multiplying them by σ_j. In the infinite-dimensional space $L^2(\mathbb{T})$ we have an infinite basis given by $e_n = \exp(int)$, $n \in \mathbb{Z}$, and the analog of a diagonal operator is an operator T such that $Tf = \Sigma_n \sigma_n c_n \exp(int)$ whenever $f = \Sigma_n c_n \exp(int)$ ($\{\sigma_n\}$ a fixed sequence independent of f), i.e., $(Tf)^\wedge(n) = \sigma_n \hat{f}(n)$. Such T is called a multiplier operator and the fixed sequence $\{\sigma_n\}$ is called the multiplier or symbol of T.

Remark 2.2. Formula (1.14) tells us that H is a multiplier operator in $L^2(\mathbb{R}^1)$ with symbol $\sigma(x) = -i \operatorname{sgn} x$.

Proposition 2.1. For a given bounded linear operator T acting on $L^2(\mathbb{R}^n)$, the following conditions are equivalent:

(1) T is a multiplier operator;

(2) T commutes with translations;

(3) $T\varphi * \psi = \varphi * T\psi$ for all $\varphi, \psi \in L^2(\mathbb{R}_n)$.

Proof. (1) implies (2): Let $(Tf)^\wedge(x) = \sigma(x)\hat{f}(x)$. For any translation τ_h, $h \in \mathbb{R}^n$, by relation (ii) of Chapter 2, Section 1, we have

$$(T(\tau_h f))^\wedge(x) = \sigma(x)\exp(-2\pi i\, x\cdot h)\hat{f}(x) = \exp(-2\pi i\, x\cdot h)\sigma(x)\hat{f}(x)$$
$$= \exp(-2\pi i\, x\cdot h)(Tf)^\wedge(x) = (\tau_h(Tf))^\wedge(x)$$

and, therefore, $T\tau_h = \tau_h T$ in L^2.

(2) implies (3): For any $g \in L^2$ and $\varphi, \psi \in L^1 \cap L^2$ it is

$$<T\varphi * \psi, g> = \int (T\varphi * \psi)(x)g(x)dx = \int(\int T\varphi(x - t)\psi(t)dt)g(x)dx$$
$$= \int\int \tau_t(T\varphi)(x)\psi(t)g(x)dxdt$$
$$= \int(\int T(\tau_t\varphi)(x)g(x)dx)\psi(t)dt \tag{2.2}$$

But by the F. Riesz representation theorem there is $g_1 \in L^2$ such that

$$<Tf, g> = <f, g_1>$$

so (2.2) becomes

$$<T\varphi * \psi, g> = \int(\int \tau_t\varphi(x)g_1(x)dx)\psi(t)dt$$

$$= \int\int \varphi(x - t)\psi(t)g_1(x)dtdx$$

$$= \int (\varphi * \psi)(x)g_1(x)dx$$

Similarly,

$$<\varphi * T\psi, g> = \int (\varphi * \psi)(x)g_1(x)dx$$

so that

$$<T\varphi * \psi, g> = <\varphi * T\psi, g> \tag{2.3}$$

for all $g \in L^2$. From (2.3) we conclude that $T\varphi * \psi = \varphi * T\psi$ for all $\varphi, \psi \in L^1 \cap L^2$ and therefore (3) holds.

<u>(3) implies (1)</u>: Let $f \in L^2$, $f(x) \neq 0$ for all x and $\psi = \hat{f}$. As $T\varphi * \psi = \varphi * T\psi$ for all $\varphi \in L^2$, it will be $(T\varphi)^\wedge \hat{\psi} = \hat{\varphi}(T\psi)^\wedge$ and $(T\varphi)^\wedge = \hat{\varphi}(T\psi)^\wedge/\hat{\psi}$ for all $\varphi \in L^2$. Let us call $(T\psi)^\wedge/\hat{\psi} = \sigma$. We claim that $\sigma \in L^\infty$. In fact for any $\varphi \in L^2$,

$$\|\sigma\hat{\varphi}\|_2 = \|(T\varphi)^\wedge\|_2 = \|T\varphi\|_2 \leq \|T\| \cdot \|\hat{\varphi}\|_2$$

and so, if $E = \{x : |\sigma(x)| > \|T\|\}$, it cannot be $|E| > 0$, since then, for $\hat{\varphi} = \chi_E$, $\|\hat{\varphi}\|_2 = |E|^{\frac{1}{2}}$, and $\|\sigma\hat{\varphi}\|_2 > \|T\| \, |E|^{\frac{1}{2}}$ leads to a contradiction. But $|E| = 0$ means that σ is essentially bounded, as we had claimed. ∇

<u>Corollary 2.2</u>. If $k \in L^1 \cap L^2$ and K is the convolution operator of kernel k, then K commutes with translations and is a multiplier operator with symbol $\hat{k}$.

Corollary 2.3. If $\{T_n\}$ is a sequence of multiplier operators with symbols $\{\sigma_n\}$ such that $\|\sigma_n\|_\infty \leq C$ for all n, and $\sigma_n(x) \to \sigma(x)$ a.e., then $\|T_n f - Tf\|_2 \to 0$ for all $f \in L^2$, for T the multiplier operator with symbol σ.

Proof. Since $\|T_n f - Tf\|_2^2 = \|(T_n f)\hat{} - (Tf)\hat{}\|_2^2 = \|\sigma_n \hat{f} - \sigma \hat{f}\|_2^2$

$$= \int_{\mathbb{R}^n} |\sigma_n(x) - \sigma(x)|^2 |\hat{f}(x)|^2 dx$$

the thesis follows taking limits and applying the Lebesgue dominated convergence theorem. ∇

The translations form a group acting in $\mathbb{R}^n$, so do the (positive) dilations. The operators that commute with both groups are closely linked with the homogeneous functions through the following result. Recall that a function f is said to be homogeneous of degree α if

$$f(\lambda x) = \lambda^\alpha f(x) \quad \text{for all} \quad \lambda > 0 \tag{2.4}$$

Proposition 2.4. A bounded linear operator T acting on $L^2(\mathbb{R}^n)$ commutes with all the translations and (positive) dilations of $\mathbb{R}^n$ if and only if T is a multiplier operator whose symbol is a homogeneous function of degree zero.

Proof. By Proposition 2.1 we have only to prove that the necessary and sufficient condition for a multiplier T to commute with dilations is that its symbol must be a homogeneous function of degree zero. This means

$$\sigma(\lambda x) = \sigma(x) \quad \text{for all} \quad \lambda > 0 \tag{2.5}$$

In fact, since by relation (iii) of Chapter 2, Section 1, we have for any $a > 0$,

$$(\delta_a(Tf))^\wedge(x) = a^{-n}(Tf)^\wedge(x/a) = a^{-n}\sigma(x/a)\hat{f}(x/a)$$

and also

$$(T(\delta_a f))^\wedge(x) = \sigma(x)(\delta_a f)^\wedge(x) = \sigma(x)a^{-n}\hat{f}(x/a)$$

it follows that $T\delta_a = \delta_a T$ if and only if $\sigma(x/a) = \sigma(x)$ for all $a > 0$, which is (2.5). ∇

With these facts in mind let us consider the Hilbert operator H defined on $L^2(\mathbb{R}^1)$ by (1.6a). Since H was shown to be a multiplier operator with symbol $\sigma(x) = -i \operatorname{sgn} x$, Proposition 2.4 yields the following two essential properties of H:

(I) H commutes with translations

(II) H commutes with dilations

and we may add that

(III) H anticommutes with the reflection: $H(\rho f) = -\rho(Hf)$, since, if $\rho : f(x) \to f(-x)$, then,

$$H(\rho f)(x) = \frac{1}{\pi} \text{P.V.} \int_{-\infty}^{\infty} \frac{f(-t)}{x-t}\, dt$$

$$= -\frac{1}{\pi} \text{P.V.} \int_{-\infty}^{\infty} \frac{f(t)}{-x-t}\, dt$$

$$= - Hf(-x) = - (\rho Hf)(x)$$

Properties (II) and (III) imply that, for every $\varepsilon \neq 0$,

$$\delta_{\varepsilon^{-1}} H \delta_\varepsilon = (\operatorname{sgn} \varepsilon) H \tag{2.6}$$

Moreover, (I), (II) and (III) characterize the Hilbert operator H up to a multiplicative constant. In fact,

Theorem 2.5. Given a bounded linear operator T acting on $L^2(\mathbb{R}^1)$, T satisfies conditions (I), (II) and (III) if and only if T is a constant times H, the Hilbert operator.

Proof. Since H satisfies (I), (II) and (III), we only have to prove the other half of the theorem. By (I) and (II) and Proposition 2.4, T is a multiplier operator with homogeneous symbol σ, so that $\sigma(x) = c$ for $x > 0$ and $\sigma(x) = c'$ for $x < 0$. But by (III), $T(\rho f)(x) = -\rho(Tf)(x)$, and therefore

$$\sigma(x)(\rho f)^{\wedge}(x) = -\rho(\sigma(x)\hat{f}(x))$$

or

$$\sigma(x)\hat{f}(-x) = -\sigma(-x)\hat{f}(-x),$$

so that $\sigma(-x) = -\sigma(x)$, $c' = -c$ and $\sigma(x) = c \operatorname{sgn} x$. The symbol corresponding to H is $-i \operatorname{sgn} x$, so the theorem holds with constant equal to ic. ∇

Remark 2.2. Note that the differential operator D given by $Df(x) = f'(x)$ (that does not act on $L^2(\mathbb{R})$ but acts on $C_0^\infty(\mathbb{R})$, for instance) has the following properties:

(I) D commutes with translations, (II') $D(\delta_a f) = a\delta_a(Df)$; (III) D anticommutes with reflection. Thus D satisfies the same properties as the Hilbert operator H with (II) replaced by (II'). The differentiation operator of order zero, D_0 (the identity operator) satisfies (I) and (II) but, instead of (III), it satisfies (III') $D_0(\rho f) = \rho(D_0 f)$, since it commutes with reflection. Thus H can be thought of as "an antisymmetric differentiation operator of order zero" and it is natural to extend the notion of differentiation operator in $\mathbb{R}^1$ so as to include H. A similar situation arises in $\mathbb{R}^n$.

We now turn to $L^2(\mathbb{R}^n)$, $n > 1$, and consider the singular integral operators given by the C-Z kernels as defined in (1.28). For these we have the following result.

Theorem 2.6. Let Ω be a function defined in $\mathbb{R}^n$ such that

(i) Ω is homogeneous of degree zero, i.e., Ω is determined by its values on the unit sphere Σ;

(ii) Ω has mean value zero on the unit sphere Σ, i.e.,

$$\int_\Sigma \Omega(x')dx' = 0$$

(iii) Ω satisfies a "Dini type" condition, i.e., if

$$\omega(\rho) = \sup\{|\Omega(x') - \Omega(y')| : x', y' \in \Sigma, |x' - y'| < \rho\}$$

then ω is a positive increasing function of $\rho > 0$ such that

$$\int_0^1 (\omega(\rho)/\rho)d\rho < \infty \tag{1.30}$$

Under these hypotheses, for each $f \in L^2(\mathbb{R}^n)$, the function

$$Kf(x) = \text{P.V.} \int_{\mathbb{R}^n} f(x - t)\frac{\Omega(t)}{|t|^n}\,dt \tag{1.28}$$

exists, is in L^2, and $K : f \to Kf$ is a multiplier operator with symbol σ given by

$$\sigma(x) = -(i\pi/2)\int_\Sigma \operatorname{sgn}(x' \cdot t')\Omega(t')dt' + \int_\Sigma \log(x' \cdot t')^{-1}\Omega(t')dt' \tag{2.7}$$

Proof. Since the given C-Z kernel $k(x) = \Omega(x)|x|^{-n}$ is not integrable, we first truncate it in order to take Fourier transforms.

Thus, for $0 < \varepsilon < \eta < \infty$, let

$$k_{\varepsilon\eta}(x) = \begin{cases} \Omega(x)|x|^{-n} & \text{if } \varepsilon < |x| < \eta, \\ 0 & \text{otherwise} \end{cases} \tag{2.8}$$

Since $k_{\varepsilon\eta} \in L^1(\mathbb{R}^n)$, for every $f \in L^2$ we have $k_{\varepsilon\eta} * f \in L^2$ and $\hat{k}_{\varepsilon\eta} \cdot \hat{f} \in L^2$. We claim that $\hat{k}_{\varepsilon\eta}$ has the two main properties:

(A) $\sup_x |\hat{k}_{\varepsilon\eta}(x)| \leq C$, independently of ε and η;

(B) $\lim_{\substack{\varepsilon \to 0 \\ \eta \to \infty}} \hat{k}_{\varepsilon\eta}(x) = \sigma(x) \in L^\infty$ for each $x \neq 0$.

In order to prove this, let us take polar coordinates $x = Rx'$, $y = ry'$, $R = |x|$, $r = |y|$, $x', y' \in \Sigma$, in the Fourier transform of $k_{\varepsilon\eta}$. Then,

$$\begin{aligned} \hat{k}_{\varepsilon\eta}(x) &= \int_{\mathbb{R}^n} k_{\varepsilon\eta}(t)\exp(-2\pi i\, x \cdot t)dt \\ &= \int_{\varepsilon<|t|<\eta} \Omega(t)|t|^{-n}\exp(-2\pi i\, x \cdot t)dt \\ &= \int_\Sigma \Omega(t')\Big(\int_\varepsilon^\eta \exp(-2\pi i\, Rr\, x' \cdot t')r^{-n}r^{n-1}dr\Big)dt' \\ &= \int_\Sigma \Omega(t')\Big(\int_{\varepsilon R}^{\eta R} \exp(-2\pi i\, s\, x' \cdot t')ds/s\Big)dt' \end{aligned}$$

Since $\int_\Sigma \Omega = 0$, we may write

$$\hat{k}_{\varepsilon\eta}(x) = \int_\Sigma \Omega(t')g_{\varepsilon,\eta,x}(t')dt' \tag{2.9}$$

where

$$g_{\varepsilon,\eta,x}(t') = \int_{R\varepsilon}^{R\eta} \frac{\exp(-2\pi i\, s x'\cdot t') - \cos 2\pi s}{s}\, ds \qquad (2.10)$$

We claim that

$$\lim_{\substack{\varepsilon\to 0\\ \eta\to\infty}} g_{\varepsilon,\eta,x}(t') \quad \text{exists for each} \quad x$$

and

$$|g_{\varepsilon,\eta,x}(t')| \le C \log|x'\cdot t'|^{-1} + C \quad \text{for some} \quad C > 1$$

The imaginary part of $g_{\varepsilon,\eta,x}$ is

$$\operatorname{Im}(g_{\varepsilon,\eta,x}(t')) = -\int_{R\varepsilon}^{R\eta} \frac{\sin 2\pi s\, x'\cdot t'}{s}\, ds$$

$$= -\operatorname{sgn}(x'\cdot t') \int_{2\pi x'\cdot t' R\varepsilon}^{2\pi x'\cdot t' R\eta} \frac{\sin u}{u}\, du$$

and the last expression is uniformly bounded, independently of the limits of integration and tends, when $\varepsilon \to 0$ and $\eta \to \infty$, to $-\pi/2\, \operatorname{sgn}(x'\cdot t')$. To study the real part of $g_{\varepsilon,\eta,x}$ let us consider first the case $\varepsilon R \le 1 \le \eta R$. Then

$$\operatorname{Re}(g_{\varepsilon,\eta,x}(t')) = \int_{R\varepsilon}^{R\eta} \frac{\cos 2\pi s\, x'\cdot t' - \cos 2\pi s}{s}\, ds$$

$$= \int_{R\varepsilon}^{1} + \int_{1}^{R\eta} = I_1 + I_2 \qquad (2.11)$$

As $|\cos 2\pi s\, x' \cdot y' - \cos 2\pi s| =$
$2|\sin \pi s(x' \cdot y' + 1)\sin \pi s(x' \cdot y' - 1)| \leq 2\pi s|x' \cdot y' + 1| 2\pi s|x' \cdot y' - 1|$
$\leq 8\pi^2 s^2$, we have

$$|I_1| \leq 8\pi^2 \int_{R\varepsilon}^{1} s\,ds = 4\pi^2(1 - (R\varepsilon)^2) \leq 4\pi^2$$

On the other hand, $|I_2| \leq |\int_{x' \cdot t'}^{(x' \cdot t')R\eta} s^{-1} \cos 2\pi s\, ds| + C.$
If $(x' \cdot t')R\eta > 1$, then

$$|I_2| \leq \int_{x' \cdot t'}^{1} \frac{ds}{s} + \int_{1}^{(x' \cdot t')R\eta} \frac{\cos 2\pi s}{s}\,ds \leq \log \frac{1}{|x' \cdot t'|} + 2C$$

where $C = \sup_R \int_1^R \frac{\cos u}{u}\,du.$

If $0 < (x' \cdot t')R\eta \leq 1$, then

$$|I_2| \leq \int_{x' \cdot t'}^{1} \frac{ds}{s} = \log \frac{1}{|x' \cdot t'|}$$

An analogous result holds for $x' \cdot t' < 0$, so that if $\varepsilon R \leq 1 \leq \eta R$,

$$\mathrm{Re}(g_{\varepsilon,\eta,x}(t') \leq 2 \log|x' \cdot t'|^{-1} + 4\pi^2 + 2C$$

Furthermore, if $R\varepsilon > 1$, then

$$\mathrm{Re}(g_{\varepsilon,\eta,x}) = \int_{R\varepsilon}^{R\eta} = \int_{1}^{R\eta} - \int_{1}^{R\varepsilon}$$

and so

$$|\mathrm{Re}(g_{\varepsilon,\eta,x})| \leq 2 \log|x' \cdot t'|^{-1} + 4C$$

Similarly, if $R\eta < 1$,

$$\mathrm{Re}(g_{\varepsilon,\eta,x}) = \int_{R\varepsilon}^{1} - \int_{R\eta}^{1}$$

and so $|\mathrm{Re}(g_{\varepsilon,\eta,x})| \leq 8\pi^2$. Thus, in all cases,

$$|\mathrm{Re}(g_{\varepsilon,\eta,x})| \leq C_1 \log|x' \cdot t'|^{-1} + C_2 \tag{2.12}$$

as claimed. Thus $g_{\varepsilon,\eta,x}(y')$ is majorized, independently of ε and η, by a function integrable on Σ. Since Ω is bounded on Σ, we get from (2.9) the estimate (A) for $\hat{k}_{\varepsilon\eta}$, with

$$C = \int_{\Sigma} \Omega(y')(1 + \log|x' \cdot t'|^{-1})dt$$

If h is a "good enough" function, then, for each pair $\lambda, \mu > 0$, we have

$$\lim_{\substack{\varepsilon \to 0 \\ \eta \to \infty}} \int_{\varepsilon}^{\eta} \frac{h(\lambda s)-h(\mu s)}{s}\, ds = h(0) \log \mu/\lambda \tag{2.13}$$

(see Exercise 2.1 below).

Applying (2.13) to $h(s) = \cos 2\pi s$, $\lambda = x' \cdot t'$ (which may be taken positive since cosine is an even function), $\mu = 1$, we get

$$\lim_{\substack{\varepsilon \to 0 \\ \eta \to \infty}} \int_{R\varepsilon}^{R\eta} (\cos 2\pi s\, x' \cdot t' - \cos 2\pi s)(ds/s) = \log|x' \cdot t'|^{-1}$$

Taking limits in (2.9) as $\varepsilon \to 0$ and $\eta \to \infty$ and applying the estimate (A) and the Lebesgue dominated convergence theorem, we obtain the existence of the limit in (B) with

$$\lim_{\substack{\varepsilon \to 0 \\ \eta \to \infty}} \hat{k}_{\varepsilon\eta}(x) = \int_{\Sigma} \Omega(t')\left(-\frac{i\pi}{2} \operatorname{sgn}(x' \cdot t') + \log|x' \cdot t'|^{-1}\right)dt' = \sigma(x)$$

as claimed.

Now, if we consider the operators $K_{\varepsilon\eta}$ given by convolution with the kernels $k_{\varepsilon\eta}$ then, for each $\varepsilon, \eta, K_{\varepsilon\eta}$ is, by Corollary 2.2, a multiplier operator of symbol $\hat{k}_{\varepsilon\eta}$ and, by (A), $\|\hat{k}_{\varepsilon\eta}\|_\infty \le C$ for all ε, η. Thus $\{K_{\varepsilon\eta}\}$ satisfy the conditions of Corollary 2.3 and, for $\varepsilon \to 0$ and $\eta \to \infty$, $K_{\varepsilon\eta}$ tends to the multiplier operator of symbol σ. On the other hand, if $f \in L^2$, then

$$K_{\varepsilon\eta}f(x) = k_{\varepsilon\eta} * f(x) = \int_{\varepsilon<|t|<\eta} f(x - t)k(t)dt$$

so that for every x,

$$\lim_{\eta\to\infty} K_{\varepsilon\eta}f(x) = \int_{\varepsilon<|t|} f(x - t)k(t)dt$$

Thus

$$\lim_{\varepsilon\to 0} (\lim_{\eta\to\infty} K_{\varepsilon\eta}f(x)) = \text{P.V.} \int_{\mathbb{R}^n} f(x - t)k(t)dt = Kf(x)$$

exists in L^2, $(Kf)^\wedge = \sigma(x)\hat{f}(x)$ and the theorem is proved. ∇

<u>Exercise 2.1</u>. Let $h \in C^1(R)$ be such that (i) $h'(u)/u$ is bounded in a neighborhood of 0, and (ii) $\int_a^\infty (h(u)/u)du$ converges absolutely if $a > 0$. Prove that

$$\lim_{\substack{\varepsilon \to 0 \\ \eta \to \infty}} \int_{\varepsilon}^{\eta} (h(\lambda\rho) - h(\mu\rho))(d\rho/\rho) = h(0)\log(\lambda/\mu)$$

for all $\lambda \geq \mu > 0$.

Remark 2.3. The proof of Theorem 2.6 holds under very general conditions on Ω. In fact, writing $\Omega = \Omega_o + \Omega_e$, where Ω_o is the odd part and Ω_e the even part of Ω, the uniform boundedness of $\operatorname{Im}(g_{\varepsilon,\eta,x})$ requires only that $\int_\Sigma |\Omega_o(y')|dy' < \infty$, i.e., $\Omega_o \in L^1(\Sigma)$. For the estimate of $\operatorname{Re}(g_{\varepsilon,\eta,x})$, the given proof requires the uniform boundedness of

$$\int_\Sigma |\Omega_e(y')| \log|x' \cdot y'|^{-1} dy' \tag{2.14}$$

Thus, if k is an odd kernel (i.e., $k(-x) = -k(x)$), homogeneous of degree $-n$, so that Ω is odd on Σ, the type (2, 2) of operator K will ge guaranteed by requiring only that $\Omega \in L^1(\Sigma)$.

For k not necessarily odd, the more general hypothesis that validates (2.14) is $\Omega \in L \log^+ L(\Sigma)$, the Zygmund class (for definitions see Chapter 0, Section 1.) In fact, by the Young convexity inequality (see Chapter 0, inequality (3.2)), which asserts in particular that for each $a, b \geq 0$, $ab \leq a\log(a+1) + e^b$, taking $a = |\Omega(y')|$, $b = \lambda \log|x' \cdot y'|^{-1}$, for $\lambda > 0$, we get

$$|\Omega(y')| \log|x' \cdot y'|^{-1} \leq \frac{1}{\lambda} |\Omega(y')| \log(1 + |\Omega(y')|) + \frac{1}{\lambda} |x' \cdot y'|^{-\lambda}$$

Since for $0 < \lambda < 1$, $|x' \cdot y'|^{-\lambda} = |\cos\varphi|^{-\lambda}$ is integrable on Σ, it is enough to insure the integrability of $|\Omega|\log(1 + |\Omega|)$. This is equivalent, Σ being compact, to $|\Omega|\log^+|\Omega| \in L^1(\Sigma)$. Hence $\Omega \in L \log^+ L$ guarantees the uniform boundedness of (2.14).

It has been proved in [1] that this hypothesis, together with the mean value zero of Ω on Σ, is sufficient to insure the L^p boundedness of operator K for $1 < p < \infty$. Furthermore, it is also the best possible hypothesis, as shown by a counterexample in [6].

Remark 2.4. The mean value zero condition on Σ, $\int_\Sigma \Omega = 0$, cannot be omitted. Since

$$\int_{\mathbb{R}^n} \Omega(y)|y|^{-n} f(x - y)dy = \int_{|y|\leq 1} + \int_{|y|>1}$$

and the second integral exists for Ω merely integrable on Σ, the treatment of the first is the main difficulty. For instance, if $\Omega \equiv 1$ and f is constant and different from zero in the neighborhood of a point x_0, the first integral diverges in that neighborhood. To get convergence, some kind of cancellation is required, hence the need for the mean value zero condition on Ω.

We now turn to considering the symbol σ of operator K.

Remark 2.5. The symbol $\sigma(x)$ is a homogeneous function of degree zero, as seen by Formula (2.7). This is merely a special case of the fact that in $\mathbb{R}^n$ the Fourier transform of any function homogeneous of degree α is homogeneous of degree $\alpha - n$.

Remark 2.6. From Formula (2.7) it follows immediately that σ has mean value zero on the unit sphere, since

$$\int_\Sigma \sigma(x')dx' = \int_\Sigma \left(\int_\Sigma \Omega(t')(-(\pi i/2)\mathrm{sgn}(x' \cdot t') + \log|x' \cdot t'|^{-1})dt'\right)dx'$$

$$= \int_\Sigma \Omega(t')\left(\int_\Sigma (-\pi i/2)\mathrm{sgn}(x' \cdot t') + \log|x' \cdot t'|^{-1}dx'\right)dt'$$

where, as is easily seen, the inner integral is independent of t' (see Chapter 0, Section 2) and Ω has mean value zero on Σ.

Remark 2.7. If $k(x)$ is an odd function the formula (2.7) reduces to

$$\sigma(x) = -(\pi i/2)\int_{\Sigma} \Omega(t')\operatorname{sgn}(x' \cdot t')dt'$$

$$= -\pi i \int_{\Sigma^+(x)} \Omega(t')dt' \tag{2.15}$$

where $\Sigma^+(x)$ denotes the hemisphere where $x' \cdot t' \geq 0$.

If we consider now the Riesz transforms as a particular case of the singular integrals of Theorem 2.6, we obtain expressions for their symbols that will enable us to see that these transforms are analogs in $\mathbb{R}^n$, $n > 1$, of the Hilbert transform in the sense that they can be characterized as H was in Theorem 2.5.

Definition 2.2. Given a function f defined in $\mathbb{R}^n$, its Riesz transforms are the n functions

$$R_j f(x) = c_n \lim_{\varepsilon \to +0} \int_{|t|>\varepsilon} f(x - t)\frac{t_j}{|t|^{n+1}}\,dt \tag{1.25}$$

$j=1, \ldots, n$, and,

$$k_j(x) = \Omega_j(x)|x|^{-n} \quad \text{where } \Omega_j(x) = c_n x_j |x|^{-1} \tag{1.24}$$

are the Riesz kernels. In these expressions,

$$c_n = \pi^{-(n+1)/2}\Gamma((n + 1)/2) \tag{2.16}$$

Since by (1.24) above, $\Omega_j(x)$ is for each $j = 1, \ldots, n$, an odd and homogeneous function of degree zero, has mean value zero on Σ, is discontinuous only at $x = 0$ and is smooth on Σ, every Ω_j satisfies the conditions of Theorem 2.6, and thus defines a multiplier operator that coincides with R_j, whose symbol σ_j is given, as indicated in (2.15), by

$$\sigma_j(x) = -\pi i c_n \int_{\Sigma+(x)} t'_j \, dt' \tag{2.17}$$

To compute σ_j, j fixed, note that $t'_j = t' \cdot x_j$, and since $t' = \Sigma_{k=1}^n (t' \cdot \ell_k)\ell_k$ for $\{\ell_1, \ldots, \ell_n\}$ any orthonormal basis in $\mathbb{R}^n$, then

$$t'_j = \sum_{k=1}^{n} (t' \cdot \ell_k)(\ell_k \cdot x_j)$$

Let us choose the ℓ_k's such that $\ell_j = x'$ for a fixed x. Then

$$t'_j = (t' \cdot x')x'_j + \sum_{k \neq j} c_k(t' \cdot \ell_k) \tag{2.18}$$

where $c_k = \ell_k \cdot x_j$. Integrating (2.18) over $\Sigma^+(x)$ we get

$$\sigma_j(x) = -\pi i c_n x'_j \int (x' \cdot t')dt' - \pi i c_n \sum_{j \neq k} c_k \int (t' \cdot \ell_k)dt'$$

Here the first integral on the right is equal to Ω_{n-1}, the volume of the unit sphere in $\mathbb{R}^{n-1}$, and the other integrals vanish. Therefore, by (2.16) and Lemma 2.1 of Chapter 0,

$$\sigma_j(x) = -\pi i c_n x'_j \Omega_{n-1} = -ix'_j \tag{2.19}$$

for every $j = 1, \ldots, n$.

From (2.19) we can deduce for the Riesz transforms a property analogous to one of the Hilbert transform in $L^2(\mathbb{R}^1)$, namely that $H^2 = -I$, I the identity operator. In fact, since for $f \in L^2(\mathbb{R}^n)$ every $R_j f$ is also in $L^2(\mathbb{R}^n)$, and since $(R_j f)^\wedge(x) = -i x_j |x|^{-1} \hat{f}(x)$, writing $R_j^2 f = R_j(R_j f)$, it follows that

$$(R_j^2 f)^\wedge(x) = -x_j^2 |x|^{-2} \hat{f}(x) \tag{2.20}$$

Thus,

$$\left(\sum_{j=1}^{n} R_j^2 f\right)^\wedge(x) = -\hat{f}(x) \tag{2.21}$$

and (2.21) can be interpreted as $\Sigma R_j^2 = -I$ in $L^2(\mathbb{R}^n)$.

Furthermore, (2.20) and Plancherel's theorem imply

$$\sum_{j=1}^{n} \|R_j f\|_2^2 = \|f\|_2^2$$

An n-tuple $(v_1(x), \ldots, v_n(x))$ of functions defined on $\mathbb{R}^n$ is said to transform like a vector under a rotation $A = (a_{jk})$, $j, k = 1, \ldots, n$, if, for each j,

$$v_j(A^{-1}x) = \sum_k a_{jk} v_k(x)$$

Another important property of the Riesz transforms is that their symbols $(\sigma_1, \ldots, \sigma_n)$ are transformed as a vector under rotations in $\mathbb{R}^n$. This fact characterizes the Riesz transform up to a constant through the following

Lemma 2.7. If $m_1(x), \ldots, m_n(x)$ are homogeneous functions of degree zero such that $(m_1, \ldots, m_n)$ transforms as a vector under every rotation of $\mathbb{R}^n$, then $m_j(x) = cx_j|x|^{-1}$, for $j = 1, \ldots, n$, and some constant c.

Proof. By hypothesis it is sufficient to consider $x \in \Sigma$. Let $e_1, \ldots, e_n$ be the canonical basis of $\mathbb{R}^n$ and let $c = m_1(e_1)$. We claim that $m_j(e_1) = 0$ for $j \neq 1$. In fact, if $A = (a_{jk})$ is a rotation that leaves e_1 fixed, we have for $j \neq 1$,

$$m_j(e_1) = m_j(A^{-1}e_1) = \sum_{k=2}^{n} a_{jk}m_j(e_1)$$

and so $(m_2(e_1), \ldots, m_n(e_1))$ is a fixed vector under every rotation of $\mathbb{R}^{n-1}$ and it must be the zero vector. Thus $m_j(e_1) = 0$ for $j = 2, \ldots, n$. If A is now a general rotation in $\mathbb{R}^n$, then

$$m_j(Ae_1) = a_{j1}m_1(e_1) + \sum_{k=2}^{n} a_{jk}m_j(e_1) = a_{j1}m_1(e_1)$$

but, if $Ae_1 = x \in \Sigma$ then $a_{j1} = x_j$, so

$$m_j(x) = cx_j$$

and the lemma is proved, since $|x| = 1$. ∇

Now we have the following analog of Theorem 2.5.

Theorem 2.8 (Geometric characterization of the Riesz Transforms). Given an n-tuple of bounded linear operators $(T_1, \ldots, T_n)$ acting on $L^2(\mathbb{R}^n)$, they satisfy the conditions

(I) every T_j, $j = 1, \ldots, n$, commutes with all translations in $\mathbb{R}^n$,

(II) every T_j, $j = 1, \ldots, n$, commutes with all dilations in $\mathbb{R}^n$,

(III) for every rotation $A = (a_{jk})$ in $\mathbb{R}^n$, $(T_1, \ldots, T_n)$ is transformed as a vector, i.e.,

$$AT_jA^{-1} = \sum_k a_{jk}T_k, \qquad j = 1, \ldots, n$$

if and only if there is a constant c such that

$$T_j = cR_j, \qquad j = 1, \ldots, n$$

where R_j is the Riesz transform operator.

<u>Proof</u>. Since R_j, $j = 1, \ldots, n$, is a multiplier operator with homogeneous symbol σ_j given by (2.19), R_j commutes with translations and dilations and satisfies (III). Conversely, assume that $(T_1, \ldots, T_n)$ satisfy (I), (II) and (III). By (I) and (II) every T_j, $j = 1, \ldots, n$, is a multiplier operator with homogeneous symbol m_j and Lemma 2.7 implies then that, by (III), $m_j(x) = cx_j|x|^{-1}$. By (2.19) the theorem is proved. ∇

The Riesz transforms are closely linked, through the Poisson integrals, to the theory of harmonic functions of several variables. We have seen in Section 1 how the Hilbert transform appears in connection with the boundary values of the harmonic conjugate function. To indicate the analogy in the multidimensional case, let us note that an analytic function of one complex variable can be thought (at least locally) as a gradient of a harmonic function in two real variables, and so it is natural to consider gradients of multidimensional harmonic functions as generalizations of the analytic functions to several variables (for more on this outlook, see [7]).

More precisely, let a <u>system of conjugate functions</u> in $\mathbb{R}^n$ be an n-tuple of functions $\{u_1(x), \ldots, u_n(x)\}$ such that $u_j \in C^2(\mathbb{R}^n)$, $j = 1, \ldots, n$, and satisfy the <u>generalized Cauchy-Riemann equations</u>

$$\sum_{j=1}^{n} \left(\frac{\partial u_j}{\partial x_j}\right) = 0 \quad \text{and} \quad \frac{\partial u_j}{\partial x_k} = \frac{\partial u_k}{\partial x_j} \quad \text{for} \quad j \neq k \tag{2.22}$$

In a simply connected domain, the second equation implies the existence of a function in n variables of which $\{u_1, \ldots, u_n\}$ is the gradient, and the first equation implies that this function is harmonic. The existence of a system of conjugate functions in $\mathbb{R}^n$ implies, at least locally, the existence of a harmonic function $h(x)$, $x \in \mathbb{R}^n$, such that $u_j(x) = \partial h/\partial x_j$ for $j = 1, \ldots, n$. Based on the study of the Poisson integrals done in Chapter 3 we have the following characterization for the systems of conjugate functions in terms of Riesz transforms.

<u>Theorem 2.9</u>. Given $n + 1$ functions $f, f_1, \ldots, f_n \in L^2(\mathbb{R}^n)$, and their Poisson integrals $u_0 = P_y * f$, $u_j = P_y * f_j$, $j = 1, \ldots, n$, the $(n + 1)$-tuple $\{u_0(x, y), u_1(x, y), \ldots, u_n(x, y)\}$ is a system of conjugate functions in $\mathbb{R}^{n+1}$ if and only if

$$f_j = R_j f, \quad j = 1, \ldots, n \tag{2.23}$$

(In $\mathbb{R}^{n+1}$, the variable y plays the role of x_0.)

<u>Proof</u>. If (2.23) holds, then as by (2.19), $\hat{f}_j(x) = -ix_j|x|^{-1}\hat{f}(x)$, Theorem 1.3 of Chapter 3, that expresses the Poisson integral of a function by the Abel means of its Fourier transform, yields

$$u_j(x, y) = P_y * f_j(x) = -i\int \hat{f}(t)t_j|t|^{-1}\exp(2\pi ix \cdot t)\exp(-2\pi|t|y)dt$$

for $j = 1, \ldots, n$. These integrals are rapidly convergent, so derivation under the integral sign is justified and the equations (2.20) are immediately verified. Conversely, for each $j = 1, \ldots, n$, let

$$u_j(x, y) = P_y * f_j(x) = \int \hat{f}_j(t)\exp(2\pi i x \cdot t)\exp(-2\pi|t|y)dt$$

and

$$u_0(x, y) = P_y * f(x) = \int \hat{f}(t)\exp(2\pi i x \cdot t)\exp(-2\pi|t|y)dt$$

Since $\partial u_0/\partial x_j = \partial u_j/\partial x_0 = \partial u_j/\partial y$ by equations (2.20),

$$2\pi i t_j \hat{f}(t)\exp(-2\pi|t|y) = 2\pi|t|\hat{f}_j(t)\exp(-2\pi|t|y)$$

Thus $\hat{f}_j(t) = it_j|t|^{-1}\hat{f}(t)$ and $f_j(x) = R_j f(x)$ for $j = 1, \ldots, n$.

∇

Remark 2.7. This result has been generalized for functions in $L^p(\mathbb{R}^n)$, $1 < p < \infty$, [8]. In the case $p = 1$ it leads to the characterization of the Hardy space $H^1(\mathbb{R}^n)$ in terms of a subclass of $L^1(\mathbb{R}^n)$ and, in view of the Remark 4.2 of Section 4, to the duality between H^1 and BMO.

3. GENERAL THEOREMS IN L^p AND BMO

In the two preceding sections we studied the Hilbert transform operator H and its n-dimensional generalizations as operators acting on L^2, and proved that they are of type (2, 2). These results extend to L^p, $1 < p < \infty$, insuring the existence of Hf for every $f \in L^p$. More precisely, there is the classical result of M. Riesz that asserts that H is of type (p, p), i.e.,

$$\int_{-\infty}^{\infty} |Hf(x)|^p dx \le M_p \int_{-\infty}^{\infty} |f(x)|^p dx \qquad (3.1)$$

for all $f \in L^p(\mathbb{R})$, $1 < p < \infty$, and M_p independent of f.

Unlike the case $p = 2$, where the results for H can be obtained by Fourier methods and so extend to $n > 1$, M. Riesz proved inequality (3.1) by means of the theory of analytic functions of one complex variable and his methods do not extend to $n > 1$. Besicovitch and Titchmarsh introduced real variable techniques to the study of the conjugate function and these were later developed by Calderón and Zygmund, permitting the treatment of the n-dimensional case.

As in other theories, even for $n = 1$ the cases $p = 1$ and $p = \infty$ are excluded in the type (p, p) result (3.1). The case $n = p = 1$ was studied by Kolmogoroff, Privalov, and others, obtaining that H is of weak type $(1, 1)$. The result for $p = \infty$ is more recent and asserts that $H : L^\infty \to BMO$ continuously. These assertions, as well as their generalizations for $n > 1$, are proven in this and the following sections.

As we deal with the type (p, p) of convolution operators we start by a general property of these.

Proposition 3.1. Given an integrable or square integrable kernel, if the corresponding convolution operator T is of type $(2, 2)$ and of weak type $(1, 1)$, then T is of type (p, p) for all p, $1 < p < \infty$.

Proof. By the hypotheses and the Marcinkiewicz interpolation theorem (Theorem 4.1 of Chapter 4), T is of type (p, p) for $1 < p \le 2$.

The case $2 < p < \infty$ is obtained by duality. Let $2 < p < \infty$ and p' be such that $1/p + 1/p' = 1$, with $1 < p' < 2$. Let k be the kernel of the convolution operator T, i.e.,

$$Tf(x) = \int_{\mathbb{R}^n} k(x - y)f(y)dy$$

for every $f \in L^1 \cap L^p$. If $\phi \in L^{p'}$, as $k \in L^1$ or L^2, then by Fubini's theorem,

$$\langle Tf, \phi\rangle = \int Tf(x)\phi(x)dx = \int (\int k(x - y)f(y)dy)\phi(x)dx$$

$$= \int (\int k(x - y)\phi(x)dx)f(y)dy \tag{3.2}$$

where $\int k(x - y)\phi(x)dx = T_-\phi(x) \in L^{p'}$, by the first part of the proposition, since T_- is the convolution operator with kernel $k_-(x) = k(-x)$, and shares the same type properties of T. Furthermore, $\|T_-\phi\|_{p'} \leq C_{p'} \|\phi\|_{p'}$, so by Hölder's inequality, (3.2) can be estimated as

$$|\langle Tf, \phi\rangle| \leq C_{p'} \|f\|_p \|\phi\|_{p'} \quad \text{for all} \quad \phi \in L^{p'}$$

By the F. Riesz representation theorem, $Tf \in L^p$, with

$$\|Tf\|_p = \sup_{\|\phi\|_{p'} \leq 1} |\langle Tf, \phi\rangle| \leq C_{p'} \|f\|_p$$

which is the type (p, p) of operator T for $2 < p < \infty$. ∇

In view of this proposition, to check the type (p, p) of a convolution operator we need only check what happens when $p = 2$ and $p = 1$. We are going to present some results leading to the weak type $(1, 1)$ of a class of convolution operators which includes the singular integrals.

Given a convolution operator,operator T of kernel k, we are going to prove as Theorem 3.3, that under certain conditions on k, T is of weak type $(1, 1)$. The idea of the proof of that proposition stems from the classical consideration of the Hilbert transform in the P.V. sense, where an integrable function is presented as the sum of a "good" function g and a "bad" function b and the cancellation of the kernel is used to deal with b. In fact, the principal problem in dealing with the integral $\int_{-L}^{L} b(t)(x - t)^{-1}dt$ comes from the logarithm resulting from the integration of $1/x$. If we can replace the given integral by $\int_{-L}^{L} b(t)((x - t)^{-1} - x^{-1})dt$ the situation is better, since $|(x - t)^{-1} - x^{-1}| \sim L/x^2$ when x is far from $[-L, L]$, and $L\int_{|x|>2L} (dx/x^2) \leq 1$. But the change can always be done if $\int_{-L}^{L} b(t)dt = 0$, so the solution of the problem involves finding a "bad" function with mean value zero on appropriate intervals. This is done in $\mathbb{R}^n$ through a lemma due to Calderón and Zygmund, that is of remarkable use also in other problems.

<u>Lemma 3.2 (The Calderón-Zygmund decomposition lemma)</u>. Let $f \in L^1(\mathbb{R}^n)$ be a positive function. Then $\mathbb{R}^n = P \cup Q$, $P \cap Q = \phi$, where $Q = \bigcup_{j=1}^{\infty} Q_j$, $\{Q_j\}$ nonoverlapping cubes, and $f = g + b$, where $g \in L^2(\mathbb{R}^n)$, $b(x) = 0$ a.e. in P, and b has mean value zero on every Q_j.

<u>Proof</u>. Given a positive $f \in L^1$, for each $\alpha > 0$, the decomposition $\mathbb{R}^n = P \cup Q$ exists by Lemma 3.5 of Chapter 5, and is such that $\Sigma_j |Q_j| \leq \|f\|_1/\alpha$, with $f(x) \leq \alpha$ a.e. in P and $\alpha < |Q_j|^{-1}\int_{Q_j} f(x)dx \leq 2^n\alpha$, for every $j = 1, 2, \ldots$. Thus, defining

$$g(x) = \begin{cases} f(x) & \text{if } x \in P \\ |Q_j|^{-1}\int_{Q_j} f(x)dx & \text{if } x \in Q_j \end{cases} \tag{3.3}$$

and $b(x) = f(x) - g(x)$, it is $b(x) = 0$ a.e. in P and $\int_{Q_j} b(x)dx = 0$, for $j = 1, 2, \ldots$. It remains only to show that $g \in L^2$. In fact,

$$\|g\|_2^2 = \int |g(x)|^2 dx = \int_P + \int_Q$$

$$\leq \alpha \int_P |f(x)|dx + \sum_{j=1}^{\infty} \int_{Q_j} \left|\frac{1}{Q_j} \int_{Q_j} f(t)dt\right|^2 dx$$

$$\leq \alpha \|f\|_1 + \sum_{k=1}^{\infty} (2^n \alpha)^2 |Q_j|$$

$$\leq (1 + 2^{2n})\alpha \|f\|_1 \tag{3.4}$$

and the lemma is proved. ∇

Remark 3.1. For every $\alpha > 0$ we get a different decomposition of f.

Theorem 3.3. Let $k \in L^2(\mathbb{R}^n)$ be such that there exists a constant $A > 0$ for which

(a) $|\hat{k}(x)| \leq A$ for all $x \in \mathbb{R}^n$

and

(b) $\int_{|x|>2|y|} |k(x - y) - k(x)|dx \leq A$ for all $y \neq 0$ (3.5)

Then the convolution operator T of kernel k is of type (2, 2) and of weak type (1, 1) with constants depending only on A and n.

Remark 3.2. The assumption $k \in L^2$ can be weakened. We will show how to do this for singular kernels in Section 4.

Proof. Let us prove first the type (2, 2) in $L^1 \cap L^2$ and then extend it to all L^2 by continuity. For $f \in L^1 \cap L^2$, $(Tf)\hat{}(x) = \hat{k}(x)\hat{f}(x)$ and Plancherel's theorem yields

$$\|Tf\|_2 = \|k * f\|_2 = \|\hat{k} \cdot \hat{f}\|_2 \leq A\|f\|_2 \tag{3.6}$$

by condition (a) on k, which is the type (2, 2) condition with constant A.

For the weak type (1, 1) of T, we remark that it is sufficient to consider positive integrable functions, since for a general $f = f^+ - f^-$, $f^+, f^- \geq 0$ it is $Tf = Tf^+ - Tf^-$ and $(Tf)_*(\alpha) \leq (Tf^+)_*(\alpha/2) + (Tf^-)(\alpha/2)$, and therefore the weak type (1, 1) of T on positive functions entails its weak type (1, 1) on all functions. Then, let fix $f \geq 0$ and $\alpha > 0$. By Lemma 3.2 we can express $f = g + b$ and

$$(Tf)_*(\alpha) \leq (Tg)_*(\alpha/2) + (Tb)_*(\alpha/2)$$

Since $g \in L^2$ by Lemma 3.2, with L^2 norm bounded as in (3.4), the type (2, 2) of operator T and Chebyshev's inequality (Proposition 3.2 (a) of Chapter 4) give us

$$(Tg)_*(\alpha/2) \leq 4\alpha^{-2}\|Tg\|_2^2 \leq 4(1 + 2^{2n})A^2\|f\|_1/\alpha \tag{3.7}$$

As for the bad part b, let $b_j(x) = b(x)\chi_{Q_j}(x)$ for each j, thus $b(x) = \Sigma_j b_j(x)$, for a.e.x. For each j, let y_j be the center of the cube Q_j and let S_j be the sphere of center y_j and radius equal to the diameter of Q_j. Let $S = \bigcup_j S_j$ and S^c be its complement. Since $|S_j| = c_n|Q_j|$, c_n a fixed constant, it is

$$|S| \leq c_n|Q| = c_n \Sigma_j |Q_j| \leq c_n\|f\|_1/\alpha \tag{3.8}$$

By Lemma 3.2,

$$Tb_j(x) = \int_{Q_j} k(x - y)b_j(y)dy$$

(continued)

$$= \int_{Q_j} (k(x - y) - k(x - y_j))b_j(y)dy \tag{3.9}$$

and so, as $Tb(x) = \Sigma_j\, Tb_j(x)$,

$$\int_{S^c} |Tb(x)|dx \le \Sigma_j \int_{S^c} |Tb_j(x)|dx$$

$$\le \Sigma_j \int_{S_j^c} |Tb_j(x)|dx$$

$$\le \Sigma_j \int_{y \in Q_j} |b_j(y)| \left(\int_{x \notin S_j} |k(x - y) - k(x - y_j)|dx \right) dy$$

Let us write $x - y = (x - y_j) - (y - y_j)$. Thus, if $x \notin S_j$ and $y \in Q_j$, then $|x - y_j| \ge 2|y - y_j|$ and, by hypothesis (b),

$$\int_{x \notin S_j} |k(x - y) - k(x - y_j)|dx \le \int_{|z| \ge 2|y - y_j|} |k(z - (y - y_j)) - k(z)|dz$$

$$\le A$$

Then

$$\int_{S^c} |Tb(x)|dx \le \Sigma_j \int_{Q_j} A|b_j(x)|dx$$

$$= A \int_Q |b(x)|dx \tag{3.10}$$

Since for $x \in Q$, $|b(x)| \le |f(x)| + 2^n\alpha$, we get

$$\int_Q |b(x)|dx \le \|f\|_1 + 2^n\alpha|Q|$$

$$\le (1 + 2^n)\|f\|_1 \tag{3.11}$$

and from (3.10) and (3.11),

$$\int_{S^c} |Tb(x)|dx \leq (1 + 2^n)A\|f\|_1 \tag{3.12}$$

Finally, as

$$(Tb)_*(\alpha/2) = |\{|Tb| > \alpha/2\}| \leq |S^c \cap \{|Tb| > \alpha/2\}| + |S|$$

from (3.8) and (3.12) we get

$$(Tb)_*(\alpha/2) \leq ((1 + 2^n)A + c_n)\|f\|_1/\alpha \tag{3.13}$$

and by (3.7) and (3.13),

$$(Tf)_*(\alpha) \leq C_{An}\|f\|_1/\alpha \tag{3.14}$$

Since this can be done for every $\alpha > 0$, (3.14) gives the weak type (1, 1) of T with constant $C_{An} = (2 + 2^n + 2^{2n})A + c_n$. ∇

As an immediate consequence of Theorem 3.3 and Proposition 3.1 we have the

Corollary 3.4. If $k \in L^2(\mathbb{R}^n)$ satisfies conditions (a) and (b) of Theorem 3.3, then the convolution operator of kernel k is of type (p, p), for all $1 < p < \infty$, and of weak type (1, 1).

As it was already mentioned, there is a substitute result for $p = \infty$. As an example in the one-dimensional case, let us observe that the Hilbert transform of the bounded function $f(x) = \operatorname{sgn} x$ is $H(\operatorname{sgn} x) = (2/\pi)\log|x|^{-1}$, which is not bounded but belongs to the space BMO (see Chapter 5, Section 3).

Proposition 3.5. If $k \in L^1(\mathbb{R}^n)$ satisfies conditions (a) and (b), of Theorem 3.3, then the convolution operator T of kernel k transforms L^∞ into BMO and, for all $f \in L^\infty$, there is a C depending only on A and n such that

$$\|Tf\|_{BMO} \le C\|f\|_\infty \tag{3.15}$$

Remark 3.3. By Young's inequality, this lemma holds trivially for any $k \in L^1$, with $C = \|k\|_1$. The fact that in (3.16) C is independent of k will enable us to relax the condition $k \in L^1$ by taking limits on a sequence of kernels. See Theorem 4.4 in next section.

Proof. Let $f \in L^\infty$ be such that $\|f\|_\infty \le 1$ and fix Q, a cube with center at the origin, and S, a sphere also centered at the origin and of radius equal twice the diameter of Q. Let $f = f\chi_S + f\chi_{S^c} = f_1 + f_2$, thus $k * f = k * f_1 + k * f_2 = u_1 + u_2$. Since $f_1 \in L^2$, by Plancherel's theorem and condition (a),

$$\|u_1\|_2 = \|k * f_1\|_2 = \|\hat{k} \cdot \hat{f}_1\|_2 \le A\|f_1\|_2 \le Ac_n|Q|^{1/2} \tag{3.16}$$

and

$$\int_Q |u_1(x)|dx \le \|u_1\|_2|Q|^{1/2} \le Ac_n|Q| \tag{3.17}$$

where c_n is a fixed constant depending only on n.

On the other hand, while

$$u_2(x) = k * f_2(x) = \int_{\mathbb{R}^n} k(x - y)f_2(y)dy$$

let

$$a_Q = \int_{\mathbb{R}^n} k(-y)f_2(y)dy \tag{3.18}$$

where $a_Q < \infty$, since $k \in L^1$. Thus

$$u_2(x) - a_Q = \int_{\mathbb{R}^n} (k(x - y) - k(-y))f_2(y)dy \tag{3.19}$$

and, by condition (b),

$$\begin{aligned} \int_Q |u_2(x) - a_Q|dx &\le \int_{x \in Q} \left(\int_{y \notin S} |k(x - y) - k(-y)|dy\right)dx \\ &\le |Q| \int_{|y|>2|x|} |k(x - y) - k(-y)|dy \\ &\le |Q|A \end{aligned} \tag{3.20}$$

By (3.17) and (3.20),

$$\begin{aligned} |Q|^{-1} \int_Q |u(x) - a_Q|dx &\le |Q|^{-1} \int_Q |u_1(x)|dx \\ &\quad + |Q|^{-1} \int_Q |u_2(x) - a_Q|dx \\ &\le A(c_n + 1) \end{aligned}$$ ∇

The hypotheses (a) and (b) of Theorem 3.3 are of two kinds: (a) corresponds to the L^2 theory and its formulation is the most general (even if it is not the easiest or simplest to apply), and (b), which is used to insure the weak type (1, 1) of operator T, also represents essentially the weakest condition that validates the arguments of the proof of Theorem 3.3. Nevertheless, in applications it is enough to check that the kernels k satisfy stronger hypotheses that imply (a) and (b). In this direction we have

Corollary 3.6. Let $k \in C^1(\mathbb{R}^n - \{0\})$ be such that (a) holds and

$$\left|\frac{\partial k}{\partial x_j}(x)\right| \leq C|x|^{-n+1} \quad \text{for} \quad |x| > 0, \quad j = 1, \ldots, n \qquad (3.21)$$

Then the convolution operator T of kernel k is of type (p, p) for $1 < p < \infty$ and of weak type $(1, 1)$.

Proof. By Corollary 3.4, it is sufficient to show that (3.21) implies (b). If $|x| > 2|y| > 0$ then the segment through the points x and $x - y$ does not go by the origin. By the mean value theorem of elementary calculus, there exists a point ξ belonging to that segment such that $|\xi| > |x|/2$ and

$$|k(x - y) - k(x)| = |y|\left|\sum_{j=1}^{n} (\partial k/\partial x_j)(\xi)\theta_j\right|$$

where θ_j, $j = 1, \ldots, n$, are the cosine directors of the segment. Thus,

$$\begin{aligned} |k(x - y) - k(x)| &\leq |y| \sum_j |(\partial k/\partial x_j)(\xi)| \\ &\leq |y| nC|\xi|^{-n-1} \\ &\leq n2^{n+1}C|y||x|^{-n-1} \end{aligned} \qquad (3.22)$$

From (3.22),

$$\begin{aligned} \int_{|x|>2|y|} |k(x - y) - k(x)|dx &\leq 2^{n+1}nC|y| \int_{|x|>2|y|} |x|^{-n-1}dx \\ &\leq C_n \int_{|t|>1} |z|^{-n-1}dz \\ &= A < \infty \end{aligned}$$

∇

With respect to hypothesis (a), it is not always easy to assess the boundedness of $\hat{k}$ and, in any case, it is of interest to replace this hypothesis by one in terms of k itself. The next result is useful in applications.

Corollary 3.7. Let $k \in L^2(\mathbb{R}^n)$ and $A > 0$, a constant, such that

(b) $\int_{|x|>2|y|} |k(x - y) - k(x)|dx \le A$ for all $|y| > 0$,

(c) $\int_{|x|<R} |x||k(x)|dx \le AR$ for all $R > 0$,

(d) $|\int_{|x|<R} k(x)dx| \le A$ for all $R > 0$.

Then hypotheses (a) and (b) and, therefore, Corollary 3.4, hold.

Proof. By assumption, (b) is granted, so we have to show that (b), (c), (d) imply (a), i.e., the uniform boundedness of $\hat{k}(x)$. To estimate $\hat{k}(x)$ we choose $w \in \mathbb{R}^n$ such that $\exp(-2\pi ix \cdot w) = -1$ (this can always be done, taking $w = x/(2|x|^2)$, $|w| = (2|x|)^{-1}$). Then, in the L^2 sense,

$$\hat{k}(x) = \int e^{-2\pi ix\cdot y}k(y)dy = \frac{1}{2}\int e^{-2\pi ix\cdot y}(k(y) - k(y - w))dy$$

$$= \frac{1}{2}\int_{|y|>1/|x|} + \frac{1}{2}\int_{|y|\le 1/|x|}$$

$$= I_1 + I_2 \tag{3.23}$$

By (b),

$$|I_1| \le \frac{1}{2}\int_{|y|>2|w|} |k(y) - k(y - w)|dy \le A/2 \tag{3.24}$$

On the other hand, as $|x|^{-1} = 2|w|$,

$$2I_2 = \int_{|y|\leq 1/|x|} (e^{-2\pi i x \cdot y} - 1)k(y)dy + \int_{|y|\leq 1/|x|} k(y)dy$$

$$- \int_{|y|\leq 1/|x|} (e^{-2\pi i x \cdot y} + 1)k(y - w)dy + \int_{|y|\leq 2|w|} k(y - w)dy$$

$$= I + II - III + IV \tag{3.25}$$

Since, while $|x| \cdot |y| < 1$, there exists a constant c such that $|\exp(-2\pi i x \cdot y) - 1| \leq c|x| \cdot |y|$,

$$|I| \leq c|x| \int_{|y|\leq 1/|x|} |y||k(y)|dy \leq cA \tag{3.26}$$

by hypothesis (c). By the choice of w, $\exp(-2\pi i x \cdot y) = - \exp(2\pi i x \cdot (y - w))$, and so $\exp(-2\pi i x \cdot y) + 1 = - (\exp(2\pi i x \cdot (y - w) - 1)$, thus $|\exp(-2\pi i x \cdot y) + 1| \leq c|x| \cdot |y - w|$ and, as in (3.26),

$$|III| \leq c|x| \int_{|y|\leq 2|w|} |y - w||k(y - w)|dy$$

$$\leq c'|x| \int_{|z|\leq 3/(2|x|)} |z||k(z)|dz$$

$$\leq c''A$$

since $|y| \leq 2|w|$ implies $|y - w| \leq 3|w| = 3/(2|x|)$.

For estimating II and IV we use hypotheses (c) and (d). First,

$$|II| = \left|\int_{|y|\leq |w|} k(y)dy\right| \leq A$$

Finally, since $\{y : |y - w| \leq 3|w|\} = \{y : |y| \leq 2|w|\} \cup \{y : |y| > 2|w|, |y - w| \leq 3|w|\}$,

$$IV = \left(\int_{|y-w|\le 3|w|} - \int_{\substack{|y-w|\le 3|w| \\ |y|>2|w|}}\right) k(y - w)dy$$

and

$$|IV| \le \left|\int_{|z|\le 3|w|} k(z)dz\right| + |w|^{-1} \int_{|w|<|z|\le 3|w|} |z|\,|k(z)|dz \le 4A$$

Adding up all the estimates for (3.24) and (3.25), we get that

$$|\hat{k}(x)| \le CA$$

for C a constant. ∇

An important class of kernels that satisfy the hypotheses of Corollary 3.7 are those $k \in L^2$ such that

(a') $|k(x)| \le B|x|^{-n}$ for $|x| > 0$

(b) $\int_{|x|>2|y|} |k(x - y) - k(x)|dx \le A$ for all $|y| > 0$

(d') $\int_{R_1<|x|<R_2} k(x)dx = 0$ for $0 < R_1 < R_2 < \infty$

Remark 3.4. If T is a convolution operator of kernel k, and δ_ε is a dilation of factor $\varepsilon > 0$, then the operator $\delta_{\varepsilon^{-1}} T \delta_\varepsilon$ is the convolution operator of kernel k_ε, where

$$k_\varepsilon(x) = \varepsilon^{-n} k(x/\varepsilon) \tag{3.27}$$

It is important to note that *if* k *satisfies any of the hypotheses stated in this section, so does* k_ε, $\varepsilon > 0$, *and with the same bounds*.

In the following we are going to concentrate our attention on the study of the convolution operators T given by kernels k that make them invariant under dilations, that is by the above remark, kernels such that $k(x) = \varepsilon^{-n}k(x/\varepsilon)$ for all $\varepsilon > 0$. Since these kernels need not belong to L^2 it will be necessary to truncate them in order to make use of the results of this section.

4. THE CALDERÓN-ZYGMUND SINGULAR INTEGRALS

Let us reconsider the C-Z kernels defined at the end of Section 1 by

$$k(x) = \Omega(x)|x|^{-n} \tag{1.27}$$

for Ω a homogeneous function of degree zero that has mean value zero on the unit sphere.

Our present aim is to prove that, under suitable smoothness conditions on Ω, some truncations of the corresponding C-Z kernel k satisfy the conditions (a) and (b) of the preceding section and thus give rise to convolution operators bounded in L^p, $1 < p < \infty$. We recall that the need for the truncation comes from the fact that $k(x)$ as in (1.27) is singular both at the origin and at infinity and can never belong to L^2.

We need first to prove two elementary lemmas.

<u>Lemma 4.1.</u> If $|x| > 2|y|$ then $\left|\frac{x-y}{|x-y|} - \frac{x}{|x|}\right| \le 2\left|\frac{y}{x}\right|$.

<u>Proof.</u> In Figure 1 we have that $\varphi = \pi/2 > \varphi'$, $2\varphi' + \theta = \pi$, $OP' = 1$. Also $Q'P'/OP' = \sin\theta/\sin\varphi'$ and $\sin\theta = |y|/|x|$, thus $Q'P' = (1/\sin\varphi')(|y|/|x|)$. But

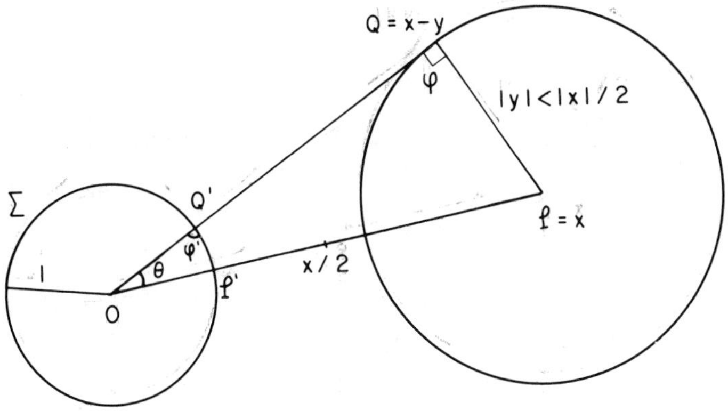

Figure 1

$$\sin \varphi' = \sin \frac{\pi - \theta}{2} = \cos \frac{\theta}{2} = \left(\frac{1+\cos\theta}{2}\right)^{1/2} = \left(\frac{1+|x-y|/|x|}{2}\right)^{1/2}$$

$$= \left(\frac{|x|+|x-y|}{2|x|}\right)^{1/2}$$

Then

$$(\sin \varphi')^{-1} = 2\left(\frac{|x|}{|x-y|+|x|}\right)^{1/2} \le 2$$

and the thesis follows since $Q'P' = |(x-y)/|x-y| - x/|x||$. ∇

<u>Lemma 4.2.</u> Let k be a kernel such that

(a'') $\int_{1\le|x|\le 2} |k(x)|dx \le B$,

(b) $\int_{|x|>2|y|} |k(x-y) - k(x)|dx \le A$ for all $|y| > 0$,

and let $k_1(x) = k(x)$ if $|x| > 1$ and zero otherwise. Then k_1 also satisfies condition (b) with constant equal to $A + 2B$.

Proof. Let $V = \{x : 1 \leq |x| \leq 2\}$ and χ_V be its characteristic function. Then we claim that

$$|k_1(x - y) - k_1(x)| \leq |k(x - y) - k(x)| + \chi_V(x - y)|k(x - y)| + \chi_V(x)|k(x)| \tag{4.1}$$

whenever $|x| > 2|y|$. In fact, if $|x| \geq 1$ and $|x - y| \geq 1$ then, $k_1(x - y) - k_1(x) = k(x - y) - k(x)$, and if $|x| \geq 1$ and $|x - y| < 1$ then $1 > |x - y| > |x| - |x|/2 = |x|/2$. So, if $x \in V$, then $|k_1(x - y) - k_1(x)| = |k_1(x)| = \chi_V(x)|k(x)|$. If $|x| < 1$ and $|x - y| \geq 1$, then $|x - y| \leq |x| + |x|/2 < 3/2$ and $x - y \in V$ so that $|k_1(x - y) - k_1(x)| = |k_1(x - y)| = \chi_V(x - y)|k(x - y)|$. Finally, if $|x| < 1$ and $|x - y| < 1$ then $k_1(x - y) - k_1(x) = 0$. Thus (4.1) is satisfied whereas $|x| > 2|y|$, hence

$$\begin{aligned}\int_{|x|>2|y|} |k_1(x - y) - k_1(x)|dx &\leq \int_{|x|>2|y|} |k(x - y) - k(x)|dx \\ &\quad + \int_{1\leq|x|\leq 2} |k(x)|dx \\ &\quad + \int_{1\leq|x-y|\leq 2} |k(x - y)|dx \\ &\leq A + 2B\end{aligned}$$

∇

We are now ready to state and prove the n-dimensional generalization of the Riesz inequality (3.1).

Theorem 4.3 (The Calderón-Zygmund theorem). Let Ω be a function defined in $\mathbb{R}^n$ such that

(i) Ω is homogeneous of degree zero, i.e., $\Omega(\lambda k) = \Omega(x)$, $\lambda > 0$,

(ii) Ω has mean value zero on the unit sphere, i.e., $\int_\Sigma \Omega = 0$,

(iii) Ω satisfies a "Dini type" condition, i.e.,

if $\omega(\rho) = \sup\{|\Omega(x') - \Omega(y')| : x', y' \in \Sigma, |x' - y'| < \rho\}$ then ω is a positive increasing function of $\rho > 0$ such that

$$\int_0^1 \omega(\rho)\rho^{-1}d\rho < \infty \tag{1.30}$$

For each $\varepsilon > 0$, let K_ε be the truncated operator defined for every $f \in L^p$, $1 < p < \infty$, by

$$K_\varepsilon f(x) = \int_{|y|>\varepsilon} \frac{\Omega(y)}{|y|^n} f(x - y)dy \tag{4.2}$$

Then the following assertions hold:

(1) For all $f \in L^p(\mathbb{R}^n)$, $K_\varepsilon f \in L^p(\mathbb{R}^n)$, and there exists a constant C_p (independent of ε and f) such that

$$\|K_\varepsilon f\|_p \leq C_p \|f\|_p$$

i.e., K_ε is of type (p, p) for all $\varepsilon > 0$, with the same constant C_p. Furthermore, K_ε is of weak type $(1, 1)$ also, with constant independent of ε.

(2) For every $f \in L^p$, the limit of $\{K_\varepsilon f\}$ when $\varepsilon \to 0$ exists in L^p, $\|K_\varepsilon f - Kf\|_p \to 0$, and the limit operator so defined is of type (p, p) with constant C_p, i.e.,

$$\|Kf\|_p \leq C_p \|f\|_p, \quad \text{for all} \quad f \in L^p$$

<u>Proof.</u> Let us remark first that it will be sufficient to prove (1) for $\varepsilon = 1$. In fact, under hypothesis (i), $K_\varepsilon = \delta_{\varepsilon^{-1}} K_1 \delta_\varepsilon$ for all $\varepsilon > 0$ (see Remark 3.4 at the end of the last section). Since for each $\varepsilon > 0$, δ_ε is a linear 1-1 operator from L^p to L^p and $\|\delta_\varepsilon f\|_p = \varepsilon^{-n/p} \|f\|_p$, property (1) for K_1 will imply that

$$\|K_\varepsilon f\|_p = \|(\delta_{\varepsilon^{-1}}K_1\delta_\varepsilon)f\|_p = \varepsilon^{n/p}\|K_1\delta_\varepsilon f\|_p$$

$$\leq \varepsilon^{n/p}C_p\|\delta_\varepsilon f\|_p = \varepsilon^{n/p}C_p\varepsilon^{-n/p}\|f\|_p$$

$$= C_p\|f\|_p$$

and (1) would be proved for all K_ε, $\varepsilon > 0$. Thus let us fix $\varepsilon = 1$ and consider the truncated kernel

$$k_1(x) = \begin{cases} \Omega(x)/|x|^n & \text{if} \quad |x| \geq 1 \\ 0 & \text{if} \quad |x| < 1 \end{cases} \tag{4.3}$$

Now $k_1 \in L^2$, so to prove (1) for K_1 it will be sufficient, by Corollary 3.7, to prove that k_1 satisfies its hypotheses (b), (c) and (d).

<u>Condition (b)</u>: Consider $k(x) = \Omega(x)|x|^{-n}$. By Lemma 4.2, to prove that k_1 satisfies Condition (b) it is enough to show that $k(x)$ does, since

$$\int_{1\leq|x|\leq 2} |k(x)|dx = \int_\Sigma |\Omega(x')|\int_1^2 r^{-n}r^{n-1}dr\, dx'$$

$$\leq \|\Omega\|_\infty \omega_n \log 2 < \infty$$

But now

$$k(x-y) - k(x) = \frac{\Omega(x-y)}{|x-y|^n} - \frac{\Omega(x)}{|x-y|^n} + \frac{\Omega(x)}{|x-y|^n} - \frac{\Omega(x)}{|x|^n}$$

$$= I + II$$

and, by (iii), Ω is bounded on Σ, so

$$\int_{|x|>2|y|} |II|dx \leq \|\Omega\|_\infty \int_{|x|>2|y|} (|x - y|^n - |x|^n)^{-1}dx$$

and the last integral is bounded independently of y.

To estimate the other factor we use Lemma 4.1. In fact, the distance between the projections on Σ of the points $x - y$ and x is

$$\left|\frac{x-y}{|x-y|} - \frac{x}{|x|}\right| \leq 2\left|\frac{y}{x}\right| \quad \text{for} \quad |x| > 2|y|$$

and, in such case, by (iii), $|\Omega(x - y) - \Omega(x)| \leq \omega(2|y|/|x|)$. Thus,

$$\begin{aligned}\int_{|x|>2|y|} |I|dx &= \int_{|x|>2|y|} |\Omega(x - y) - \Omega(x)|\,|x - y|^{-n}dx \\ &\leq \int_{|x|>2|y|} \omega(2|y|/|x|)|x - y|^{-n}dx \\ &\leq 2^n \int_{|x|>2|y|} \omega(2|y|/|x|)|x|^{-n}dx \\ &= 2^n\omega_n \int_{2|y|}^{\infty} \omega\left(\frac{2|y|}{r}\right) \frac{dr}{r} \\ &= 2^n\omega_n \int_0^1 \omega(\rho) \frac{d\rho}{\rho}\end{aligned}$$

and the last integral is bounded, by (iii), independently of $|y|$. So, k satisfies condition (b), i.e.,

$$\int_{|x|>2|y|} |k(x - y) - k(x)|dx \leq A \quad \text{for all} \quad |y| > 0$$

Condition (c): for every given $R > 0$,

$$\int_{|x|<R} |x|\,|k_1(x)|dx = \int_{1<|x|<R} |x|\,|\Omega(x)|\,|x|^{-n}dx$$

$$= \int_{\Sigma} |\Omega(x')| \int_1^R r^{1-n} r^{n-1} dr\, dx'$$

$$\le \|\Omega\|_\infty \omega_n (R - 1) \le AR$$

Condition (d): for a given $R > 0$,

$$\int_{|x|\le R} k_1(x)dx = \int_{1<|x|\le R} \Omega(x)|x|^{-n}dx$$

$$= \int_{\Sigma} \Omega(x') \int_1^R r^{-n} r^{n-1} dr\, dx'$$

$$= \log R \int_{\Sigma} \Omega(x')dx' = 0$$

by (ii), so

$$\left|\int_{|x|\le R} k_1(x)dx\right| = 0 \le A$$

Thus $K_1 : f \to k_1 * f$ is of type (p, p), $1 < p < \infty$, and of weak type $(1, 1)$, by Corollary 3.4, and (1) is satisfied for all $\varepsilon > 0$ uniformly on ε.

(2) We are going to prove that for each $f \in L^p$, $1 < p < \infty$, $\{K_\varepsilon f\}$ is a Cauchy sequence in L^p and, thus, that its limit exists in L^p.

Let us first show that this is the case for functions in the class $C_0^1(\mathbb{R}^n)$. If $g \in C_0^1(\mathbb{R}^n)$ and $0 < \varepsilon_1 < \varepsilon_2$,

$$K_{\varepsilon_1} g(x) - K_{\varepsilon_2} g(x) = \int_{\varepsilon_1<|y|<\varepsilon_2} \Omega(y)|y|^{-n} g(x - y)dy$$

$$= \int_{\varepsilon_1 < |y| < \varepsilon_2} \Omega(y)|y|^{-n}(g(x-y) - g(x))dy$$

by Condition (d) and, by Minkowski's integral inequality,

$$\|K_{\varepsilon_1} g - K_{\varepsilon_2} g\|_p \leq \int_{\varepsilon_1 < |y| < \varepsilon_2} |\Omega(y)|\,|y|^{-n}\|g(.\, - y) - g\|_p dy \tag{4.4}$$

But since $g \in C_0^1$, $\|g(.\, - y) - g\|_p$ is bounded by a constant times $|y|$, and (4.4) becomes

$$\|K_{\varepsilon_1} g - K_{\varepsilon_2} g\|_p \leq C_p \|\Omega\|_\infty \int_{\varepsilon_1 < |y| < \varepsilon_2} |y|^{-n+1} dy \tag{4.5}$$

which tends to zero as $\varepsilon_1, \varepsilon_2 \to 0$.

Let f be a general element of L^p. Then for every $\eta > 0$, there is a $g \in C_0^1$ such that $f = g + h$ with $\|h\|_p < \eta$. Thus

$$\begin{aligned}\|K_{\varepsilon_1} f - K_{\varepsilon_2} f\|_p &\leq \|K_{\varepsilon_1} g - K_{\varepsilon_2} g\|_p + \|K_{\varepsilon_1} h - K_{\varepsilon_2} h\|_p \\ &\leq \|K_{\varepsilon_1} g - K_{\varepsilon_2} g\|_p + 2C_p\|h\|_p \\ &< \|K_{\varepsilon_1} g - K_{\varepsilon_2} g\|_p + 2C_p \eta\end{aligned}$$

by part (1) of the theorem. Therefore, by (4.5) there exists $\lim_{\varepsilon \to 0} K_\varepsilon f = Kf$ in L^p and, by the bound proved in part (1) for all $\varepsilon > 0$,

$$\|Kf\|_p \leq C_p \|f\|_p$$

for every $f \in L^p$, $1 < p < \infty$. ∇

We now give the substitute result for $p = \infty$ ([9], [10], [5]).

Theorem 4.4. Under the hypotheses of Theorem 4.3, the C-Z operator K given by

$$Kf(x) = \text{P.V.} \int_{\mathbb{R}^n} k(x - y)f(y)dy$$

transforms continuously L^∞ into BMO, so that

$$\|Kf\|_{BMO} \leq C\|f\|_\infty \tag{4.6}$$

holds for all $f \in L^\infty$, where C is independent of f.

Remark 4.1. We have already proved, as Proposition 3.5, a similar result under the additional hypothesis that $k \in L^1$. This hypothesis also insures the existence of Kf.

Proof. Given $f \in L^\infty$ and the C-Z kernel k, let $k_\varepsilon(x) = k(x)$ for $|x| > \varepsilon$ and zero otherwise,

$$u_\varepsilon(x) = \int (k_\varepsilon(x - y) - k_1(-y))f(y)dy \tag{4.7}$$

and

$$C_\varepsilon = \int (k_\varepsilon(-y) - k_1(-y))f(y)dy \tag{4.8}$$

By (4.7), $u_\varepsilon(x) - u_\eta(x) = f * k_{\varepsilon\eta}(x)$ with $k_{\varepsilon\eta} = k_\varepsilon - k_\eta$. Since $k_{\varepsilon\eta} \in L^1$ for any $0 < \varepsilon < \eta < \infty$ and satisfies, by Lemma 4.2, the same hypothesis as k does, Proposition 3.5 implies that

$$\|k_{\varepsilon\eta} * f\|_{BMO} \leq C\|f\|_\infty$$

Thus, for every cube Q,

$$|Q|^{-1}\int_Q |u_\varepsilon(x) - u_\eta(x) - (u_\varepsilon)_Q + (u_\eta)_Q|dx \le C\|f\|_\infty \quad (4.9)$$

where g_Q is the mean value of the function $g(x)$ over Q.

Since $u_\eta(x) - C_\eta = \int(k_\eta(x - y) - k_\eta(-y))f(y)dy$ tends to zero as $\eta \to \infty$ uniformly in Q, by the truncation of k_η and the conditions k satisfies, both

$$u_\eta(x) \to C_\eta \quad \text{and} \quad (u_\eta)_Q \to C_\eta \quad \text{as} \quad \eta \to \infty \quad (4.10)$$

By (4.9)

$$|Q|^{-1}\int_Q |u_\varepsilon(x) - (u_\eta(x) - C_\eta) - (u_\varepsilon)_Q + (u_\eta)_Q - C_\eta|dx \le C\|f\|_\infty$$

so, letting $\eta \to \infty$ and then $\varepsilon \to 0$, we get

$$|Q|^{-1}\int_Q |u(x) - a_Q|dx \le C\|f\|_\infty$$

for a_Q a constant, and the theorem is proved. ∇

Remark 4.2. Theorem 4.4 has a converse for the case of the Riesz transforms that is equivalent to the duality of the Hardy space H^1 and BMO (for this fact and the proof of the theorem see [11]). Theorem: Every $f \in$ BMO may be written as $f = g_0 + \Sigma_{j=1}^n R_j g_j$, where $g_0, g_1, \ldots, g_n \in L^\infty(\mathbb{R}^n)$. If $n = 1$, every $f \in$ BMO is equal to $g + Hh$ where g, $h \in L^\infty$ and Hh is the Hilbert transform of h.

Remark 4.3. As shown in this and the preceding sections, the homogeneity of the C-Z kernels plays an important role in the theory of singular integrals. In connection with the study of parabolic differential equations, B. F. Jones [12] introduced the study of kernels for which homogeneity is replaced by

$$k(\lambda x, \lambda^m t) = \lambda^{-n-m} k(x, t), \quad \lambda > 0 \tag{4.11}$$

for $x \in \mathbb{R}^n$, $t \in \mathbb{R}_+$.

A theory of "parabolic" singular integrals was developed, including most results of this chapter, for kernels satisfying

$$k(\lambda^{\alpha_1} x_1, \ldots, \lambda^{\alpha_n} x_n) = \lambda^{-|\alpha|} k(x), \quad \lambda > 0 \tag{4.11a}$$

for $\alpha_1, \ldots, \alpha_n \geq 1$ fixed, and $\alpha_1 + \ldots + \alpha_n = |\alpha|$. For details see [13], [14], [15].

Remark 4.4. Let a C-Z kernel be given by an Ω defined in Σ satisfying (i), (ii) and (iii) of Theorem 4.3. For $\mu \in \mathcal{M}(\mathbb{R}^n)$ it can be proved that if

$$K_\varepsilon(d\mu)(x) = \int_{|x-y|>\varepsilon} \Omega(x - y)|x - y|^{-n} d\mu(y)$$

then $\lim_{\varepsilon \to 0} K_\varepsilon(d\mu)(x)$ exists a.e. [1]. Furthermore, if $\varphi \in C_0^\infty(\mathbb{R}^n)$, then its Riesz transforms $R_j\varphi$, $j = 1, \ldots, n$, are continuous functions (since then $R_j\varphi = C_n x_j |x|^{-n-1} * \varphi = C_n(1 - n)|x|^{1-n} * (\partial\varphi/\partial_{x_j})$). Therefore, given $\mu \in \mathcal{M}$, its Riesz transforms $R_j\mu = \nu_j$, $j = 1, \ldots, n$, may be defined by $\int \varphi d\nu_j = -\int (R_j\varphi) d\mu$ for all $\varphi \in C_0^\infty$. Thus $\nu_1, \ldots, \nu_n$ are given by tempered distributions that in general are not finite measures. The theory of H^p spaces in several variables leads to the following generalization of a classical theorem of F. and M. Riesz on analytic measures. Theorem: If μ and all its Riesz transforms $\nu_1, \ldots, \nu_n$ belong to $\mathcal{M}$, then they are all absolutely continuous, i.e., there exist $f_0, f_1, \ldots, f_n \in L^1$ such that $d\mu = f_0 dx$, $d\nu_j = f_j dx$, $j = 1, \ldots, n$. [7]

5. POINTWISE CONVERGENCE OF SINGULAR INTEGRALS

The Calderón-Zygmund theorem asserts the convergence in L^p of the singular integral sequence $\{K_\varepsilon f\}$ for every $f \in L^p$, $1 < p < \infty$ (part (2) of Theorem 4.3). It is natural, then, to ask if under the conditions given on singular kernels there is pointwise convergence of $\{K_\varepsilon f(x)\}$, i.e., if $\lim_{\varepsilon \to 0} K_\varepsilon f(x) = Kf(x)$ exists a.e. In fact, in the classical case $(n = 1)$, the pointwise convergence of the Hilbert transform was obtained through the complex method (by Lusin, Privalov, Plessner) earlier than the convergence in L^p. As that method is essentially linked--through the connection with the theory of analytic functions of one complex variable--with the one-dimensional case, the methods used for $n > 1$ need to be different and we shall rely on the maximal theory developed in Chapter 5.

Again, for every $\varepsilon > 0$ and $f \in L^p(\mathbb{R}^n)$, $1 \le p < \infty$, $n \ge 1$, let

$$K_\varepsilon f(x) = \int_{|y|>\varepsilon} \Omega(y)|y|^{-n} f(x - y)dy \tag{4.2}$$

so that the maximal operator of the family $\{K_\varepsilon\}_{\varepsilon > 0}$ is given by $K^* : f \to K^* f$ where

$$K^* f(x) = \sup_\varepsilon |K_\varepsilon f(x)| \tag{5.1}$$

In order to apply the maximal method given in Theorem 4.1 of Chapter 5, that will insure the pointwise convergence a.e. of singular integrals, we have to prove that $\{K_\varepsilon\}$ satisfies the hypotheses (A) and (B) of that theorem. Hypothesis (B), i.e., the existence of a class D, dense in L^q for some q, $1 \le q < \infty$, where the pointwise limit exists, was already proved in part (2) of Theorem 4.3, for $D = C_0^1$, which is dense in every L^p, $1 \le p < \infty$. Hypothesis (A), i.e., the weak type (p, p) of the maximal operator for all p, $1 \le p < \infty$, will be proved in what follows, separately for $p > 1$ and for $p = 1$.

To do this, let us first recall an estimate for the convolution operator given by a convolution unit in terms of the Hardy-Littlewood maximal operator Λ, namely,

Lemma 5.1 (Exercise 2.1 of Chapter 5). Let $\{\varphi_\varepsilon\}$ a convolution unit given by $\varphi \in L^1$, $\int\varphi = 1$, $\varphi_\varepsilon(x) = \varepsilon^{-n}\varphi(x/\varepsilon)$, be such that its least decreasing radial majorant $\psi(x) = \sup_{|y|\ge|x|} |\varphi(y)| \in L^1$. Then $\sup_{\varepsilon>0} |\varphi_\varepsilon * f(x)| \le C\Lambda f(x)$ for every $f \in L^p$, $1 \le p \le \infty$, where $C = \int\psi$. (Observe that the convolution unit generated by any positive radial decreasing φ with $\int\varphi = 1$ satisfies the hypotheses of the lemma).

Theorem 5.2. The maximal operator K^* defined in (5.1) is of type (p, p), $1 < p < \infty$, under the hypotheses of Theorem 4.3.

Proof. The existence and the type (p, p), $1 < p < \infty$, of the limit operator K of the sequence $\{K_\varepsilon\}_{\varepsilon>0}$ were already proved in Theorem 4.3. This fact, together with the type (p, p), $1 < p \le \infty$ of the Hardy-Littlewood operator Λ (proved in Theorem 1.2 of Chapter 5), will give the thesis, through the subordination estimate

$$K^*f(x) \le C_1\Lambda(Kf)(x) + C_2\Lambda f(x) \qquad (5.2)$$

where C_1 and C_2 are constants independent of f.

The aim, thus, is to prove (5.2). As in (4.3), let $k_\varepsilon(x) = \Omega(x)|x|^{-n}$ for $|x| > \varepsilon$ and zero otherwise and consider $\varphi \in C_0^1(\mathbb{R}^n)$, a positive function with support in the unit sphere, such that $\int\varphi = 1$, and φ is a radial function decreasing in $|x|$. Let $k * \varphi = \lim_{\varepsilon\to 0} k_\varepsilon * \varphi$, where the limit exists in the pointwise sense and let $\Phi = \varphi * k - k_1$. As the operator $\varphi \to k * \varphi$ is a convolution operator and k is homogeneous of degree $-n$, the operator commutes with dilations, so

$$\Phi_\varepsilon = \varphi_\varepsilon * k - k_\varepsilon \tag{5.3}$$

where $\Phi_\varepsilon(x) = \varepsilon^{-n}\Phi(x/\varepsilon)$, $\varphi_\varepsilon(x) = \varepsilon^{-n}\varphi(x/\varepsilon)$. Convolving (5.3) with $f \in L^p$, $1 < p < \infty$, we get

$$K_\varepsilon f = k_\varepsilon * f = (\varphi_\varepsilon * k) * f - \Phi_\varepsilon * f \tag{5.4}$$

Let us observe that

$$(\varphi_\varepsilon * k) * f(x) = (Kf) * \varphi_\varepsilon(x) \tag{5.5}$$

for every $x \in \mathbb{R}^n$. In fact, for every $\delta > 0$ and every x,

$$(\varphi_\varepsilon * k_\delta) * f(x) = \varphi_\varepsilon * (k_\delta * f)(x)$$

$$= \varphi_\varepsilon * (K_\delta f)(x)$$

and, as $\varphi_\varepsilon \in L^{p'}$, $\varphi_\varepsilon * k_\delta \to \varphi_\varepsilon * k$ in the $L^{p'}$ norm, as $\delta \to 0$, while $K_\delta f \to Kf$ in the L^p norm, (5.5) holds. Thus (5.4) becomes

$$K_\varepsilon f(x) = (Kf) * \varphi_\varepsilon(x) - \Phi_\varepsilon * f(x) \tag{5.6}$$

By the hypotheses on φ, Lemma 5.1 applies, and for every $g \in L^p$, $1 \leq p \leq \infty$,

$$\sup_{\varepsilon} |\varphi_\varepsilon * g(x)| \le C_1 \Lambda g(x) \tag{5.7}$$

where $C_1 = \int \varphi = 1$. We have to prove that the same is true for Φ. In fact this will be true if we check that the radial majorant $\Psi(x) = \sup_{|y| \ge |x|} |\Phi(y)|$ is an integrable function of x.

For $|x| < 1$, $\Phi(x) = \varphi * k(x) = \int \varphi(x - y)k(y)dy = \int (\varphi(x - y) - \varphi(x))k(y)dy$, and so is bounded, since $k(y) = \Omega(y)|y|^{-n}$ and $\varphi \in C_0^1(\mathbb{R}^n)$. $\Phi(x)$ is also bounded, and for the same reasons, for $1 \le |x| \le 2$, since there $\Phi(x) = \varphi * k(x) - k(x)$. For $|x| > 2$, $\Phi(x) = \int_{\mathbb{R}^n} k(x - y)\varphi(y)dy - k(x) = \int_{|y| \le 1} (k(x - y) - k(x))\varphi(y)dy$, since φ has integral one and support in the unit sphere. Since it was proved in Theorem 4.3 that k satisfies condition (b),

$$\begin{aligned}\int_{|x|>2} |\Phi(x)|dx &\le \int_{|x|>2} \int_{|y|\le 1} |k(x - y) - k(x)|\varphi(y)dy\, dx \\ &\le \int_{|y|\le 1} \int_{|x|>2|y|} |k(x - y) - k(x)|\varphi(y)dx\, dy \\ &= \int_{|y|\le 1} \int_{|x|>2|y|} |k(x - y) - k(x)|dx\, \varphi(y)dy \\ &\le A\end{aligned}$$

Thus, for every $g \in L^p$, $1 \le p \le \infty$,

$$\sup_{\varepsilon>0} |\Phi_\varepsilon * g(x)| \le C_2 \Lambda g(x) \tag{5.7a}$$

where $C_2 = \int \Psi$.

Taking supremum in expression (5.6), by (5.7) and (5.7a), we get the inequality (5.2), and the thesis follows. ∇

The proof of the weak type (1, 1) of the maximal operator K^* follows closely that of the weak type of the convolution operator T given in Theorem 3.3, as it relies mainly on the use of the Calderón-

Zygmund decomposition Lemma 3.2. In the following proof we use the same notations as in Theorem 3.3.

<u>Theorem 5.3</u>. Under the hypotheses of Theorem 4.3, the maximal operator K^* is of weak type (1, 1).

<u>Proof</u>. As in Theorem 3.3, it is enough to consider a positive integrable function f. We use the notation in that proof. For $\alpha > 0$ let $f = g + b$ be the decomposition given in Lemma 3.2. Again $(K^*f)_*(\alpha) \leq (K^*g)_*(\alpha/2) + (K^*b)_*(\alpha/2)$. As by (3.4), $\|g\|_2^2 \leq C\alpha\|f\|_1$, and K^* is of type (2, 2) by Theorem 5.2,

$$(K^*g)_*(\alpha/2) \leq C'\|K^*g\|_2^2\,\alpha^{-2} \leq C''\|f\|_1\,\alpha^{-1} \tag{5.8}$$

The claim that a similar result holds for K^*b will be sustained, as in Theorem 3.3, by estimating the measure of the set $\{x \in S^c : K^*b(x) > \alpha/2\}$, as less than or equal to $C\|f\|_1/\alpha$, since then

$$\begin{aligned}(K^*b)_*(\alpha/2) &\leq |S| + |\{x \in S^c : K^*b(x) > \alpha/2\}| \\ &\leq C_1\|f\|_1/\alpha \end{aligned} \tag{5.9}$$

Since (5.8) and (5.9) add up to the weak type (1, 1) inequality for K^*, the goal is to prove the estimate that yields (5.9).

Let us fix $x \in S^c$ and $\varepsilon > 0$ and consider, as in (3.9),

$$K_\varepsilon b(x) = \sum_j \int_{Q_j} k_\varepsilon(x - y)b_j(y)dy \tag{5.10}$$

There are three possibilities for a cube Q_j: (i) that for all $y \in Q_j$, $|x - y| < \varepsilon$; (ii) that for all $y \in Q_j$, $|x - y| > \varepsilon$, or (iii) that there is a $y \in Q_j$ such that $|x - y| = \varepsilon$. If Q_j is in case (i) then

$k_\varepsilon(x - y) = 0$ and the integral over the cube Q_j in (5.10) is zero. If Q_j is in the case (ii) then $k_\varepsilon(x - y) = k(x - y)$ and the corresponding term in (5.10) is bounded by $\int_{Q_j} |k(x - y) - k(x - y_j)||b_j(y)|d$ where y_j is the center of Q_j. If Q_j is as in case (iii) then

$$|\int_{Q_j} k_\varepsilon(x - y)b_j(y)dy| \le \int_{Q_j} |k_\varepsilon(x - y)||b_j(y)|dy = \int_{Q_j \cap S(x, r)}$$

where $r = c_n \varepsilon$, c_n depending only on dimension n, since then $S(x, r) \supset Q_j$. Now,

$$|k_\varepsilon(x - y)| \le |\Omega(x - y)||x - y|^{-n} \le \|\Omega\|_\infty |x - y|^{-n}$$
$$= \|\Omega\|_\infty \varepsilon^{-n}$$

for $|x - y| = \varepsilon$,

$$|\int_{Q_j} k_\varepsilon(x - y)b_j(y)dy| \le Cr^{-n} \int_{S(x, r) \cap Q_j} |b_j(y)|dy$$

Adding over all cubes Q_j, (5.10) yields

$$|K_\varepsilon b(x)| \le \sum_j \int_{Q_j} |k(x - y) - k(x - y_j)||b_j(y)|dy$$
$$+ \frac{c}{|S(x, r)|} \int_{S(x, r)} |b(y)|dy \qquad (5.11)$$

Since the first term of the right side of (5.11) is independent of ε and the second term depends on $r = c_n \varepsilon$, taking supremum over $\varepsilon > 0$ in (5.11) yields

$$K^* b(x) = \sup_\varepsilon |K_\varepsilon b(x)| \le \sum_j \int_{Q_j} \ldots + c\Lambda b(x) \qquad (5.12)$$

for every $x \in S^c$. Thus

$$|\{x \in S^c : K^* b(x) > \alpha/2\}| \leq |\{x \in S^c : \sum_j \int_{Q_j} \ldots > \alpha/4\}|$$

$$+ |\{x \in S^c : c\Lambda b(x) > \alpha/4\}| \qquad (5.13)$$

The first term of the right of (5.13) is bounded by a constant times $\|f\|_1/\alpha$, as proved in (3.12), and the second term is also bounded by a constant times $\|f\|_1/\alpha$, since Λ is a weak type (1,1) operator. Thus (5.13) implies (5.9) and the thesis. ∇

Theorems 5.2 and 5.3, as well as the existence of the pointwise limit for the singular integrals of C_0^1 functions, enable us to use the method of the maximal function for the family $\{K_\varepsilon\}$, and thus obtain

<u>Corollary 5.4 (The pointwise convergence for singular integrals)</u>. Let Ω be defined in $\mathbb{R}^n$ such that conditions (i), (ii) and (iii) of Theorem 4.3 hold. If $K_\varepsilon f$ is defined as in (4.2) for every $f \in L^p$, $1 \leq p < \infty$, then $\lim_{\varepsilon \to 0} K_\varepsilon f(x) = Kf(x)$ a.e.

<u>Corollary 5.5</u>. The limit operator K is of weak type (1,1) under the same hypothesis of Theorem 4.3.

*6. EXTENSIONS TO LEBESGUE SPACES WITH WEIGHTED MEASURES

One of the basic properties of the Hilbert transform is that it satisfies the Riesz inequality

$$\int_{\mathbb{R}} |Hf(x)|^p dx \leq M_p \int_{\mathbb{R}} |f(x)|^p dx \qquad (3.1)$$

for all $f \in L^p(\mathbb{R})$, $1 < p < \infty$, or similarly,

$$\int_{\mathbb{T}} |Hf(x)|^p dx \leq M_p \int_{\mathbb{T}} |f(x)|^p dx \tag{3.1a}$$

for all $f \in L^p(\mathbb{T})$, where dx stands for the ordinary Lebesgue measure in $\mathbb{R}$ or $\mathbb{T}$. Hardy and Littlewood proved in [16] (and their result was rediscovered by Babenko [17]) that the Riesz inequality (3.1a) remains valid if the Lebesgue measure is replaced by the "weighted measure" $d\mu = |x|^{\gamma p} dx$, provided that $-1/p < \gamma < 1/p'$. This result was generalized by Stein to $\mathbb{R}^n$, $n > 1$, [18]. We give now a proof for the theorem in $\mathbb{R}^n$, $n \geq 1$ (taken from [19], where the result is proved for the more general case of parabolic singular integrals), based on the properties of the Riesz potential operators introduced in Chapter 4, Section 5.

The remainder of this section deals with more general weighted measures.

Theorem 6.1. Let $k(x) = \Omega(x)|x|^{-n}$ be a C-Z kernel (that is, $\Omega(x)$ is a homogeneous function of degree zero, bounded on Σ and with mean value zero) and K be the operator given by $K : f \to$ P.V. $k * f$. Then K is continuous in $L^p_\mu = L^p(\mathbb{R}^n, d\mu)$, for $1 < p < \infty$, $d\mu = |x|^{\gamma p} dx$ and $-n/p < \gamma < n/p'$. Furthermore, K is of weak type $(1, 1)$, with respect to the measure $d\mu = |x|^\gamma dx$, whenever $-n < \gamma \leq 0$.

Proof. Under our assumptions, K is continuous on $L^p = L^p(\mathbb{R}^n, dx)$, for $1 < p < \infty$, by Theorem 4.3. Therefore,

$$\int_{\mathbb{R}^n} |Kf|^p dx \leq C_p \int_{\mathbb{R}^n} |f|^p dx \tag{6.1}$$

holds for every $f \in L^p(\mathbb{R}^n)$, $1 < p < \infty$. Given $f \in L^p_\mu$, i.e.,

$$\|f\|^p_{p,\mu} = \int_{\mathbb{R}^n} |f|^p |x|^{\gamma p} dx < \infty$$

let $g(x) = f(x)|x|^\gamma$. Then $g \in L^p$ and, by (6.1),

$$\|Kg\|^p_p = \int |Kg|^p dx \le C_p \int |g|^p dx = C_p \|f\|^p_{p,\mu} \qquad (6.2)$$

To insure the continuity of K in L^p_μ we have to prove that

$$\int |Kf|^p |x|^{\gamma p} dx \le A_p \int |f|^p |x|^{\gamma p} dx \qquad (6.3)$$

holds, but since

$$Kf(x)|x|^\gamma = Kf(x)|x|^\gamma - Kg(x) + Kg(x)$$

it will be sufficient, by (6.2), to prove that

$$\int |Kf(x)|x|^\gamma - Kg(x)|^p dx \le C_p \int |f(x)|^p |x|^{\gamma p} dx \qquad (6.4)$$

Consider

$$\begin{aligned} |x|^\gamma Kf(x) - Kg(x) &= |x|^\gamma Kf(x) - K(|x|^\gamma f(x)) \\ &= |x|^\gamma \text{P.V.} \int k(x-y) f(y) dy \\ &\quad - \text{P.V.} \int |y|^\gamma f(y) k(x-y) dy \\ &= \text{P.V.} \int k(x-y)(|x|^\gamma - |y|^\gamma) f(y) dy \qquad (6.5) \end{aligned}$$

By (6.5) and the fact that $k(x)$ is a C-Z kernel,

$$\left| |x|^{\gamma} Kf(x) - Kg(x) \right| \le \text{P. V.} \int |\Omega(x - y)| |x - y|^{-n} \left| |x|^{\gamma} - |y|^{\gamma} \right| |f(y)| dy$$
$$\le \|\Omega\|_{\infty} \int \Gamma(x, y) |f(y)| |y|^{\gamma} dy$$

where

$$\Gamma(x, y) = \left|1 - \left(\frac{|x|}{|y|}\right)^{\gamma}\right| |x - y|^{-n} \tag{6.6}$$

for γ fixed.

We claim that the operator U, defined by

$$U\varphi(x) = \int_{\mathbb{R}^n} \Gamma(x, y)\varphi(y)dy \tag{6.7}$$

is of type (p, p), for $1 < p < \infty$, and of weak type $(1, 1)$.

In fact, let us begin with the case $\underline{\gamma = -\beta < 0}$. Then,

$$\begin{aligned} \Gamma(x, y) &= \left|1 - (|y| / |x|)^{\beta}\right| |x - y|^{-n} \\ &= \left| |x|^{\beta} - |y|^{\beta} \right| |x|^{-\beta} |x - y|^{-n} \end{aligned} \tag{6.8}$$

(1) Let $|y| \le 2|x - y|$. If $z = x - y$ then

$$|x|^{\beta} = |y + z|^{\beta} \le (|y| + |z|)^{\beta} \le 2^{\beta}(|y|^{\beta} + |z|^{\beta})$$

and

$$\begin{aligned} |y + z|^{\beta} - |y|^{\beta} &\le (2^{\beta} - 1)|y|^{\beta} + 2^{\beta}|z|^{\beta} \\ &\le (2^{\beta} - 1)2^{\beta}|x - y|^{\beta} + 2^{\beta}|x - y|^{\beta} \end{aligned}$$

therefore

$$|x|^{\beta} - |y|^{\beta} \le 2^{2\beta}|x - y|^{\beta} \tag{6.9}$$

If $z = y - x$ we obtain, as in (6.9),

$$\left| |x|^\beta - |y|^\beta \right| \le 2^{2\beta} |x - y|^\beta \tag{6.10}$$

Thus, for $|y| \le 2|x - y|$,

$$\Gamma(x, y) \le \frac{2^{2\beta} |x-y|^\beta}{|x|^\beta |x-y|^n} = 2^{2\beta} |x|^{-\beta} |x - y|^{-n+\beta} \tag{6.11}$$

and, if H_γ is the operator defined in (5.10) of Chapter 4, i.e., $H_\gamma f = |x|^{-\gamma} f * |x|^{\gamma-n}$, (6.11) yields the subordination estimate

$$|U\varphi(x)| \le 2^{2\beta} |H_\beta \varphi(x)| \tag{6.12}$$

By Proposition 5.6 of Chapter 4, H_β is of type (p, p) for $p < n/\beta$ or, equivalently, for $\gamma(=-\beta) > -n/p$, and so is U. Furthermore, H_β is of weak type $(1, 1)$, and so is U, for $-n < \gamma < 0$, since then $0 < \beta < n$.

(2) Let $|y| \ge 2|x - y|$. Then $|y| \ge 2|x| - 2|y|$ and $|y| \ge 2|y| - 2|x|$. Therefore, $3|y| \ge 2|x| \ge |y|$, and, by the mean value theorem,

$$\left| |x|^\beta - |y|^\beta \right| = |x - y| \beta |\xi|^{\beta-1}$$

where ξ is an intermediate point between x and y, $\xi = x + \theta(y - x)$, $0 < \theta < 1$. Thus $|\xi| \le |x| + |y - x| \le |x| + |y| + |x| \le 4|x|$ and

$$\left| |x|^\beta - |y|^\beta \right| \le 4^{\beta-1} \beta |x - y| |x|^{\beta-1} \tag{6.13}$$

Thus, for $|y| \ge 2|x - y|$,

$$\Gamma(x, y) \leq 4^{\beta-1}\beta \frac{|x-y||x|^{\beta-1}}{|x|^{\beta}|x-y|^{n}}$$

$$= 4^{\beta-1}\beta|x|^{-1}|x - y|^{-n+1} \tag{6.14}$$

and

$$|U\varphi(x)| \leq 4^{\beta-1}\beta|H_1\varphi(x)| \tag{6.15}$$

Now, H_1 is of type (p, p) only when $n > p > 1$. By hypothesis, $-\gamma = \beta < n/p$, so this will hold whenever $\beta \geq 1$ or $\gamma \leq -1$.

For the case $-1 < \gamma < 0$, or $0 < \beta < 1$, we consider

$$|x - y| \leq |x| + |y| \leq |x| + 2|x| \leq 3|x|$$

Then by (6.14),

$$\Gamma(x, y) \leq 4^{\beta-1}\beta \frac{|x-y||x|^{\beta-1}}{|x|^{\beta}|x-y|^{n}}$$

$$= 4^{\beta-1}\beta\frac{|x-y|^{1-\beta}}{|x|^{\beta}|x-y|^{n-\beta}|x|^{1-\beta}}$$

$$= 4^{\beta-1}\beta|x|^{-\beta}|x - y|^{\beta-n}\left(\frac{|x-y|}{|x|}\right)^{1-\beta} \tag{6.16}$$

Since $1 - \beta > 0$ and $|x - y|/|x| \leq 3$, (6.16) is bounded by

$$(3/4)^{1-\beta}\beta|x|^{-\beta}|x - y|^{\beta-n}$$

and

$$|U\varphi(x)| \leq (3/4)^{1-\beta}\beta|H_{\beta}\varphi(x)| \tag{6.17}$$

In any case, by (6.12), (6.15) and (6.17), U is subordinate to $H_1 + H_\gamma$ for $\gamma < 0$. Thus, U is of type (p, p) and weak type $(1, 1)$ whenever $0 > \gamma > -n/p$.

Let us consider now the case $\underline{\gamma > 0}$. For any $\varphi \in L^p$, $\psi \in L^{p'}$ we have

$$\begin{aligned} \langle U\varphi, \psi\rangle &= \int U\varphi(x)\psi(x)dx \\ &= \int \left(\int \Gamma(x, y)\varphi(y)dy\right)\psi(x)dx \\ &= \int \left(\int \Gamma(x, y)\psi(x)dx\right)\varphi(y)dy \\ &= \int U\psi(y)\varphi(y)dy \end{aligned} \tag{6.18}$$

From (6.6) it follows that

$$\begin{aligned} U\psi(y) &= \int \left|1 - \left(\frac{|x|}{|y|}\right)^{\gamma}\right| |x - y|^{-n}\psi(x)dx \\ &= \int \left|1 - \left(\frac{|y|}{|x|}\right)^{-\gamma}\right| |x - y|^{-n}\psi(x)dx \end{aligned} \tag{6.19}$$

and U is of type (q, q) by the first part of the proof, whenever $0 > -\gamma > -n/q$. With $q = p'$, and $0 < \gamma < n/p'$, we deduce from (6.19) that $U\psi \in L^{p'}$, and that

$$\|U\psi\|_{p'} \leq C_{p'} \|\psi\|_{p'} \tag{6.20}$$

Thus Hölder's inequality, (6.18) and (6.20) yield

$$|\langle U\varphi, \psi\rangle| \leq C_{p'} \|\varphi\|_p \|\psi\|_{p'}$$

By the F. Riesz representation theorem, $U\varphi \in L^p$ and $\|U\varphi\|_p \leq C_{p'} \|\varphi\|_{p'}$ so U is of type (p, p) for all γ such that $0 < \gamma < n/p'$.

Summing up the above results, U is of type (p, p) for all γ, $-n/p < \gamma < n/p'$ and of weak type $(1, 1)$ for $-n < \gamma < 0$. By (6.5) and (6.6) this leads through (6.4) and (6.2) to the continuity of the operator K from L^p_μ into L^p_μ, when $-n/p < \gamma < n/p'$, and, similarly, from L^1_μ into $(L^1_\mu)_*$ when $-n < \gamma \leq 0$, since for $\gamma = 0$ $d\mu = dx$ and K is also of weak type $(1, 1)$ (see Corollary 5.5). ∇

Theorem 6.1 has several applications: to the basis problem (whether the exponentials can be a basis in $L^2(|x|^\gamma dx)$), to the theory of differential equations in unbounded domains and to the theory of singular integrals on general curves. These problems, as well as others in probability theory, lead to the consideration of general weighted measures $d\mu = \omega(t)dt$, $0 \leq \omega \in L^1$, satisfying the Riesz inequality

$$\int_0^{2\pi} |Hf(t)|^2 \omega(t)dt \leq M \int_0^{2\pi} |f(t)|^2 \omega(t)dt \qquad (3.1b)$$

for some $M > 0$ and all $f \in C(\mathbb{T})$.

Since every $f \in C(\mathbb{T})$ is a uniform limit of trigonometric polynomials of the form $g = \Sigma_{k=-N}^{N} c_k \exp(ikt)$, it is sufficient to require that (3.1b) hold for $f \in \mathscr{P}$, the set of all trigonometric polynomials.

Gaposhkin proved that if $\tilde{\omega} = H\omega$ is the conjugate function of ω, the condition $|\tilde{\omega}| \leq k$ is sufficient for (3.1b) to hold. While studying prediction theory, Helson and Szegö [20] gave the following remarkable necessary and sufficient condition: (3.1b) holds for <u>some</u> M if and only if the weight $0 \leq \omega \in L^1$ is of the form

$$\omega = \exp(u + \tilde{v}) \quad \text{with } u, v \in L^\infty, \ \|v\|_\infty < \pi/2 \qquad (6.21)$$

The proof of this fact relies heavily on the theory of analytic functions in the disc. In [21] we presented the following variant of the Helson-Szegő theorem: (3.1b) holds for a fixed M if and only if there exists a function $h \in H^1$ (i.e., $h \in L^1$ and $\hat{h}(n) = 0$ for $n < 0$) such that

$$-4M\omega^2(t) - 2\mathrm{Re}(M+1)\omega(t)h(t) - |h(t)|^2 \geq 0 \qquad (6.22)$$

for a.e. $t \in \mathbb{T}$.

We shall see now that condition (6.22) satisfied for some M is equivalent to condition (6.21), so that the former gives more precise information on ω than the latter.

Lemma 6.2. If ω satisfies condition (6.21), then it satisfies also condition (6.22) for some M.

Proof. Let ω be of the form (6.21). Since $v + i\tilde{v}$ extends to an analytic function whose negative Fourier coefficients vanish (see Section 1, formula (1.21a)), $v + i\tilde{v} \in H^1$ and $\tilde{v} - iv \in H^1$. Letting $h = -d\exp(\tilde{v} - iv)$ for $d > 0$ a constant, we have also $h \in H^1$. By (6.21),

$$\omega = e^u e^{\tilde{v}} = e^u(1/d)|h| \quad \text{and so} \quad |h| = d\omega e^{-u}$$

and

$$\mathrm{Re}\, h = -de^{\tilde{v}}\cos v = -d\omega e^{-u}\cos v$$

where $\cos v \geq c_1 > 0$ since $\|v\|_\infty < \pi/2$. Thus the left side of (6.22) will be nonnegative if

$$-4M\omega^2 + 2(M+1)\omega^2 d\cos v e^{-u} - d^2\omega^2 e^{-2u}$$
$$\geq \omega^2(-4M + 2(M+1)dc_1 e^{-u} - d^2 e^{-2u}) \geq 0$$

Since $c_1 > 0$ is a fixed constant, and e^{-u} is essentially bounded, we can take d such that $dc_1 e^{-u(t)} > 3$ and, therefore,

$$- 4M + 2(M + 1)dc_1 e^{-u(t)} > M$$

for a.e. t. Taking M larger than $d^2 e^{-2u}$, (6.22) is satisfied.

∇

Condition (6.22) associates a function $h \in H^1$ to each positive integrable function ω (there may be many of these). From (6.22) it follows that the discriminant of the equation

$$4Mu^2 + 2(M + 1)(\operatorname{Re} h(t))u + |h(t)|^2 = 0$$

is nonnegative and that $\omega(t)$ lies between the two roots of this equation. Hence, any such h must satisfy the two estimates

$$|\pi - \arg h| \leq \arctan \frac{M-1}{2\sqrt{M}} \tag{6.23}$$

and

$$C_1 |h| \leq \omega \leq C_2 |h| \tag{6.24}$$

for some positive constants C_1 and C_2. As a consequence of these facts, we obtain the following (see [22]).

Lemma 6.3. If ω satisfies condition (6.22) then it satisfies also condition (6.21).

Proof. Let $h \in H^1$ be the function of condition (6.22). Then, $\log h = \log |h| + i \arg h$, and $\log |h|$, $\arg h$ are conjugate functions. Define

$$u = \log(\omega\,|h(0)|/|h|)$$

It follows from (6.24) that $u \in L^\infty$. If we consider

$$v = \pi - \arg h$$

then, by (6.23), $\|v\|_\infty \leq \arctan((M-1)/2\sqrt{M}) < \pi/2$ and furthermore, its conjugate function is

$$\tilde{v} = \log(|h|/|h(0)|)$$

Thus, $\exp(u + \tilde{v}) = \omega$, and condition (6.21) holds. ∇

From Lemmas 6.2 and 6.3 it follows that the Helson-Szegö characterization of the weighted measures satisfying the Riesz inequality (3.16) is included in a sharper result. In fact, the measures $d\mu = \omega(t)dt$, $0 \leq \omega \in L^1$, that satisfy (3.1b) for a <u>fixed</u> M are precisely those given by a weight ω for which condition (6.22) (and, therefore, also condition (6.21)) holds. We devote the rest of this section to a direct proof of this theorem. This, in particular, yields a proof of the Helson-Szegö theorem that does not require any further knowledge of function theory.

Let $\mathscr{P}$ be the set of all trigonometric polynomials of the form $f = \Sigma_{-N}^{N} c_k \exp(ikt)$, $t \in [0, 2\pi)$, $\mathscr{P}_+$ (and $\mathscr{P}_-$) the set of all analytic (antianalytic) polynomials of the form $f_+ = \Sigma_0^N c_k \exp(ikt)$ (or $f_- = \Sigma_{-N}^{-1} c_k \exp(ikt)$). Every $f \in \mathscr{P}$ is of the form $f = f_+ + f_-$ and $Hf = -if_+ + if_-$, so that condition (3.1b) can be rewritten as

$$M\int_0^{2\pi} (f_+ + f_-)(\bar{f}_+ + \bar{f}_-)\omega\, dt - \int_0^{2\pi} (f_+ - f_-)(\bar{f}_+ - \bar{f}_-)\omega\, dt \geq 0$$

or, naming

$$\omega_{11} = \omega_{22} = (M-1)\omega \quad \text{and} \quad \omega_{12} = \omega_{21} = (M+1)\omega \tag{6.25}$$

as

$$\int_0^{2\pi} (f_+\overline{f}_+\omega_{11} + f_+\overline{f}_-\omega_{12} + f_-\overline{f}_+\omega_{21} + f_-\overline{f}_-\omega_{22})dt \geq 0 \qquad (6.26)$$

for any pair $f_+ \in \mathscr{P}_+$, $f_- \in \mathscr{P}_-$.

For any system $W = \{\omega_{11}, \omega_{12}, \omega_{21}, \omega_{22}\}$ of four integrable functions and for any pair $F = \{f_1, f_2\}$ of polynomials in $\mathscr{P}$, we set

$$W(F) = W(f_1, f_2)$$

$$= \int_0^{2\pi} (f_1\overline{f}_1\omega_{11} + f_1\overline{f}_2\omega_{12} + f_2\overline{f}_1\omega_{21} + f_2\overline{f}_2\omega_{22})dt \qquad (6.27)$$

Then condition (3.1b) is equivalent to the following condition (3.1c if $W = \{\omega_{11}, \omega_{12}, \omega_{21}, \omega_{22}\}$ is given by (6.25), then $W(f_+, f_-) \geq 0$ for all $f_+ \in \mathscr{P}_+$, $f_- \in \mathscr{P}_-$.

This suggests the introduction of the classes Γ_0 and Γ.

Definition 6.1. The class Γ_0 is the set of all $W = \{\omega_{11}, \omega_{12}, \omega_{21}, \omega_{22}\}$ where $\omega_{11}, \omega_{12}, \omega_{21}, \omega_{22}$ are any four (complex valued) integrable functions satisfying

$$W(f_1, f_2) \geq 0 \quad \text{for all pairs} \quad f_1 \in \mathscr{P}_+, \ f_2 \in \mathscr{P}_- \qquad (6.28)$$

and Γ is the set of all such W which satisfy

$$W(f_1, f_2) \geq 0 \quad \text{for all pairs} \quad f_1, f_2 \in \mathscr{P} \qquad (6.28a)$$

Obviously, $\Gamma \subset \Gamma_0$. In Theorem 6.7 we shall establish another relation between Γ and Γ_0 and use it to deduce the equivalence between (3.1b) and (6.22). Let us first characterize Γ.

Lemma 6.4. Let $V = \{v_{11}, v_{12}, v_{21}, v_{22}\}$, v_{ij}, $i, j = 1, 2$, integrable functions. Then $V \in \Gamma$ if and only if

$$\lambda_1\overline{\lambda}_1 v_{11}(t) + \lambda_1\overline{\lambda}_2 v_{12}(t) + \lambda_2\overline{\lambda}_1 v_{21}(t) + \lambda_2\overline{\lambda}_2 v_{22}(t) \geq 0 \qquad (6.29)$$

for all $\lambda_1, \lambda_2 \in \mathbb{C}$ and almost all $t \in \mathbb{T}$.

Proof. Suppose (6.29) holds. Then for every pair $f_1, f_2 \in \mathscr{P}$,

$$f_1(t)\overline{f_1(t)}v_{11}(t) + f_1(t)\overline{f_2(t)}v_{12}(t) + \overline{f_1(t)}f_2(t)v_{21}(t) + f_2(t)\overline{f_2(t)}v_{22}(t) \geq 0$$

a.e. and therefore, $V(f_1, f_2) \geq 0$. Conversely, suppose that $V(f_1, f_2) \geq 0$ for all $f_1, f_2 \in \mathscr{P}$, hence for all $f_1, f_2 \in C$ and also for every pair of characteristic functions χ_A, χ_B (taking sequences $f_1^n \downarrow \chi_A$, $f_2^n \downarrow \chi_B$). Letting $f_1 = \lambda_1\chi_A/|A|$, $f_2 = \lambda_2\chi_A/|A|$ we obtain

$$V(f_1, f_2) = |A|^{-1}\int_A (v_{11}\lambda_1\overline{\lambda_1} + v_{12}\lambda_1\overline{\lambda_2} + v_{21}\overline{\lambda_1}\lambda_2 + v_{22}\lambda_2\overline{\lambda_2})dt \geq 0$$

Taking A as an interval with center at t and radius δ, and letting $\delta \to 0$, we obtain

$$v_{11}(t)\lambda_1\overline{\lambda_1} + v_{12}(t)\lambda_1\overline{\lambda_2} + v_{21}(t)\overline{\lambda_1}\lambda_2 + v_{22}(t)\lambda_2\overline{\lambda_2} \geq 0$$

for all $\lambda_1, \lambda_2 \in \mathbb{C}$ and almost all t, which proves the lemma. ∇

Definition 6.2. Given $V = \{v_{11}, v_{12}, v_{21}, v_{22}\}$ and $W = \{\omega_{11}, \omega_{12}, \omega_{21}, \omega_{22}\}$, two systems of integrable functions, we say that V and W are equivalent, and write $V \sim W$, whenever

$$v_{11} = \omega_{11}, \quad v_{22} = \omega_{22}, \quad v_{12} = \omega_{12} + h_1, \quad v_{21} = \omega_{21} + \overline{h}_2 \qquad (6.30)$$

for $h_1, h_2 \in H^1 = \{f \in L^1 : \hat{f}(n) = 0 \text{ for } n < 0\}$.

In the following, all integrals are taken from 0 to 2π, unless otherwise stated.

Lemma 6.5. (a) If $W \in \Gamma_0$ then $\omega_{11} \geq 0$, $\omega_{22} \geq 0$,

$$\int f_+ e^{it} \omega_{12} dt = \int f_+ e^{it} \overline{\omega_{21}} dt \tag{6.31}$$

for all $f_+ \in \mathscr{P}_+$ and

$$\left| \int f_+ \overline{f}_- \omega_{12} dt \right|^2 \leq \left(\int |f_+|^2 \omega_{11} dt \right) \left(\int |f_-|^2 \omega_{22} dt \right) \tag{6.32}$$

for all $f_+ \in \mathscr{P}_+$, $f_- \in \mathscr{P}_-$.

(b) If $W \in \Gamma$, then the stronger estimate

$$\left| \int f_1 f_2 \omega_{12} dt \right|^2 \leq \left(\int |f_1|^2 \omega_{11} dt \right) \left(\int |f_2|^2 \omega_{22} dt \right) \tag{6.32a}$$

holds for all $f_1, f_2 \in \mathscr{P}$.

(c) If $V = (v_{11}, v_{12}, v_{21}, v_{22})$ is another system such that $V(f_+, f_-) = W(f_+, f_-)$ for all $f_+ \in \mathscr{P}_+$, $f_- \in \mathscr{P}_-$, then $V \sim W$.

Proof. (a) Let $W \in \Gamma_0$. If $f_- = 0$, we obtain from (6.28) that $\int f_+ \overline{f}_+ \omega_{11} \geq 0$ for all $f_+ \in \mathscr{P}_+$. If $f = \Sigma_{-N}^{N} c_k e^{ikt} \in \mathscr{P}$, then $f\overline{f} = (fe^{iNt})(\overline{fe^{iNt}}) = g_+ \overline{g_+}$ where $g_+ \in \mathscr{P}_+$. Therefore $\int f\overline{f}\omega_{11} \geq 0$ for every $f \in \mathscr{P}$ and, since $\mathscr{P}$ is dense in C, we get that $\int F\omega_{11} \geq 0$ for all positive $F \in C$, hence $\omega_{11} \geq 0$. Similarly $\omega_{22} \geq 0$. From (6.27) and (6.28) we get then that $\int f_+ \overline{f}_- \omega_{12} = \int f_+ \overline{\overline{f}_- \omega_{21}}$ and, letting $f_-(t) = e^{-it}$, we obtain (6.31). Now, to prove (6.32), let $g_+ = \lambda_1 f_+$, $g_- = \lambda_2 f_-$ for λ_1, λ_2 constants. Then from (6.28) we deduce that for all $\lambda_1, \lambda_2 \in \mathbb{C}$,

$$a\lambda_1\bar{\lambda}_1 + b\lambda_1\bar{\lambda}_2 + c\lambda_2\bar{\lambda}_1 + d\lambda_2\bar{\lambda}_2 \geq 0$$

where $a = \int |f_+|^2\omega_{11}$, $d = \int |f_-|^2\omega_{22}$, $b = \int f_+\bar{f}_-\omega_{12} = \bar{c}$, and (6.32) follows.

(b) Same proof as for (6.32), letting $g_1 = \lambda_1 f_1$, $g_2 = \lambda_2 f_2$ for $f_1, f_2 \in \mathscr{P}$, $\lambda_1, \lambda_2 \in \mathbb{C}$.

(c) If $f_- = 0$ again, we have from $V(f_+, f_-) = W(f_+, f_-)$ that $\int f_+ f_+ v_{11} = \int f_+ f_+ \omega_{11}$ for all $f_+ \in \mathscr{P}_+$ and, as in the proof of (6.31), we see that this implies $\int Fv_{11} = \int F\omega_{11}$ for all positive $F \in C$. Hence $v_{11} = \omega_{11}$. Similarly $v_{22} = \omega_{22}$. Then

$$2\text{Re}\int f_+\bar{f}_-\omega_{12} = 2\text{Re}\int f_+\bar{f}_- v_{12}$$

for all $f_+ \in \mathscr{P}_+$, $f_- \in \mathscr{P}_-$, and letting $f_- = e^{-it}$ and $f_- = ie^{-it}$ we get

$$\int f_+ e^{it}\omega_{12} = \int f_+ e^{it} v_{12}$$

for all $f_+ \in \mathscr{P}_+$. For $f_+ = e^{nit}$, $n \geq 0$, this gives $\hat{\omega}_{12}(-n-1) = \hat{v}_{12}(-n-1)$ for all $n \geq 0$. Therefore, $(\omega_{12} - v_{12})^\wedge(k) = 0$ for $k < 0$ and $\omega_{12} - v_{12} = h_1 \in H^1$. Similarly, (6.31) yields $\omega_{21} - v_{21} = \bar{h}_2$, $h_2 \in H^1$. Collecting these results, we get $V \sim W$ as claimed.

∇

Definition 6.3. Let $E = L^1 \times L^1 \times L^1 \times L^1$ (so that $\Gamma_0 \subset E$) and let $E^* = C \times \mathscr{P}_+ \times \mathscr{P}_- \times C$ be the space of the systems of continuous functions $\Phi = (\varphi_{11}, \varphi_{12}, \varphi_{21}, \varphi_{22})$, where $\varphi_{11}, \varphi_{22} \in C(\mathbb{T})$, $\varphi_{12} \in \mathscr{P}_+$, $\varphi_{21} \in \mathscr{P}_-$.

For any pair $W \in E$ and $\Phi \in E^*$ we set

$$\langle W, \Phi\rangle = \int \varphi_{11}\omega_{11}dt + \int \varphi_{12}\omega_{12}dt + \int \varphi_{21}\omega_{21}dt + \int \varphi_{22}\omega_{22}dt \tag{6.33}$$

and we endow the linear space E with the topology defined by this duality, i.e., $W_\alpha \to W$ in E if and only if $\langle W_\alpha, \Phi\rangle \to \langle W, \Phi\rangle$ for all $\Phi \in E^*$. E becomes a topological linear space in this topology, and E^* is the dual of E, which is itself the dual of E^*. Of course, two different elements W and V of E may define the same functional in E^*, but by (c) of Lemma 6.5 this will happen if and only if $W \sim V$.

From this topology notion and (6.28), it is clear that

$$\Gamma_0 \text{ is a closed cone in } E \tag{6.34}$$

since $f_+\bar{f}_- \in \mathscr{P}_+$, $f_-\bar{f}_+ \in \mathscr{P}_-$, whenever $f_+ \in \mathscr{P}_+$, $f_- \in \mathscr{P}_-$.

<u>Lemma 6.6</u>. If $V_\alpha \in \Gamma$ and $V_\alpha \to W \in \Gamma_0$ (in the topology of E) then there exists $V \in \Gamma$ such that $V_\alpha \to V$ and $V \sim W$. Thus, Γ is closed modulo the equivalence relation $\sim$.

<u>Proof</u>. Let $V_\alpha = (v_{11}^\alpha, v_{12}^\alpha, v_{21}^\alpha, v_{22}^\alpha) \in \Gamma$ and $V_\alpha \to W$. Then, $v_{11}^\alpha \geq 0$ and $\int fv_{11}^\alpha dt \to \int f\omega_{11}dt$ for all $f \in C(\mathbb{T})$, so that the measures $d\nu_{11}^\alpha = v_{11}^\alpha dt$ converge weakly-*. Since then $\nu_{11}^\alpha(1)$ converges to $\omega_{11}(1)$ and $\nu_{11}^\alpha(1) = \|\nu_{11}^\alpha\|$, we may assume that, for all α, there is a fixed constant a such that $\int fv_{11}^\alpha dt \leq a\|f\|_\infty$ for all $f \in C$. Similarly, $\int fv_{22}^\alpha dt \leq a\|f\|_\infty$, for all $f \in C$. If $f_1 = f_2 = f \in \mathscr{P}$, (6.32a) implies $|\int |f|^2 v_{12}^\alpha dt| \leq 4\|f\|_\infty^2$ for all $f \in C$, hence also $|\int fv_{12}^\alpha dt| \leq a\|f\|_\infty$ for all $f \in C$. By the theorem of Alaoglu-Bourbaki there exists a subsequence V_β (which we continue to call informally V_α) such that $d\nu_{12}^\alpha = v_{12}^\alpha dt$ converges weakly-*

to a measure $\nu_{12} = \nu$. Let us show that ν is absolutely continuous. If $E \subset \mathbb{T}$ is any compact set of Lebesgue measure zero, we shall prove that $\nu(E) = 0$. Let $f_n \in C$ be such that $f_n^2 \downarrow \chi_E$. Then it suffices to prove that $\int f_n^2 d\nu \to 0$. Since $\int f_n^2 v_{11}^{\alpha} dt \to \int_E v_{11}^{\alpha} dt = 0$, for large n we have that $\int f_n^2 v_{11}^{\alpha} dt = \int |f_n|^2 v_{11}^{\alpha} dt < \varepsilon$ and, similarly, that $\int |f_n|^2 v_{22}^{\alpha} dt < \varepsilon$. From (6.32), we get $|\int |f_n|^2 v_{12}^{\alpha} dt| \leq \varepsilon$, as desired. Thus, there exists $v_{12} \in L^1$ such that $\int f v_{12}^{\alpha} dt \to \int f v_{12} dt$ for all $f \in C$. Similarly, there exists $v_{21} \in L^1$, such that $\int f v_{21}^{\alpha} dt \to \int f v_{21} dt$ for all $f \in C$. Therefore, letting $V = (\omega_{11}, v_{12}, v_{21}, \omega_{22})$, we have that $V_{\alpha}(f_1, f_2) \to V(f_1, f_2)$ for all $f_1, f_2 \in C$ (and not only for $f_1 = f_+$, $f_2 = f_-$). Therefore $V(f_1, f_2) \geq 0$ for all $f_1, f_2 \in \mathscr{P}$ and $V \in \Gamma$. Since $V_{\alpha}(f_+, f_-) \to W(f_+, f_-)$, we get $V(f_+, f_-) = W(f_+, f_-)$ for all $f_+ \in \mathscr{P}_+$, $f_- \in \mathscr{P}_-$, and hence $V \sim W$.

∇

<u>Theorem 6.7</u>. For every $W \in \Gamma_0$ there is a $V \in \Gamma$ such that $V \sim W$.

<u>Proof</u>. By the preceding lemma, all we have to prove is that every $W \in \Gamma_0$ is the limit of a sequence in Γ. Since Γ is a cone in $E = L^1 \times L^1 \times L^1 \times L^1$, by the polar theorem (Chapter 0, Proposition 4.5 and Corollary 4.6) it is sufficient to show that if $\Phi \in E^* = C \times \mathscr{P}_+ \times \mathscr{P}_- \times C$ is such that

$$\langle V, \Phi \rangle \geq 0 \quad \text{for all} \quad V \in \Gamma \tag{6.35}$$

then

$$\langle W, \Phi \rangle \geq 0 \quad \text{for any} \quad W \in \Gamma_0 \tag{6.35a}$$

Let Φ satisfy (6.35). Since Γ contains all the elements V of the form $\{v_{11}, v_{12}, v_{21}, v_{22}\}$ with $v_{11} = \lambda_1\bar{\lambda}_1 f$, $v_{12} = \lambda_1\bar{\lambda}_2 f$, $v_{21} = \lambda_2\bar{\lambda}_1 f$, $v_{22} = \lambda_2\bar{\lambda}_2 f$ where $0 \leq f \in L^1$ and $\lambda_1, \lambda_2 \in \mathbb{C}$, taking $f = |A|^{-1}\chi_A$, for A an interval with center at t, we see, as in Lemma 6.4, that (6.35) implies that $\Phi \in \Gamma$. By (6.29), $\Phi \in \Gamma$ is equivalent to

$$\varphi_{11} \geq 0, \; \varphi_{22} \geq 0, \; \varphi_{12} = \bar{\varphi}_{21} \tag{6.36}$$

and

$$|\varphi_{12}(t)|^2 \leq \varphi_{11}(t)\varphi_{22}(t) \tag{6.37}$$

Since the elements $|g_+|^2$, $g_+ \in \mathscr{P}_+$, are dense in the positive part of C (see proof of Lemma 6.5), we may assume that $\varphi_{11} = g_+\bar{g}_+$ and $\varphi_{22} = g_-\bar{g}_-$, where $g_+ \in \mathscr{P}_+$, $g_- \in \mathscr{P}_-$. Furthermore g_+ (and similarly g_-) can be chosen without zeros, by taking first $g_+ = \exp(u + i\tilde{u})$ for u the Poisson integral of $(1/2)\log \varphi_{11}$, and then approximating this analytic function by analytic polynomials. Thus (6.37) becomes

$$|\varphi_{12}(t)| \leq |g_+(t)\overline{g_-(t)}| \tag{6.37a}$$

where φ_{12} and $g_+\overline{g_-}$ belong to $\mathscr{P}_+$ and $g_+\overline{g_-}$ never vanishes. Therefore,

$$\varphi_{12}(t) = \varepsilon(t)g_+(t)\overline{g_-(t)} \tag{6.38}$$

where ε is an analytic function such that $|\varepsilon(t)| \leq 1$. We can now rewrite the components of Φ as

$$\varphi_{11} = \psi_+\overline{\psi}_+ + \gamma$$

$$\varphi_{22} = \psi_-\overline{\psi}_-$$

$$\varphi_{12} = \overline{\varphi_{21}} = \psi_+\overline{\psi}_-$$

where $\psi_+ = \varepsilon g_+$ is an analytic function (the uniform limit of polynomials in $\mathscr{P}_+$), $\psi_- = g_-$ is an antianalytic polynomial in $\mathscr{P}_-$, and

$$\gamma = g_+\overline{g}_+(1 - \varepsilon\overline{\varepsilon}) \geq 0$$

Now, for every $W \in \Gamma_0$, we have

$$<W, \Phi> = \int \gamma\omega_{11}dt + \int (\psi_+\overline{\psi_+}\omega_{11} + \psi_+\overline{\psi_-}\omega_{12} + \psi_+\overline{\psi_-}\omega_{21} + \psi_-\overline{\psi_-}\omega_{22})dt$$

Since $\gamma \geq 0$ and $\omega_{11} \geq 0$ (by Lemma 6.5), the first integral is nonnegative, and so is the second, as follows from the definition of Γ_0. Hence (6.35a) holds for every Φ that satisfies (6.35) and the theorem is proved. ∇

As a corollary to this theorem, we deduce the equivalence of conditions (3.1b) and (6.22).

Corollary 6.8. Let M be a fixed constant and $0 \leq \omega \in L^1$. The measure $d\mu = \omega(t)dt$ satisfies the Riesz inequality (3.1b) (or (3.1c)) for all $f \in \mathscr{P}$ if and only if there exists a function $h \in H^1(\hat{h}(n) = 0$ for $n < 0)$ which satisfies (6.22) for a.e. $t \in \mathbb{T}$.

Proof. Let $W = \{\omega_{11}, \omega_{12}, \omega_{21}, \omega_{22}\}$ with $\omega_{11} = \omega_{22} = (M - 1)\omega$, $\omega_{12} = \omega_{21} = (M + 1)\omega$. Then condition (3.1c) is equivalent to $W \in \Gamma_0$. By the above theorem and Definition 6.2, there exists $V \in \Gamma$ such that $v_{11} = \omega_{11}$, $v_{22} = \omega_{22}$, $v_{12} = \omega_{12} + h$ and $v_{21} = \overline{v_{21}} = \omega_{12} + \overline{h}$ for

$h \in H^1$. By Lemma 6.4, the condition $V \in \Gamma$ amounts to $v_{11} = v_{22} = (M - 1)\omega \geq 0$ and to the positiveness of the determinant

$$\begin{vmatrix} v_{11} & v_{12} \\ v_{21} & v_{22} \end{vmatrix} = \begin{vmatrix} \omega_{11} & \omega_2 + h \\ \omega_{12} + \bar{h} & \omega_{22} \end{vmatrix} = \begin{vmatrix} (M - 1)\omega & (M + 1)\omega + h \\ (M + 1)\omega + \bar{h} & (M - 1)\omega \end{vmatrix}$$

This last fact can be rewritten as condition (6.22) and thus the proof is completed. ∇

Remark 6.1. The preceding results generalize easily to $\mathbb{R}$. As already noted in Remark 4.2, $g \in$ BMO if and only if $g = u + \tilde{v}$ for u, $v \in L^\infty$ and, by the John-Nirenberg theorem (Theorem 3.11 of Chapter 5) this is equivalent to

$$\sup_Q \int_Q \exp(\lambda |g(x) - g_Q|)dx < \infty$$

for some $\lambda > 0$. The Helson-Szegö theorem says that $\omega(t)dt$ satisfies the Riesz inequality (3.1a) for $p = 2$ if and only if $\omega = e^g$, where $g = u + \tilde{v}$ is a special BMO function with $\|v\|_\infty < \pi/2$. The John-Nirenberg characterization suggests that the Helson-Szegö condition (6.21) implies a certain condition on the mean value $|Q|^{-1} \int_Q \omega$. In fact, Hunt, Muckenhoupt and Wheeden [23] proved the following important result: $\omega(t)dt$ satisfies the Riesz inequality (3.1) for $p = 2$ if and only if ω satisfies the A_2 condition, i.e.,

$$\left(\frac{1}{|Q|} \int_Q \omega\right)\left(\frac{1}{|Q|} \int_Q \frac{1}{\omega}\right) \leq A$$

for all intervals Q (in $\mathbb{T}$ or $\mathbb{R}$).

While the Helson-Szegö theorem works only for L^2, the A_2 condition extends to L^p, $p \neq 2$: $\omega(t)dt$ satisfies the Riesz inequality (3.1) for a given p, $1 < p < \infty$ if and only if ω satisfies the A_p condition, i.e.,

$$\left(\frac{1}{|Q|}\int_Q \omega\right)\left(\frac{1}{|Q|}\int_Q \frac{1}{\omega^{1/(p-1)}}\right)^{p-1} \leq A$$

for all intervals Q, and this result was generalized for $\mathbb{R}^n$, $n > 1$ [24]. While the Helson-Szegö condition (6.21) does not extend to $n > 1$, with the same estimate, condition (6.22) does [25]. (Recently Garnett and Jones [26] have extended condition (6.21) to the n-dimensional case, but the corresponding estimates are different for necessity and sufficiency.)

Remark 6.2. Observe that if $f = f_+ + f_-$ is as in (6.25) and if $\hat{f}(n)$, $\hat{f}_\pm(n)$ are the Fourier coefficients of f, $f_\pm$, then $\hat{f}_\pm(n) = \chi_\pm(n)\hat{f}(n)$ with $\chi_+(n) = 1$ if $n \geq 0$ and zero otherwise and $\chi_-(n) = 1$ if $n < 0$ and zero otherwise. Therefore, if $\gamma_n = \int e^{-inx}\omega(x)dx$ and $\lambda_n = \int e^{-inx} f(x)dx$ are the Fourier transforms of ω and f, then condition (6.25) can be rewritten as

$$\sum_{n,k} (M-1)\gamma_{n-k}\chi_+(n)\lambda_n\chi_+(k)\overline{\lambda}_k + (M+1)\gamma_{n-k}\chi_+(n)\lambda_n\chi_-(n)\overline{\lambda}_k$$

$$+ (M+1)\gamma_{n-k}\chi_-(n)\lambda_n\chi_+(k)\overline{\lambda}_k + (M-1)\gamma_{n-k}\chi_-(n)\lambda_n\chi_-(k)\overline{\lambda}_k \geq 0 \tag{6.39}$$

or, equivalently,

$$\sum_{n,k} K_{nk}\lambda_n\overline{\lambda}_k \geq 0 \tag{6.40}$$

where

$$K_{nk} = \begin{cases} (M - 1)\gamma_{n-k} & \text{if } n \geq 0,\ m \geq 0 \text{ or } n < 0,\ m < 0 \\ (M + 1)\gamma_{n-k} & \text{otherwise} \end{cases} \tag{6.41}$$

From Corollary 6.8 we obtain the following

Corollary 6.9. A given sequence $\{\gamma_n\}$ is the Fourier transform of a measure $\omega(t)dt$ that satisfies the Riesz condition (3.1b) if and only if the associated kernel K_{nk} given by (6.41) is positive definite. In this case we have the integral representation for the kernel:

$$K_{nk} = \int_0^{2\pi} e^{-i(n-k)t} v_{\alpha\beta}(t)dt \tag{6.42}$$

where $\alpha = \operatorname{sgn} n$, $\beta = \operatorname{sgn} k$ and

$$\lambda_1\bar{\lambda}_1 v_{++}(t) + \lambda_1\bar{\lambda}_2 v_{+-}(t) + \lambda_2\bar{\lambda}_1 v_{-+}(t) + \lambda_2\bar{\lambda}_2 v_{--}(t) \geq 0$$

for all $\lambda_1, \lambda_2 \in \mathbb{C}$ and almost all $t \in \mathbb{T}$.

Observe that the Herglotz-Bochner theorem (see Chapter 2, Section 4) asserts that $\{\gamma_n\}$ is the Fourier transform of a positive measure μ if and only if the kernel $K_{nk} = \gamma_{n-k}$ is positive definite, and in such case, the integral representation

$$K_{nk} = \int_0^{2\pi} e^{-i(n-k)t} d\mu(t)$$

holds.

Thus Corollary 6.9 can be considered as the analog of Bochner's theorem for measures satisfying the Riesz inequality (cfr. [21] and [25]).

Remark 6.3. Unlike the Helson-Szegö characterization (6.21) or the A_2 condition, condition (6.22) generalizes for the case of two different measures μ, ν. The measures $d\mu = u\,dt$, $d\nu = v\,dt$ satisfy

$$\int_{\mathbb{T}} |Hf(t)|^2 d\mu \leq M \int_{\mathbb{T}} |f(t)|^2 d\nu$$

for all $f \in \mathscr{P}$, if and only if $u(t) \leq Mv(t)$ and there exists a function $h \in H^1$ such that

$$-4Mu(t)v(t) - 2\,\mathrm{Re}\, h(t)(u(t) + Mv(t)) - |h(t)|^2 \geq 0$$

for a.e. $t \in \mathbb{T}$. (See [27].)

REFERENCES

1. A. P. Calderón and A. Zygmund, Acta Math., 88:85 (1952).
2. A. P. Calderón and A. Zygmund, Amer. J. Math., 78:310 (1956).
3. A. Zygmund, Rend. di Mat., 16:468 (1957).
4. A. P. Calderón, Bull. A.M.S., 72:426 (1966).
5. E. M. Stein, Proc. Symp. Pure Math., 10:316 (1967).
6. Mary Weiss and A. Zygmund, Studia Math., 26:101 (1966).
7. E. M. Stein and G. Weiss, Acta Math., 103:25 (1960).
8. J. Hórvath, Indag. Math., 15:17 (1953).
9. S. Spanne, Ann. Scuola Norm. Sup. Pisa, 20:625 (1966).
10. J. Peetre, Ann. Mat. Pura Appl., 72:295 (1966).
11. C. Fefferman and E. M. Stein, Acta Math., 129:137 (1972).
12. B. F. Jones, Jr., Amer. J. Math., 86:441 (1964).
13. E. B. Fabes and C. Sadosky, Studia Math., 26:75 (1966).
14. C. Sadosky, Studia Math., 27:73 (1967).
15. E. B. Fabes and N. M. Rivière, Studia Math., 27:19 (1966).
16. G. H. Hardy and J. E. Littlewood, Duke Math. J., 2:351 (1936).

17. K. I. Babenko, Dokl. Akad. Nauk SSSR, 62:157 (1948).

18. E. M. Stein, Proc. A. M. S., 8:250 (1958).

19. Cora Sadosky, Studia Math., 26:327 (1966).

20. H. Helson and G. Szegö, Ann. Mat. Pura Appl., 51:107 (1960).

21. M. Cotlar and C. Sadosky, C. R. Acad. Sci. Paris, A, 285: 433 (1977).

22. R. Arocena, C. R. Acad. Sci. Paris, A, 228:721 (1979).

23. R. Hunt, B. Muckenhoupt and R. L. Wheeden, Trans. A. M. S., 176:227 (1973).

24. R. R. Coifman and C. Fefferman, Studia Math., 51:241 (1974).

25. M. Cotlar and C. Sadosky, C. R. Acad. Sci. Paris, A, 285: 611 (1977).

26. J. B. Garnett and P. W. Jones, Ann. Math., 108:373 (1978).

27. M. Cotlar and C. Sadosky, Proc. Symp. Pure Math., 35 (1979).

Appendix A

SINGULAR INTEGRALS AND PARTIAL DIFFERENTIAL EQUATIONS

Let us start with some considerations on the algebra of singular integral operators and its applications to the study of partial differential equations.

The references are from Chapter 6, and we use the notations introduced there. Let K be a singular integral operator given by convolution with a C-Z kernel k (and assume $k \in C^{\infty}(\mathbb{R}^n - \{0\})$. We have seen that K can be expressed as a multiplier given by $\hat{k}$ (e.g., in L^2), where $\hat{k}$ is also a C-Z kernel (a function homogeneous of degree zero and with mean value zero on Σ) by Theorem 2.6 and Remarks 2.4 and 2.5, and furthermore $\hat{k} \in C^{\infty}(\mathbb{R}^n - \{0\})$. It can be proved (see [1]) that the converse holds, namely the second part of the following

<u>Proposition A.1</u>. Let $k \in C^{\infty}(\mathbb{R}^n - \{0\})$ be a homogeneous function of degree zero such that $\int_{\Sigma} k(x')dx' = 0$. Then its Fourier transform $\hat{k}$ also belongs to $C^{\infty}(\mathbb{R}^n - \{0\})$, is homogeneous of degree zero and $\int_{\Sigma} \hat{k}(x')dx' = 0$. Conversely, if $h \in C^{\infty}(\mathbb{R}^n - \{0\})$, is homogeneous of degree zero and $\int_{\Sigma} h(x')dx' = 0$, then $h = \hat{k}$ for $k \in C^{\infty}(\mathbb{R}^n - \{0\})$, homogeneous of degree zero and such that $\int_{\Sigma} k(x')dx' = 0$.

The C-Z kernels that belong to the class $C^{\infty}(\mathbb{R}^n - \{0\})$ give rise to convolution operators that do not preserve their class through composition: if $T_j : f \to k_j * f$ for $j = 1, 2$, then

$$((T_1 \circ T_2)f)^{\wedge} = (k_1 * k_2 * f)^{\wedge} = \hat{k}_1 \cdot \hat{k}_2 \cdot \hat{f}$$

where $\hat{k}_1 \cdot \hat{k}_2$ is a function homogeneous of degree zero that is indefinitely differentiable outside the origin as $\hat{k}_1$ and $\hat{k}_2$ are, but $\hat{k}_1 \cdot \hat{k}_2$ has not necessarily mean value zero on Σ. But by Proposition A.1, if we substract to $\hat{k}_1 \cdot \hat{k}_2$ its mean value on Σ, we obtain a $\hat{k}_3$, thus

$$\hat{k}_1 \cdot \hat{k}_2 = \hat{k}_3 + c \tag{A.1}$$

where

$$c = \int_{\Sigma} \hat{k}_1 \cdot \hat{k}_2(x')dx'$$

From (A.1), (at least if $f \in L^2$) then

$$k_1 * k_2 * f = k_3 * f + cf \tag{A.2}$$

(A.2) suggests the possibility of defining a class of singular integral operators such that it will be an algebra under composition.

Definition A.1. The operator T is a generalized singular integral operator if

$$T : f \to cf + Kf = cf + k * f \tag{A.3}$$

where $c \in \mathbb{C}$ is a constant and $k \in C^{\infty}(\mathbb{R}^n - \{0\})$ is a C-Z kernel.

For the class of generalized singular integral operators the composition is commutative:

$$(T_1 \circ T_2)f = T_1(T_2 f) = T_1(c_2 f + k_2 * f)$$

$$= c_1 c_2 f + c_1 k_2 * f + k_1 * c_2 f + k_1 * k_2 * f$$

$$= c_2 c_1 f + c_2 k_1 * f + c_1 k_2 * f + k_2 * k_1 * f$$

$$= T_2(T_1 f) = (T_2 \circ T_1)f$$

Definition A. 2. Given a generalized singular integral operator T given by $Tf = cf + k * f$, the symbol of T is the function

$$\sigma(T) = c + \hat{k} \tag{A. 4}$$

Observe that $\sigma(T)$ is a homogeneous function of degree zero and that $\sigma(T) \in C^{\infty}(\mathbb{R}^n - \{0\})$.

By (A. 4) it is

$$(Tf)\hat{} = c\hat{f} + \hat{k}.\hat{f} = \sigma(T).\hat{f} \tag{A. 5}$$

and there is a 1-1 correspondence between the operators and their symbols.

The generalized singular integral operators form an algebra, and if $\sigma(T)$ does not vanish, T is an invertible element of the algebra.

With these facts in mind, let us approach the study of partial differential equations.

Given $f \in \mathscr{S}(\mathbb{R}^n)$, we know from Chapter 2, Section 1, that $(\partial f/\partial x_j)\hat{} = -2\pi i x_j \hat{f}(x)$ and that $(\Delta f)\hat{}(x) = -4\pi^2 |x|^2 \hat{f}(x)$. Let us define the operator Λ by

$$(\Lambda f)^\wedge(x) = 2\pi|x|\hat{f}(x) \tag{A.6}$$

so that

$$(\Lambda^m f)^\wedge(x) = (2\pi|x|)^m \hat{f}(x) \tag{A.7}$$

for m integer and, in particular for $m = 2$,

$$\Lambda^2 = -\Delta$$

and we may formally write

$$\Lambda = (-\Delta)^{1/2} \tag{A.8}$$

Thus, if $\partial^\alpha f = \partial^\alpha f/\partial x^\alpha$, then

$$(\partial^\alpha f)^\wedge(x) = (-2\pi i x)^\alpha \hat{f}(x) = (-i)^{|\alpha|} \frac{x^\alpha}{|x|^{|\alpha|}} (2\pi|x|)^{|\alpha|}\hat{f}(x)$$

$$= (-i)^{|\alpha|}\left(\frac{x}{|x|}\right)^\alpha (\Lambda^{|\alpha|} f)^\wedge(x) \tag{A.9}$$

But since $x^\alpha|x|^{-|\alpha|}$ is a homogeneous function of degree zero that belongs to C^∞ outside the origin, by Proposition A.1 it must be the symbol $\sigma(K_\alpha)$ of a generalized singular integral operator K_α. From this and (A.9) we get

$$\partial^\alpha f = (-i)^{|\alpha|} K_\alpha \Lambda^{|\alpha|} f \tag{A.10}$$

thus the application of the operator derivation of order α is reduced to the application of one "bad" operator (Λ is unbounded and defined only on a dense set) followed by a known continuous operator (the singular integral operator K_α).

Given a homogeneous polynomial of degree m

$$P(x) = \sum_{|\alpha|=m} a_\alpha x^\alpha \tag{A.11}$$

the corresponding differential polynomial is

$$P(\partial) = \sum_{|\alpha|=m} a_\alpha \partial^\alpha \tag{A.12}$$

that by (A.10) can be written as

$$P(\partial) = (-i)^m K_m \Lambda^m \tag{A.13}$$

where

$$K_m = \sum_{|\alpha|=m} a_\alpha K_\alpha \tag{A.14}$$

is a generalized singular integral operator, whose symbol is

$$\sigma(K_m) = \sum_{|\alpha|=m} a_\alpha \frac{x^\alpha}{|x|^m} = \frac{P(x)}{|x|^m} \tag{A.15}$$

If $P(x) \neq 0$ for all $x \neq 0$, $P(\partial)$ is called an <u>elliptic</u> operator and, in such case $(\sigma(K_m))^{-1}$ is a homogeneous function of degree zero that belongs to C^∞ outside the origin, i.e., $(\sigma(K_m))^{-1}$ is the symbol of a generalized singular integral operator K_m^{-1}, that is the inverse of K_m. Furthermore, if $P(x) \neq 0$ for $x \neq 0$ and the coefficients of P are real, then m must be an even number, $m = 2r$, and we may write (A.13) as

$$P(\partial) = K\Delta^r \tag{A.16}$$

where $K = i^m K_m$ is a generalized singular integral operator and Δ is the laplacian.

Therefore, the elliptic partial differential equation with constant real coefficients, homogeneous of degree $m = 2r$

$$P(\partial)f = g \tag{A.17}$$

is transformed into

$$K\Delta^r f = g \tag{A.18}$$

and, as K is invertible for $P(\partial)$ elliptic, to solve (A.18) is reduced to solve

$$\Delta^r f = K^{-1} g = h \tag{A.19}$$

In connection with the study of the Riesz transforms done in Chapter 6, we give the following <u>a priori</u> estimate that is typical.

<u>Proposition A.2.</u> Let $f \in C_0^2(\mathbb{R}^n)$ and $1 < p < \infty$. Then

$$\left\|\frac{\partial^2 f}{\partial x_j \partial x_k}\right\|_p \leq A_p \|\Delta f\|_p \tag{A.20}$$

where A_p is independent of f.

<u>Remark A.1.</u> Given the Laplace equation $\Delta f = g$, where $g \in L^p$ is a known function, (A.20) yields an a priori estimate (i.e., an estimate obtained without solving the equation) for all the second derivatives of the (unknown) f in terms of a fixed multiple of $\|g\|_p$.

<u>Proof.</u> The estimate (A.20) follows from applying twice the Calderón-Zygmund theorem (Theorem 4.3 of Chapter 6) to the identity

$$\frac{\partial^2 f}{\partial x_j \partial x_k} = - R_j R_k \Delta f \tag{A. 21}$$

and (A. 21) holds since transforming Fourier both sides we obtain the identity

$$- 4\pi^2 x_j x_k \hat{f}(x) = - (i \frac{x_j}{|x|})(i \frac{x_k}{|x|})(- 4\pi^2 |x|^2)\hat{f}(x)$$

∇

The above considerations indicate a general approach that deals in the first place with differential polynomials that are nonhomogeneous and/or have variable coefficients. In fact, let be

$$P(\partial) = \sum_{|\alpha|=m} a_\alpha(x)\partial^\alpha \tag{A. 22}$$

Again $\partial^\alpha f = (-i)^{|\alpha|} T_\alpha \Lambda^{|\alpha|} f$, so taking $T_\alpha = c_\alpha + K_\alpha$ as in (A. 3),

$$P(\partial)f = (-i)^m (\sum_{|\alpha|=m} (a_\alpha(x)c_\alpha + a_\alpha(x)K_\alpha))\Lambda^m f \tag{A. 23}$$

Denoting $\Lambda^m f = \varphi$, (A. 23) becomes

$$P(\partial)f = A(x)\varphi(x) + \int k(x, x - y)\varphi(y)dy \tag{A. 24}$$

where

$$A(x) = (-i)^m \sum_{|\alpha|=m} c_\alpha a_\alpha(x)$$

and

$$k(x, z) = \sum_{|\alpha|=m} (-i)^m a_\alpha(x)k_\alpha(z) \tag{A. 25}$$

This is a reason for the interest in the study of the singular integrals with variable kernels $k(x, z)$ developed in [2].

The applications that we have just outlined are the simplest examples of the use of singular integrals in the study of partial differential equations. The theory of singular integrals has developed in that direction and has increased its importance through the introduction of pseudo-differential operators, that include in a same class differential and singular operators and thus constitute one of the most powerful tools of modern analysis. For the theory of pseudo-differential operators see [3] and [4].

REFERENCES

1. A. P. Calderón and A. Zygmund, Amer. J. Math., 78:310 (1956).
2. A. P. Calderón and A. Zygmund, Amer. J. Math., 78:289 (1956).
3. A. P. Calderón, Lecture Notes on Pseudo-Differential Operators and Elliptic Boundary Value Problems, I, Inst. Argentino de Matem., Buenos Aires, 1976.
4. L. Boutet de Monvel, A Course on Pseudo-Differential Operators and their Applications, Duke Univ. Math. Series II, Durham, NC, 1976.

Appendix B

THE COMPLEX METHOD OF INTERPOLATION

Let V be a (Hausdorff) topological vector space and A^0, A^1 two Banach spaces continuously imbedded in V. We may consider $A^0 \cap A^1$ and $A^0 + A^1$ as Banach spaces by introducing in them the norms

$$\|a\|_{A^0 \cap A^1} = \max\{\|a\|_0, \|a\|_1\} \tag{B.1}$$

$$\|a\|_{A^0 + A^1} = \inf\{\|a_0\|_0 + \|a_1\| : a_0 \in A^0, \ a_1 \in A^1, \ a_0 + a_1 = a\} \tag{B.2}$$

An <u>intermediate space</u> between A^0 and A^1 is any Banach space A such that

$$A^0 \cap A^1 \subset A \subset A^0 + A^1 \tag{B.3}$$

As an example, if $A^0 = L^{p_0}$ and $A^1 = L^{p_1}$, $1 < p_0, p_1 < \infty$, and p_t is given by

$$1/p_t = (1 - t)/p_0 + t/p, \quad 0 < t < 1 \tag{B.4}$$

then L^{p_t} is an intermediate space between L^{p_0} and L^{p_1}.

Let us assume that an intermediate space A between A^0 and A^1 is invariant under every linear operator $T : A^0 + A^1 \to A^0 + A^1$ with restrictions to A^0 and A^1 that leave A^0 and A^1 invariant, i.e., $T : A_0 \to A_0$ and $T : A_1 \to A_1$. If furthermore $T : A \to A$ boundedly, A is said to be a space of linear interpolation between A^0 and A^1.

The basic problems of the interpolation theory of linear operators are:

1) Given A^0 and A^1, characterize all the spaces of linear interpolation between A^0 and A^1.

2) How to construct spaces of linear interpolation between two given spaces A^0 and A^1.

3) Given a space A of linear interpolation between A^0 and A^1, and another pair of Banach spaces, B^0 and B^1, is there a space of linear interpolation between B^0 and B^1, say B, with the following property: every linear transformation from $A^0 + A^1$ into $B^0 + B^1$ that transforms continuously A^j into B^j, for $j = 0, 1$, also transforms A into B?

4) If A is a space of linear interpolation between A^0 and A^1, constructed by a certain method, and B is a space of linear interpolation between B^0 and B^1, constructed by the same method, does it hold that every linear transformation from $A^0 + A^1$ to $B^0 + B^1$ such that transforms A^j into B^j, $j = 0, 1$ also transforms A continuously into B?

Gagliardo [1] answered Question 1 and also Question 3, positively. There are two constructive methods of interpolation (that enable us to answer Questions 2 and 4): the complex method developed by Calderón [2] and the real method, developed by Lions and Peetre [3]. These methods are closely related to the methods of proof we gave for the Riesz-Thorin and the Marcinkiewicz theorems, respectively. (See the remarks that follow both theorems in Chapter 4.) In this appendix we present the basic properties of the complex interpolation method, as given in [2].

Let V be a topological vector space (Hausdorff) and let B^0 and B^1 be two Banach spaces continuously imbedded in V. B^0 and B^1 are then called <u>compatible</u>.

<u>Lemma B.1.</u> Let B^0 and B^1 be two compatible Banach spaces. Then $B^0 \cap B^1$ and $B^0 + B^1 = \{b \in V : b = b_0 + b_1,\ b_0 \in B^0,\ b_1 \in B^1\}$ are also Banach spaces, respectively under the norms

$$\|b\|_{B^0 \cap B^1} = \max\{\|b\|_{B^0}, \|b\|_{B^1}\} \tag{B.1}$$

and

$$\|b\|_{B^0+B^1} = \inf\{\|b_0\|_{B^0} + \|b_1\|_{B^1};\ b = b_0 + b_1,\ b_0 \in B^0,\ b_1 \in B^1\} \tag{B.2}$$

<u>Exercise B.1.</u> Prove Lemma B.1.

In what follows we write $\|\cdot\|_j = \|\cdot\|_{B^j}$, $j = 0, 1$.

Our aim is to construct intermediate spaces B, $B^0 \cap B^1 \subset B \subset B^0 + B^1$, such that they are spaces of linear interpolation between B^0 and B^1. To do so, we rely on the use of families of vector-valued analytic functions with values in the Banach space $B^0 + B^1$.

Let us consider in the complex plane the closed strip $D = \{z \in \mathbb{C} : 0 \leq \operatorname{Re} z \leq 1\}$, with interior $\mathring{D} = \{z \in \mathbb{C} : 0 < \operatorname{Re} z < 1\}$ and boundary $\Delta_0 \cup \Delta_1$, $\Delta_0 = \{z \in \mathbb{C} : z = iy,\ y \in \mathbb{R}\}$, $\Delta_1 = \{z \in \mathbb{C} : z = 1 + iy,\ y \in \mathbb{R}\}$.

Definition B.1. Given a fixed couple of compatible Banach space B^0, B^1 let $\mathscr{F} = \mathscr{F}(B^0, B^1)$ be the space of functions $f : D \to B^0 + B^1$ such that:

1) f is analytic in $\mathring{D}$,
2) f is continuous in D and $\|f(z)\|_{B^0+B^1}$ is bounded in D,
3) $f(iy) \in B^0$ for all $y \in \mathbb{R}$, the function $y \to f(iy)$ is continuous in $\mathbb{R}$, and the function $y \to \|f(iy)\|_0$ is bounded,
4) $f(1 + iy) \in B^1$ for all $y \in \mathbb{R}$, the function $y \to f(1 + iy)$ is continuous in $\mathbb{R}$, and the function $y \to \|f(1 + iy)\|_1$ is bounded.

Exercise B.2. Prove that $\mathscr{F} = \mathscr{F}(B^0, B^1)$ is a vector space.

Definition B.2. Given $f \in \mathscr{F} = \mathscr{F}(B^0, B^1)$ let

$$\|f\|_{\mathscr{F}} = \max\{\sup_y \|f(iy)\|_0, \sup_y \|f(1 + iy)\|_1\}$$

It is immediate that $\|\cdot\|_{\mathscr{F}}$ provides a norm for $\mathscr{F}$. Moreover, since for any $b \in B^j$, $j = 0, 1$, $b \in B^0 + B^1$ and $\|b\|_{B^0+B^1} \leq \|b\|_j$, then,

$$\|f\|_{\mathscr{F}} \geq \max\{\sup_y \|f(iy)\|_{B^0+B^1}, \sup_y \|f(1 + iy)\|_{B^0+B^1}\}$$

and, by the maximum principle,

$$\|f\|_{\mathscr{F}} \geq \|f(z)\|_{B^0+B^1} \tag{B.3}$$

for each $z \in D$.

Lemma B.2. $\mathscr{F}$ is a Banach space.

Proof. Let $\{f_n\} \in \mathscr{F}$ be such that $\Sigma_n \|f_n\|_{\mathscr{F}} < \infty$. By (B.3), $\|f_n(z)\|_{B^0+B} \leq \|f_n\|_{\mathscr{F}}$ for every $z \in D$. Since $B^0 + B^1$ is a Banach space by Lemma B.1, $\Sigma_n f_n(z)$ converges uniformly in D to a function in $B^0 + B^1$. Furthermore, $\|f_n(j + iy)\|_j \leq \|f_n\|_{\mathscr{F}}$ and $\Sigma_n f_n(j + iy)$ converges uniformly in Δ_j to a function in B^j, $j = 0, 1$, and the limits coincide in $B^0 + B^1$. Thus the limit $f \in \mathscr{F}$, and $\Sigma_n f_n$ converges to f in $\mathscr{F}$. ∇

<u>Definition B.3</u>. Let be $s \in \mathbb{R}$, $0 \leq s \leq 1$, and a couple of compatible Banach spaces B^0 and B^1. Then the space $B_s = [B^0, B^1]_s \subset B^0 + B^1$ is defined by

$$B_s = \{b \in B^0 + B^1 : b = f(x),\ f \in \mathscr{F}(B^0, B^1)\} \tag{B.4}$$

<u>Definition B.4</u>. Given B_s as in Definition B.3, let us define the norm

$$\|b\|_s = \|b\|_{B_s} = \inf\{\|f\|_{\mathscr{F}} : f(s) = b\} \tag{B.5}$$

Then $B_s \subset B^0 + B^1$ as a continuous imbedding.

Let $\mathscr{N}_s = \{f \in \mathscr{F} : f(s) = 0\}$. $\mathscr{N}_s$ is a closed subspace of $\mathscr{F}$ and the quotient space $\mathscr{F}(B^0, B^1)/\mathscr{N}_s$ is a Banach space isomorphic and isometric to B_s. In fact, B_s is the image of $\mathscr{F}$ under the linear mapping into $B^0 + B^1$, $f \to f(s)$. This mapping is continuous since by (B.3) $\|f(s)\|_{B^0+B^1} \leq \|f\|_{\mathscr{F}}$, and $\mathscr{N}_s$ is its kernel. The norm in B_s defined in (B.5) is the norm corresponding to the quotient space. We now study some properties of the space B_s.

<u>Proposition B.3</u>. For every $s \in \mathbb{R}$, $0 \leq s \leq 1$,

$$B^0 \cap B^1 \subset B_s \tag{B.6}$$

and the norm of B_j, the B_s space for $s = j$, coincides in $B^0 \cap B^1$ with the norm of B^j, $j = 0, 1$. In particular, since $B_s \subset B^0 + B^1$, B_s is an intermediate space between B^0 and B^1.

Proof. Let $b \in B^0 \cap B^1$. We claim that for a fixed s there exists an $f \in \mathcal{F} = \mathcal{F}(B^0, B^1)$ such that $f(s) = b$. In fact, let be $f(z) = b \exp(z - x)^2$, so that $f \in \mathcal{F}$ and $f(s) = b$. In particular, for $s = 0$ there exists $f \in \mathcal{F}$ such that $f(0) = b$. Since $\|b\|_{B_0} = \inf_f\{\|f\|_{\mathcal{F}} : f(0) = b\}$, it is $\|f\|_{\mathcal{F}} \le \|b\|_{B_0} + \varepsilon$. Moreover, $\|f\|_{\mathcal{F}} \ge \|f(0)\|_{B^0} = \|b\|_{B^0}$ and so $\|b\|_{B^0} \le \|b\|_{B_0}$. If $f_n(z) = b \exp(z^2 - nz)$, $f_n \in \mathcal{F}$ and $f_n(0) = b$. Since $\|f_n\|_{\mathcal{F}} \le \max\{\|b\|_0, \|b\|_1 Ce^{-n}\}$ that tends to $\|b\|_{B^0}$ as $n \to \infty$ (where C stands for the maximum of $\exp z^2$ on Δ_1, that is bounded), then $\|b\|_{B_0} = \|b\|_{B^0}$ for all $b \in B^0 \cap B^1$. The same argument holds for B^1 and B_1. ∇

Proposition B. 4. For every $s \in \mathbb{R}$, $0 \le s \le 1$, and $b \in B^0 \cap B^1$,

$$\|b\|_{B_s} \le \|b\|_0^{1-s} \|b\|_1^s \tag{B. 7}$$

Proof. Given b and s, let

$$f_\varepsilon(z) = b\|b\|_0^{z-s} \|b\|_1^{s-z} \exp(z - s)^2 \tag{B. 8}$$

Then $f_\varepsilon \in \mathcal{F}$, $f_\varepsilon(s) = b$ and

$$\|f_\varepsilon\|_{\mathcal{F}} = C_\varepsilon \|b\|^{1-s} \|b\|_1^s, \quad C_\varepsilon = \max_s(\exp \varepsilon s^2, \exp \varepsilon(1 - s^2)) \tag{B. 9}$$

As $C_\varepsilon \to 1$ when $\varepsilon \to 0$ and as $\|f_\varepsilon\|_{\mathcal{F}} \ge \|b\|_{B_s}$, letting $\varepsilon \to 0$ in (B. 9) we obtain (B. 7). ∇

Let now be $\mathcal{F}_0 = \{f \in \mathcal{F} : f$ is an entire function, $f(z) \in B^0 \cap B^1$ for all $z \in D\}$.

Exercise B.3. Prove that $\mathscr{F}_0$ is a dense subspace of $\mathscr{F}$. (Hint: Prove first that $\mathscr{F}_0$ is dense in $\{f \in \mathscr{F} : f = (\exp \varepsilon z^2)g, g \in \mathscr{F}\}$).

Proposition B.5. $B^0 \cap B^1$ is dense in B_s, for all $s \in \mathbb{R}$, $0 \leq s \leq 1$.

Proof. Let $b \in B_s$. Then there is an $f \in \mathscr{F}$ such that $f(s) = b$. For this f and $\varepsilon > 0$ there is an $f_0 \in \mathscr{F}_0$ such that $\|f - f_0\|_{\mathscr{F}} < \varepsilon$. By definition of $\mathscr{F}_0$, $f_0(z) \in B^0 \cap B^1$ and $\|f_0(s) - b\|_{B_s} = \|f_0(s) - f(s)\|_{B_s} \leq \|f_0 - f\|_{\mathscr{F}} < \varepsilon$. ∇

Corollary B.6. $B_j = B^j$ if and only if $B^0 \cap B^1$ is dense in B^j, $j = 0, 1$.

Proof. For $j = 0, 1$, by Proposition B.5, $B^0 \cap B^1$ is dense in B_j, and by Proposition B.3, the norms of B_j and B^j coincide in $B^0 \cap B^1$. ∇

Proposition B.7. If $0 < s < 1$ then

$$[B^0, B^1]_s = [B^1, B^0]_{1-s}$$

Proof. Observe that $f \in \mathscr{F}(B^0, B^1)$ if and only if $g \in \mathscr{F}(B^1, B^0)$ for $g(z) = f(1 - z)$. ∇

The following result gives an answer to Question 4.

Theorem B.8. Let B^0, B^1 and C^0, C^1 be two pairs of compatible Banach spaces and $T : B^0 + B^1 \to C^0 + C^1$ be a continuous linear operator such that $T : B^j \to C^j$, $j = 0, 1$, continuously with norms M_j. Then, if $B_s = [B^0, B^1]_s$ and $C_s = [C^0, C^1]_s$,

$$\|Tb\|_{C_s} \le M_0^{1-s} M_1^s \|b\|_{B_s} \tag{B.10}$$

for all $b \in B_s$, $0 \le s \le 1$.

<u>Proof</u>. Given $b \in B_s$ and $\varepsilon > 0$, there is an $f \in \mathscr{F}(B^0, B^1)$ such that $f(s) = b$ and $\|f\|_{\mathscr{F}} \le \|b\|_{B_s} + \varepsilon$. The function

$$g(z) = M_0^{z-1} M_1^{-z} T(f(z)) \tag{B.11}$$

is such that $g \in \mathscr{F}(C^0, C^1)$ and, by the hypothesis on T,

$$\|g\|_{\mathscr{F}} \le \max\{\sup_y \|f(iy)\|_{B^0}, \sup_y \|f(1+iy)\|_{B^1}\}$$

$$= \|f\|_{\mathscr{F}} \le \|b\|_{B_s} + \varepsilon$$

Thus,

$$\|b\|_{B_s} + \varepsilon \ge \|g\|_{\mathscr{F}} > \|g(s)\|_{C_s}$$

$$= \|M_0^{s-1} M_1^{-s} T(f(s))\|_{C_s}$$

$$= M_0^{s-1} M_1^{-s} \|Tb\|_{C_s}$$

for every $\varepsilon > 0$ and (B.10) follows. ∇

To deal with problems of duality, Calděron introduced yet another construction of intermediate spaces, given by a family $\mathscr{G}$ of analytic functions.

<u>Definition B.5</u>. Given a couple of compatible Banach spaces B^0, B^1, let $\mathscr{G} = \mathscr{G}(B^0, B^1)$ be the space of function $f : D \to B^0 + B^1$ such that

1) f is analytic in $\mathring{D}$, the interior of D,

2) f is continuous in D,

3) $\|f(z)\|_{B^0+B^1} \leq C(1 + |z|)$,

4) $\Delta_h f(j + iy) = f(j + i(y + h)) - f(j + iy) \in B^j$ for $j = 0, 1$ and

$$\|f\|_{\mathcal{G}} = \max\{\sup_{y,h} \|\frac{\Delta_h f(iy)}{h}\|_0, \sup_{y,h} \|\frac{\Delta_h f(1+iy)}{h}\|_1\} \qquad (B.11)$$

is finite.

Lemma B.9. The space $\mathcal{G}(B^0, B^1)$, modulo constant functions, is a Banach space provided with the norm given in (B.11).

Proof. From Definition B.5, if $\Delta_h f(z) = (f(z + ih) - f(z))/ih$ and $h \neq 0$, then $\|\Delta_h f(z)\|_{B^0+B^1} \leq \|f\|_{\mathcal{G}}$. Thus, $\|f'(z)\|_{B^0+B^1} \leq \|f\|_{\mathcal{G}}$ for all $z \in D$. Therefore, if $\|f\|_{\mathcal{G}} = 0$ then f = constant, and so $\mathcal{G}$ modulo the constant functions is a normed space. Furthermore, $\|f(z) - f(0)\|_{B^0+B^1} \leq |z| \, \|f\|_{\mathcal{G}}$ for all $z \in \mathring{D}$.

Suppose $\Sigma_n \|f_n\|_g < \infty$. Then $\Sigma_n (f_n(z) - f(0))$ converges uniformly on every compact in $\mathring{D}$. The limit f(z) satisfies (1), (2) and (3). Moreover, $\Sigma_n (f_n(j + i(y + h)) - f_n(j + iy))$ converges in B^j, $j = 0, 1$. Thus $f(j + i(y + h) - f(j + iy) \in B^j$ and therefore $f \in \mathcal{G}$ and $\mathcal{G}$ is complete. ∇

Definition B.6. If $s \in \mathbb{R}$, $0 < s < 1$, and B^0 and B^1 are a couple of compatible Banach spaces, then the space $B^s = [B^0, B^1]^s \subset B^0 + B^1$ is defined by

$$B^s = \{b \in B^0 + B^1 : b = \frac{df}{dz}(s), \ f \in \mathcal{G}(B^0, B^1)\} \qquad (B.12)$$

Definition B.7. Given B^s as in Definition B.6, let us define the norm

$$\|b\|_{B^s} = \inf\{ \|f\|_{\mathcal{G}} : f'(s) = b\} \tag{B.13}$$

Proposition B.10. B^s is a Banach space with respect to the norm given in (B.13). Furthermore, B^s is an intermediate space between B^0 and B^1.

Proof. Since $\|f'(z)\|_{B^0+B^1} \leq \|f\|_{\mathcal{G}}$ for all $z \in D$, the mapping $f \to f'(z)$ is continuous from $\mathcal{G}$ to $B^0 + B^1$. The kernel $\mathcal{N}^s$ of this mapping is closed, and B^s is the image of $\mathcal{G}$ under the mapping. Thus $B^s \sim \mathcal{G}/\mathcal{N}^s$ and the norm (B.13) is precisely the norm induced by $\mathcal{G}$ on the quotient space. Thus B^s is a Banach space, as $\mathcal{G}$ is.

Furthermore $B^s \subset B^0 + B^1$ continuously and, if $b \in B^0 \cap B^1$, by taking $f(z) = bz$ we conclude that $B^0 \cap B^1 \subset B^s$. Thus, $B^0 \cap B^1 \subset B^s \subset B^0 + B^1$ and B^s is intermediate between B^0 and B^1. ∇

Proposition B.11. Given a couple of compatible Banach spaces B^0 and B^1 then, for $0 < s < 1$,

$$B_s \subset B^s \quad \text{and} \quad \|b\|_{B^s} \leq \|b\|_{B_s} \tag{B.14}$$

for all $b \in B^0 \cap B^1$.

Proof. Let $b \in B_s$ and take $f \in \mathcal{F} = \mathcal{F}(B^0, B^1)$ such that $f(s) = b$ and $\|f\|_{\mathcal{F}} \leq \|b\|_{B_s} + \varepsilon$. Set $g(z) = \int_0^z f(u)du$. Then $g \in \mathcal{G} = \mathcal{G}(B^0, B^1)$ and $\|g\|_{\mathcal{G}} \leq \|f\|_{\mathcal{F}}$. Furthermore, $g'(s) = f(s) = b$ and thus, for any $\varepsilon > 0$,

$$\|b\|_{B^s} \leq \|g\|_{\mathscr{G}} \leq \|f\|_{\mathscr{F}} \leq \|b\|_{B_s} + \varepsilon$$

and (B.14) is proved. ∇

Remark B.1. If at least one of the two spaces B^0, B^1 is reflexive then

$$B_s = B^s \quad \text{and} \quad \|b\|_{B^s} = \|b\|_{B_s} \tag{B.15}$$

The proof of this equivalence of the two complex methods is deep and we shall not give it here. Let us note that this result, together with the Duality theorem asserting that if $B^0 \cap B^1$ is dense in B^0 and in B^1 then the dual space $[B^0, B^1]'_s = [B^{0\prime}, B^{1\prime}]^s$, immediately imply the following

Corollary. Given a compatible pair B^0, B^1 such that at least one of the spaces is reflexive and with $B^0 \cap B^1$ dense in B^j, $j = 0, 1$, then the dual space is

$$[B^0, B^1]'_s = [B^{0\prime}, B^{1\prime}]_s$$

with equal norms.

Another (nonobvious) consequence of the equivalence and the duality theorem is the Reiteration theorem: Given a pair of compatible spaces B^0, B^1, then, if $0 < s_0 < s_1 < 1$,

$$[[B^0, B^1]_{s_0}, [B^0, B^1]_{s_1}]_s = [B^0, B^1]_{s'}$$

where $s' = (1 - s)s_0 + ss_1$, with equality for the norms.

Applications

(I) Let $(\mathscr{X}, \mu)$ be a measure space and $B^j = L^{p_j}(\mathscr{X})$, $1 \leq p_j \leq \infty$ for $j = 0, 1$. We can consider $B^j \subset L_{loc}$, $j = 0, 1$, so that B^0, B^1 is a compatible pair of Banach spaces. There is a norm preserving isomorphism between B_s and $L^{p_s}(\mathscr{X})$, $1/p_s = (1 - s)/p_0 + s/p_1$, $0 \leq s \leq 1$, whenever $p_s < \infty$, and we write $B_s \sim L^{p_s}(\mathscr{X})$ in such case. (If $p_s = \infty$ it is because p_0 or $p_1 = \infty$ and then $B_j \sim B^j = L^{\infty}(\mathscr{X})$ does not hold.)

Given another measure space $(\mathscr{Y}, \nu)$ and $C^j = L^{q_j}(\mathscr{Y})$, $1 \leq q_j \leq \infty$, $j = 0, 1$, Theorem B.8 is a reformulation of the M. Riesz-Thorin convexity theorem (see Chapter 4).

(II) Let $(\mathscr{X}, \mu)$ be a measure space, μ_j, $j = 0, 1$, be two measures absolutely continuous with respect to μ and $B^j = L^{p_j}(\mathscr{X}, \mu_j)$, $j = 0, 1$. Then $B_s \sim L^{p_s}(\mathscr{X}, \mu_s)$, with $0 < 1/p_s = (1 - s)/p_0 + s/p_1$, $0 \leq s \leq 1$, and $d\mu_s = \omega_0^{1-s}\omega_1^{s}\,d\mu$, where ω_j is the Radon-Nikodym derivative of μ_j with respect to μ, $j = 0, 1$.

Through the complex method we can set also interpolation theorems for the Hardy spaces H^p, the Lipschitz spaces Λ_α^p, the Besov and Solobev spaces, etc., that were not mentioned in this book, so we are not giving the statements of those results here. Furthermore, the classical M. Riesz-Thorin interpolation theorem can be thus extended to operators transforming vector valued functions, in particular for the case of L^p functions with values in L^q. For these and other applications, and also for the real method, see [4].

REFERENCES

1. E. Gagliardo, C. R. Acad. Sci. Paris, A, 248:1912, 3388, 3517 (1959).
2. A. P. Calderón, Studia Math., 24:113 (1964).
3. J. L. Lions and J. Peetre, Inst. Hautes Etudes Sci. Publ. Math., 19:5 (1964).
4. J. Bergh and J. Löfström, Interpolation Spaces, Springer-Verlag, Berlin-Heidelberg-New York, 1976.

BIBLIOGRAPHY

All chapters conclude with a Reference section, where books and monographs are listed as they appear quoted in the text. From Chapter 4 on, the References include also some original papers, where the referred results were originally presented.

The following list includes books and monographs, most of which were not referred before, that should be consulted for further study on the material given in this text.

1. M. Cotlar, Condiciones de continuidad de operadores potenciales y de Hilbert, Cursos y seminarios de Matemáticas, Universidad de Buenos Aires, fasc. 2, Buenos Aires, 1959.
2. U. Neri, Singular Integrals, Lecture Notes in Math., # 200, Springer-Verlag, Berlin-Heidelberg-New York, 1971.
3. G. O. Okikiolu, Aspects of the Theory of Bounded Integral Operators in L^p spaces, Academic Press, London-New York, 1971.
4. E. T. Oklander, Interpolación, espacios de Lorentz y teorema de Marcinkiewicz, Cursos y seminarios de Matemáticas, Universidad de Buenos Aires, fasc. 20, Buenos Aires, 1965.

5. E. M. Stein, Singular Integrals and Differentiability Properties of Functions, Princeton University Press, Princeton, 1970.
6. E. M. Stein and G. Weiss, Introduction to Fourier Analysis on Euclidean Spaces, Princeton University Press, Princeton, 1971.
7. G. Weiss, Análisis armónico en varias variables: teoría de los espacios H^p, Cursos y seminarios de Matemáticas, Universidad de Buenos Aires, fasc. 9, Buenos Aires, 1960.
8. A. Zygmund, Intégrales singulières, Lecture Notes in Math., # 204, Springer-Verlag, Heidelberg-Berlin-New York, 1971.
9. A. Zygmund, Trigonometric Series, 2nd. edition, vols. I and II, Cambridge University Press, Cambridge, 1959.

GLOSSARY OF PRINCIPAL SYMBOLS

$\mathbb{R}$	the real line
$\mathbb{R}_+$	the positive real half-line
$\mathbb{Z}$	the integers
$\mathbb{C}$	the complex numbers
$\mathbb{T}$	the unit circle
$\mathbb{R}^n$	the n-dimensional Euclidean space
$\mathbb{R}^{n+1}_+ = \mathbb{R}^n \times \mathbb{R}_+$	the upper half-space
E^c	complement of a set E
$\lvert E\rvert$	Lebesgue measure of a set E
$\lvert x\rvert$	distance from the point x to the origin
$x.t$	scalar product in $\mathbb{R}^n$
χ_E	characteristic function of a set E
dx	Lebesgue measure in $\mathbb{R}^n$
a.e.	almost everywhere, almost every
P.V.	principal value (of an integral)
$\Sigma = \Sigma_n$	the unit sphere in $\mathbb{R}^n$
ω_n	its surface area
Ω_n	its volume
x'	a point on Σ

dx'	the surface area element on Σ
$D, \partial D, \mathring{D}, \overline{D}$	a domain (open, convex set) in $\mathbb{R}^n$, its boundary, its interior, its closure
Q	a cube with sides parallel to the axes
$Q(x, r)$	a cube of side r centered at x, with sides parallel to the axes
$S(x, r)$	a sphere of radius r centered at x
$\Gamma_\alpha(x)$	a cone in $\mathbb{R}^{n+1}_+$ with vertex in $x \in \mathbb{R}^n$ and aperture α
τ_h	translation in $\mathbb{R}^n$
δ_h	dilation in $\mathbb{R}^n$
Δ	laplacean
C_0	the class of continuous functions with compact support
C^k	the class of functions with continuous derivatives up to and including total order k
C^∞	the class of indefinitely differentiable functions
C_0^k, C_0^∞	same, with compact support
$\mathcal{S}$	the Schwartz class of indefinitely differentiable functions all whose derivatives remain bounded when multiplied by any polynomial
$C_\infty(\mathbb{R}^n)$	the class of continuous functions in $\mathbb{R}^n$ that vanish at infinity
$\mathcal{M}(\mathbb{R}^n)$	its dual, the class of finite Borel measures in $\mathbb{R}^n$
$\mathcal{M}^+$	the subclass of positive finite Borel measures
L_{loc}	the class of locally integrable functions
$L^1 = L$	the class of integrable functions
L^2	the class of square integrable functions
L^p	the class of functions integrable when raised to the power p
L^∞	the class of (essentially) bounded functions
L^p_*	the Marcinkiewicz class of order p; weak L^p

L_{pq}	the Lorentz spaces
$L^p(\log^+ L)^q$	the Zygmund classes of functions
BMO	the class of functions of bounded mean oscillation
$\mathscr{L}_f$	the Lebesgue set of f
p'	the conjugate index for p, $1/p + 1/p' = 1$
p_t	the intermediate index between p_0 and p_1, $1/p_t = (1 - t)/p_0 + t/p_1$, $0 < t < 1$
$\mathscr{A}$	the group algebra of $\mathbb{R}^n$
$\mathscr{P}$	the class of positive linear continuous functionals acting on $\mathscr{A}$, also the class of trigonometric polynomials on $\mathbb{T}$
$\mathscr{P}_1$	the subclass of functionals with norm equal to one
$P_y(x)$	the Poisson kernel in $\mathbb{R}^n$
$P_r(t)$	the Poisson kernel in $\mathbb{T}$
$Q_y(x)$	the conjugate Poisson kernel in $\mathbb{R}^n$
$Q_r(t)$	the conjugate Poisson kernel in $\mathbb{T}$
$(f, g) = \int f\bar{g}\, dx$	the scalar product in L^2
$I_f(g) = \langle f, g\rangle = \int fg\, dx$; $I_\mu(g) = \langle \mu, g\rangle = \int g\, d\mu$	functionals on function spaces
$f * g$	convolution
$\hat{f} = \mathscr{F} f$	Fourier transform of f
$f_\lambda = f$ if $\lvert f\rvert \le \lambda$	and zero otherwise
$f^\lambda = f$ if $\lvert f\rvert > \lambda$	and zero otherwise
$E_\alpha(f) = \{\lvert f\rvert > \alpha\}$	
$f_*(\alpha) = \lvert E_\alpha(f)\rvert$	the distribution function of f
f^*	the nonincreasing rearrangement of f
f^{**}	its mean value function
f_Q	the mean value of f in Q
$f_Q^\# = (\lvert f - f_Q\rvert)_Q$	the mean oscillation of f in Q

$f^{\#} = \sup_Q f_Q^{\#}$	the sharp function
Λ	the Hardy-Littlewood maximal operator
Λ'	same, "non centered"
$\Lambda^{\#}$	the sharp maximal operator
$\Lambda^{\#}{}'$	same, "non centered"
I_γ	the Riesz potential operator of order γ
H	the Hilbert transform operator
R_j	the j-Riesz transform operator
K	singular integral operator
K^*	maximal singular integral operator

INDEX